D0203666

Agricultural Systems

Agricultural Systems: Agroecology and Rural Innovation for Development

Sieglinde Snapp
Department of Crop and Soil Science
and W.K. Kellogg Biological Station
Michigan State University
East Lansing, MI

Barry Pound
Natural Resources Institute
University of Greenwich
Kent, U.K.

ELSEVIER

AMSTERDAM • BOSTON • HEIDELBERG • LONDON
NEW YORK • OXFORD • PARIS • SAN DIEGO
SAN FRANCISCO • SINGAPORE • SYDNEY • TOKYO
Academic Press is an imprint of Elsevier

Academic Press is an imprint of Elsevier
30 Corporate Drive, Suite 400
Burlington, MA 01803, USA
525 B Street, Suite 1900
San Diego, California 92101-4495, USA
84 Theobald's Road
London WC1X 8RR, UK

This book is printed on acid-free paper. ♾

Cover images © iStockphoto
Design Direction: Joanne Blank
Cover Design: Riezebos Holzbaur Design Group

Library of Congress Cataloging-in-Publication Data

Snapp, Sieglinde S.
Agricultural systems: agroecology and rural innovation for development/Sieglinde Snapp, Barry Pound.
p. cm.
Includes bibliographical references and index.
ISBN-13: 978-0-12-372517-2 (hard cover: alk. paper) 1. Agricultural systems. 2. Agricultural ecology. 3. Sustainable development. 4. Rural development. I. Pound, Barry. II. Title. III. Title: Agroecology and rural innovation for development.
S494.5.S95S58 2008
630–dc22 2007045787

British Library Cataloguing-in-Publication Data
A catalogue record for this book is available from the British Library.

ISBN: 978-0-12-372517-2

For information on all Academic Press publications
visit our Web site at www.books.elsevier.com

Printed in the United States of America
08 09 10 9 8 7 6 5 4 3 2 1

Contents

Contributors

Rachel Bezner Kerr
Department of Geography
University of Western Ontario
London, Ontario, Canada

Malcolm Blackie
University of East Anglia
Cringleford
Norwich, United Kingdom

Czech Conroy
Resources Institute
Chatham Maritime
Kent, United Kingdom

Laurie Drinkwater
Department of Horticulture
Cornell University
Ithaca, New York

George Kanyama-Phiri
Professor of Agroforestry
University of Malawi
Liongwe, Malawi

Vicki Morrone
Department of Community Agriculture
Recreation and Resource Studies
Michigan State University
East Lansing, Michigan

Barry Pound
Natural Resources Institute
University of Greenwich
Kent, United Kingdom

Sieglinde Snapp
Department of Crop and Soil Science
W. K. Kellogg Biological Station
Michigan State University
East Lansing, Michigan

Robert Tripp
Overseas Development Institute
Chiddingfold
Surrey, United Kingdom

Eva Weltzien
International Center for Research in the Semi-Arid Tropics
Bamako, Mali

CHAPTER 1

Introduction

George Kanyama-Phiri, Kate Wellard, and Sieglinde Snapp

Summary

Agricultural systems are at the heart of developing countries' economies and family livelihoods. A rapidly transforming world poses considerable challenges to rural smallholders with limited resources, from globalization and poverty to pandemics and climate change. Agricultural development is explored within this dynamic context. The evolution of agricultural research is traced from simplistic commodity-focused improvement to innovation systems approaches. The goal of this book is to transform and update a "farming systems approach" through integrating the latest science. This chapter introduces principles and practices in agroecology, livelihood systems, and participatory research. We review the role of interdisciplinary and multistakeholder inquiry to support knowledge generation and local research capacity. Rural innovation, market linkages, and education are drivers of agricultural change, and scientists can support this transformation process by working with farmers and extension advisors to test and adapt new plants, animals, technologies, and information.

Agricultural Systems: Agroecology and Rural Innovation for Development

AGRICULTURAL SYSTEMS IN CONTEXT

Most economies of the developing world are predominantly agrarian. In sub-Saharan Africa, for example, agriculture employs around 70 to 90% of the population and contributes between 30 and 60% of the gross domestic product and over 60% of exports. Agricultural systems are vital to tackling poverty and malnutrition, which remain widespread and severe, with 70% or more of the population living below $1 per day and children under the age of 5 and the elderly being highly vulnerable. Agricultural productivity gains are occurring in developing countries. This is shown by staple grain yields, which have increased by 70 kg/ha/year over the last decade in Latin America and southeast Asia to an average yield of 2800 kg/ha. Sub-Saharan Africa remains the exception, where grain yields have stagnated at around 1000 kg/ha over the last 2 decades.

High population pressure with limited land holdings leads to continuous arable cultivation on the same piece of land or extension of cultivation on fragile ecosystems such as steep slopes and river banks. These in turn bring about biological, chemical, and physical land degradation. Many African countries that were self-sufficient are now net importers of food. Food production has not kept pace with population growth in the face of shrinking land holdings. At the same time, government contingency planning is largely inadequate in the face of adverse weather conditions caused by climate change. The past 60 years are widely reported to have been the warmest in the last 1000 years while temperatures continue to climb. Associated with these global climatic changes are increasing risks of epidemics and invasive species such as weeds. Taken together, the need for rural innovation and adaptation to rapid change is more critical than ever.

Globalization[1] and liberalization of many developing economies of the world, especially Africa, have not brought about commensurate agricultural economic growth and prosperity. Later chapters in this book consider this essential context to development; however, the primary focus of the book is on working with smallholder farmers and rural stakeholders, where educators, researchers, and extension advisors can make a difference. We recognize the critical need to engage with policy makers and consider fully the context for equitable development. Trade barriers and tariffs, including subsidies, cause considerable disparities and tend to favor Northern hemisphere investors in agricultural trade and related intellectual property rights. The uneven sequencing of liberalization is impoverishing and widening the gap between rich and poor countries, resulting in limited competitive capacity among developing countries. Conflict and wars have further impacted negatively on food production and have led to loss of property and life, displacements, and misery throughout much of developing world (Table 1.1).

[1]Globalization refers to increasing global connectivity, information flow, and interdependence, which lead to convergence across markets, ideas, and ecologies.

TABLE 1.1 Development Barriers and Promoters for Agricultural Systems and Livelihoods of Rural Smallholders and Other Stakeholders

Development barriers	Development promoters
Population growth	Population growth, with investment in education and human development
Poor infrastructure	Investment in transport, communication, finance, and market access
Lack of governance: civic unrest and corruption	Commitment to civic structures, governance, and transparency
Poor agricultural productivity and limited value-added opportunities	Support for innovations in productivity enhancement and postharvest processing
Epidemics, human and animal diseases	Health care
Erratic and poorly resourced environment (climate, soils, water access, genetics)	Buffered environment with resources of light, water, nutrients, and genetic materials

The development of agricultural systems of the south is being undermined by the HIV/AIDS pandemic and by emerging epidemics such as the bird flu H5-N1 virus. The productive workforce, rural families, and research, extension, and education staff have been devastated. Gender inequality is another major social challenge. Despite contributing 70% of agricultural labor in many developing economies, women rarely have access to requisite resources and technologies compared to their male counterparts. The consequence is a vicious cycle of poverty and food insecurity, especially in households headed by women and children.

Agricultural development depends to a great extent on investment in human capacity and education for a successful generation and application of knowledge. It is a conundrum that increasing human population density can exhaust resources and impoverish an area or lead to innovation and prosperity through education and human capacity building. Investments in knowledge, especially science and technology, have featured prominently and consistently in most national agricultural strategies. In a number of countries, particularly Asia, these strategies have been highly successful. Research on food crop technologies, especially genetic improvements, has resulted in average grain yields doubling since the mid-1960s (Fig. 1.1).

Gains in agricultural productivity and ingenuity in devising superior storage and postharvest processing have directly contributed to enhanced food security around the globe. Time and again the predictions that population growth will outstrip food supply have been disproved. New disease-resistant crop varieties and integrated crop management have provided measurable gains for farmers, from the adoption of disease-resistant cassava varieties to high-yielding, maize-based systems. Agricultural scientists in developing countries are innovators in genetic improvement, including partnering with farmers to develop new varieties of indigenous

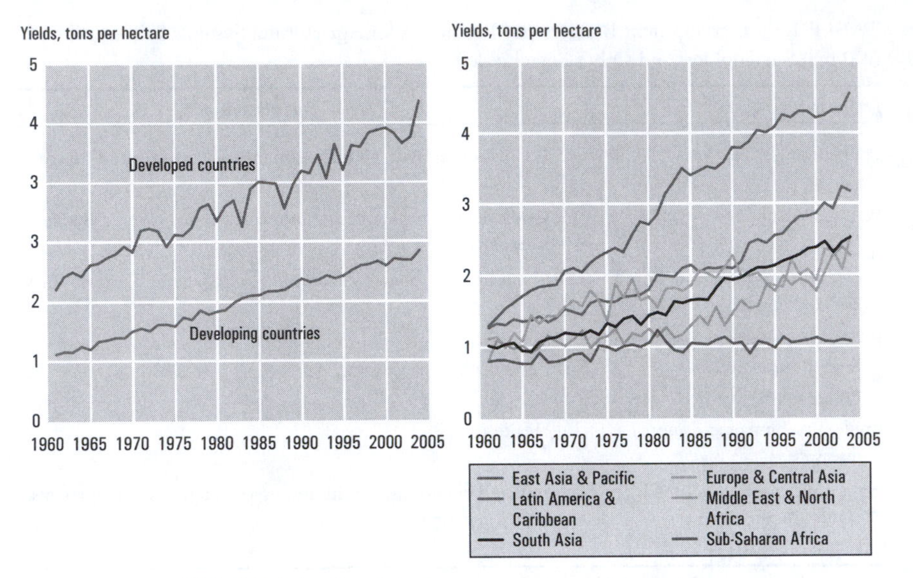

FIGURE 1.1 Cereal yields rose, except in sub-Saharan Africa.

crop plants (Fig. 1.2). Complementary technology innovations have allowed farmers to protect gains in productivity, such as biological control practices to suppress pests and postharvest storage improvements (Fig. 1.3).

The green revolution[2] is an example of widespread and rapid transformation through new varieties and technologies that provided substantial, and often remarkable, increases in the productivity of rice and wheat cropping systems. Productivity gains, however, do not necessarily ensure equitable accrual of benefits. A review of over 300 studies of the green revolution found that over 80% produced unbalanced benefits and increased income inequity associated with the adoption of high-yield potential varieties and production technologies (Freebairn, 1995).

The varieties produced by the green revolution provided a new architectural plant type that could respond to high rates of nutrient inputs with heavier yields in the presence of sufficient water and productive soils. These were widely adopted by farmers on irrigated lands, in some cases displacing indigenous varieties and the biodiversity of landraces. In other locales the new varieties were adopted judiciously, not replacing but supplementing the diversity of varieties grown to provide one more option among the many plant types managed by smallholders.

[2]The green revolution of the 1960s describes the adoption of high-yield potential varieties, primarily in irrigated cropping systems, which doubled agricultural production in many areas.

FIGURE 1.2 Improvement of the indigenous bambara groundnut crop is underway in South Africa, where rapid gains in productivity and quality traits have been achieved.

An example is the development of early maturing rice varieties with a high harvest index. These plant types allocate to grain, with limited stover production, and not necessarily produce the tasty or storable grain, which were still valued by Sierra Leone farmers. Interestingly, the new high-yield potential varieties were integrated

FIGURE 1.3 Biological control is being practiced on a large scale in Thailand, where farmers are supported by innovative field stations and extension educators that demonstrate health-promoting composts and integrated pest management practices.

into both "swamp" rice (informal irrigation) and upland, rain-fed rice production systems in Sierra Leone. These "green revolution" rice varieties supplemented but did not replace long and medium duration varieties, which were moderate in yield potential, but had many other desirable properties. The new varieties allowed smallholder women and men to exploit specific soil types and land forms for rice production and to develop a wider range of intercrop systems of early and late duration rice varieties (Richards, 1986). This illustrates the adaptive and innovative nature of smallholder farming in the face of new technologies and genetic materials.

There are numerous critiques of the green revolution. A common focus is the limited adoption of high-yield potential varieties within agroecologies that have an unreliable water supply or an inadequate market infrastructure. A lack of nuanced understanding of local conditions (which vary widely in time and space and provide limited system-buffering capacity) and misconceptions of farmer priorities are key contributors to failures in some green revolution varieties and input management technologies developed for intensified production in the irrigated tropics that were inadvisably promoted in rain-fed and extensive agricultural systems.

The relevance of agricultural technologies that require substantial investment in labor and external inputs is particularly suspect for extensive agriculture where farmers often prioritize minimal investment. In a variable environment, replanting is not uncommon, so low-cost seeds and minimal labor for seedbed preparation may be a goal often not recognized by agricultural scientists. Optimizing return to small doses of inputs rather than optimizing return overall requires different types of technologies. Stable production that reduces risk is another common goal of farmers, particularly in sub-Saharan Africa, with different criteria for success than simply yield potential. The changeable and low resource environments experienced by smallholder farmers in much of the region require careful attention to technologies with high resilience.

Poor soil fertility, low and variable rainfall, underdeveloped institutions, markets, and infrastructure are realities facing the rain-fed tropics. Typically, farmer knowledge systems have been tested over many years and in a wide range of environments. The fine-tuned modifications that occur over a long period lead to resilient and relevant technologies. Agricultural researchers have only periodically been fully cognizant of this valuable resource—local knowledge systems. Recently, renewed importance has been given to valuing both indigenous and scientific understanding of the world. These knowledge systems need to be integrated rather than be seen as competing. The two worldviews can be complementary, as shown by the example of integrated nutrient management. Here, organic nutrient sources (such as residues and compost) can enhance returns from a judicious use of nutrient inputs from purchased fertilizers and herbicides that reduce crop competition for nutrients (de Jager *et al.*, 2004).

The context for agriculture is changing rapidly and the process of knowledge generation is undergoing transformation as well. Agricultural development has moved beyond a technology transfer model to one that recognizes farmers and rural inhabitants as full partners, central to change efforts. Participatory approaches that are fully cognizant of the necessity for collaborative efforts are being tried around the globe: from participatory action research (PAR) on soil fertility in Uganda (Fig. 1.4) to community watershed improvement efforts in India. Other exciting examples include dairy farmers in The Netherlands participating in research circles, land care groups in Australia, and potato growers in Peru involved in participatory integrated pest management (see these examples and more in Pound *et al.*, 2003).

The long-term goal of sustainable development is to enhance capacity and to promote food security, livelihoods, and resource conservation for all. Tremendous adaptability and understanding are required to manage a biocomplex and rapidly changing world. This is a pressing reality for the more than 3 billion people living in rural areas with limited resources. In these often risky, heterogeneous environments, access to food and income depends on a wealth of detailed knowledge evolved over generations and a capacity to integrate new findings. This book presents a research and development approach that seeks to engage fully with local knowledge producers: primarily smallholder farmers and rural innovators.

Agricultural research has historically often suffered from an oversimplistic view of development and a top-down approach toward rural people. This was one of the major critiques that led to the rise of the farming systems movement in the 1970s. The technologies developed through a reductionist understanding of agricultural problems did not take into account farmers' holistic and systems-based management and livelihood goals (Norman, 1980) (Box 1.1).

FIGURE 1.4 Participatory action research underway with Ugandan farmers interested in soil fertility improvement.

BOX 1.1 Farming System Definition

A farming system is defined as a complex, interrelated matrix of soils, plants, animals, power, labor, capital, and other inputs, controlled—in part—by farming families and influenced to varying degrees by political, economic, institutional, and social factors that operate at many levels (Dixon *et al.*, 2001).

Our goal is to bring farming systems research into the 21st century and to provide a new synthesis incorporating advances in systems analysis, participatory methodologies, and the latest understanding of agroecology and biological processes. This will improve the tools available to researchers and change agents and provide a synthesis of information directly relevant to smallholder farmers and rural inhabitants. Ensuring access to food and increasing that access depend on the broad shoulders and innovative capacity of farmers that tend 1 or 2 ha of land or less. This book sets out theory and practice linking advanced science with local knowledge, with the ultimate goal being to help catalyze rural innovation.

EVOLVING AGRICULTURAL SYSTEMS RESEARCH

Agricultural sciences are seen by some as naturally interdisciplinary: a "quasidiscipline" defined by real-life multidimensional phenomena. As such, a multidisciplinary approach is needed to address them adequately. Different integrations have occurred since the mid-1970s. By the early 1970s, crop ecology had evolved, including disciplines such as physiology, pathology, entomology, genetics, and agronomy. From the mid-1970s to 1980s, farming systems research was prominent, including biophysical and economic components. By 1985, a focus on sustainable production had become dominant. Now, worries about food production and global hunger have been modified by increased public concern about the rapid deterioration of the Earth's ecosystem, especially since the 1992 Rio "Earth Summit."[3] Thus sustainable agricultural management has been redefined as sustainable natural ecosystem management, including disciplines such as geography, meteorology, ecology, hydrology, and sociology (Janssen and Goldsworthy, 1996). These have been combined with new thinking on sustainability and poverty alleviation so

[3]The United Nations Conference on Environment and Development, also known as the Earth Summit, was held in Rio de Janeiro in June of 1992. Over 170 governments participated, and the conference focused world attention on human impacts, emerging environmental issues, and sustainable development. The Climate Change Convention moved forward, along with discussions of indigenous rights and biodiversity.

that international agricultural research centers (IARC) have altered their focus on agricultural productivity and commodity research to a more integrated natural resource management (NRM) perspective (Probst *et al.*, 2003). NRM aims to take into account issues beyond classical agronomy: spatial and temporal interdependency, on-site and off-site effects, trade-offs of different management options, and the need to involve a wider range of stakeholders in joint activities (Probst, 2000).

These evolving approaches are gradually being seen in the work of researchers on the ground, including IARCs, national agricultural research systems, extension services, nongovernmental organizations, development agencies, the private sector, and, in particular, farmers' groups (Box 1.2).

There is increasing recognition recently of farmers' ability to adapt technologies to their own purposes. This was one of the instigating factors in developing farming systems research approaches in the 1980s. Another driving factor in developing

BOX 1.2 Testing "Best-Bet Options" in Mixed Farming Systems in West Africa

The contributions of livestock to natural resource management take place within a complex of biophysical, environment, social, and economic interactions. To better understand and optimize the contribution of livestock, novel approaches have been developed that integrate these multiple aspects and consider the implications from household to regional levels. An example of such an approach is mixed farming systems in West Africa where international institutions—International Institute of Tropical Agriculture (IITA), the International Livestock Research Centre (LRI), and the International Crops Research Institute for the Semi-Arid Tropics (ICRISAT)—have been working together with farmers to increase productivity while maintaining environmental stability through integrated NRM. The process began with prioritization of the most binding constraints that research can respond to (competition for nutrients and the need to increase productivity of crops and livestock without mining the soil). The introduced technologies—the best of everything that research has produced—were presented as "best-bet" options, which were tested by farmers against current practices. The implications and impacts of introducing best-bet options were assessed, taking into account not only grain and fodder yields, but also nutrient cycling, economic/social benefits or disadvantages, and farmers' perceptions. A further step would be to capture environmental implications, such as methane emissions, construction of wells, and availability of fresh water.

Source: Tarawali *et al.*, 2000 (www.inrm.org/Workshop2000/abstract/Tarawali/Tarawali.htm)

farming systems and, more recently, participatory research methodologies has been the perceived lack of relevance and relative failures associated with monocultural, green revolution technologies. Farming in semiarid and subhumid rain-fed production areas and across the vast majority of sub-Saharan Africa has remained at low levels of productivity, often less than 800 kg of grain/ha. There are recent exceptions related to market opportunities or educational and extension reforms, which are explored in depth in this book. In areas where agricultural research and extension systems have remained stuck in a commodity-oriented mode, there have been failures to understand the complex interactions between social and biophysical processes, resulting in impractical agricultural technologies and policies that did not address farmer priorities.

There is a vast literature both for and against the case for focusing on genetic improvement of a few crop species. Many consultative groups for international agricultural research centers and many development projects are still focused primarily around improvements in monocultural, high input, and high return (to land) cropping systems. There are numerous cases where the genetic outputs and technologies have been used by farmers from diverse socioeconomic, gender, and age groups, if they provide adequate returns to their labor and investment and support improvements in their livelihoods.[4]

Addressing and understanding the complexity of goals associated with a whole farming system were the main focuses of the farming systems movement that attempted to improve client orientation and to develop a more multidisciplinary approach to agricultural research and extension (R&E). The farming systems approach shifted R&E from a commodity focus to a holistic approach that included crops, livestock, off-farm income generation, and cultural goals, as well as economic returns of the entire farm.

Farmers continually make complex trade-offs of time and labor with multiple returns from diverse farm and off-farm enterprises that address the whole farming system and livelihoods within a rapidly changing environment. Diagnostics of system complexity and understanding farmer priorities in order to develop relevant technologies and interventions led to farming systems teams that bridged social science and biological science inquiry. Collaborative endeavors among social scientists, biologists, educators, and rural community members have been growing over many years, which has led to and strengthened recognition of the whole farming system and livelihood strategies within which varieties and other technologies are assessed and adopted, discarded, or temporarily adopted.

The value of interdisciplinary inquiry has been heralded by many, but the challenges are tremendous and many whole systems approaches have devolved into a single focus or dispersed efforts over time. Communication across disciplines is a huge challenge, requiring long time frames and a commitment to working together.

[4]Livelihoods here encompass the multiple strategies used to sustain self and family, a concept developed in depth in Chapter 2.

Institutional reward structures that focus on individual achievements and changing donor priorities appear to have marginalized farming systems teams in some organizations and projects. Potential returns from a committed, enduring farming systems approach are seen in the steady enhancement of farmer livelihoods in regions of Brazil where farming systems teams have labored for 2 decades. Here a range of germplasm and technologies has been introduced: a long-season legume (pigeon pea) providing nutrition for poultry while enhancing soil productivity and linked to new maize varieties and an integrated use of poultry manure and fertilizer are components of more sustainable farming systems (Fig. 1.5).

Over time, an ecologically based understanding has informed a farming systems approach to enhance the diagnostic and descriptive aspects of R&E. A rigorous understanding of the biological and physical landscape and processes in the ecosystem can greatly improve the technical insights and knowledge that scientists bring to agricultural development. Instead of empirical trial and error, the crop types and management practices suited to a given agroecology can be predicted more accurately. This will lead to the identification of more promising options that farmers and local extension advisors can then test for performance within a given locale

FIGURE 1.5 Pigeon pea introduced on smallholder farms in Brazil. Note soil fertility-enhancing residues accumulating in the front.

and social context. Agroecology is the science of applying ecological concepts and principles to the design, development, and management of sustainable agricultural systems. The key principle is to manage biological processes, including reestablishing ecological relationships that can occur naturally on the farm, instead of managing through reliance on high doses of external inputs.

Improved understanding of plant interactions with soil microbial and insect communities is contributing to systems-oriented management practices (Chapter 3). Through carefully chosen plant combinations and integration of plants with livestock, an agroecologically informed design can improve the inherent resilience of a farming system. Indigenous practices often rely on agroecological principles such as diversity of plant types and strategic planting of accessory, or helper plants, to reduce pest problems and protect soils. This is shown in the remarkably similar plant combinations used by farmers around the world. For example, in hillside vegetable production systems from Korea to the upper Midwest in North America and the Andes in South America, farmers plant strips of winter cereals (rye in Korea and the United States, barley in Peru) along the contour across slopes where onions, potatoes, and other tubers are grown to confuse pests and prevent erosion while building soil organic matter. In more tropical zones, vetiver grass strips can play a similar role (Fig. 1.6).

DIFFERENT PATHS TAKEN

Farming system characterization and understanding livelihood strategies are at the foundation of agricultural development. It is a challenging process, one that is addressed from different perspectives in the following chapters. Factors to consider

FIGURE 1.6 Vetiver grass planted along bunds for soil conservation in Malawi.

BOX 1.3 Cowpea Variety Development and Farmer Adoption in West Africa

An illuminating example of multiple collaborative endeavors is the intercountry African Cowpea Project (PRONAF) of the International Institute of Tropical Agriculture (IITA) in West Africa. The initial focus of cowpea breeders on determinant, short-statured varieties was not successful, as cowpea is used by many farmers not only for grain and leaf production (e.g., as a vegetable) but also for livestock fodder, products that require some indeterminate, viney traits in cowpea. This adoption story [documented by Inaizumi *et al.* (1999) and Kristjansen *et al.* (2005)] shows how livestock researchers worked with plant breeders and social scientists over a number years, while extensionists, geographers, and agricultural economists were involved in dissemination and evaluation.

Because losses due to pests were evidently a major constraint, IITA established the Ecologically Sustainable Cowpea Protection (PEDUNE) project to find alternatives to the use of toxic pesticides and to promote integrated pest management (IPM) as the standard approach to cowpea pest management in the dry savannah zone. The project identified botanical pesticides such as extract of neem leaf (*Azidirachta indica*), papaya, and *Hyptis*; introduced new aphid- and striga-tolerant cowpea varieties; and promoted the use of solar drying. The program has worked with the West and Central Africa Cowpea Research Network and the Bean/Cowpea Collaborative Research Program. It uses farmer field schools (FFS), a learner-centered approach where farmers' groups conduct field experiments to test and learn about technology options under realistic conditions, improving their crop management decision-making skills in the process. FFS represent an exciting extension–farmer partnership for catalyzing evaluation of new agricultural technologies (Nathaniels, 2005).

include environmental aspects, such as the agroecology and resource base, and socioeconomic aspects, including population density and community goals, levels of technological complexity, and market orientation.

Let us take the crops and animals present on a farming system as an example of the complexity involved. A mixture of crops is grown, including intercropped cereals and legumes, where there is competition for the available land, labor, and capital resources. Where land is a limiting factor, farmers can maximize usage of land through intercropping of legumes and cereals or doubling up of legumes to both increase yields and improve soil fertility. However, to identify the best-bet cereal/legume combinations, researchers must partner with farmers to ensure that their preferences are embedded in the development process (Box 1.3). Research in western and southern Africa (Kitch *et al.*, 1998; Snapp *et al.*, 2002) has demonstrated that food legumes are preferred over nonfood legumes and that most small-scale farmers choose new varieties of legumes primarily for food and cash income security rather than for soil fertility enhancement.

There are exceptions where plant species are adopted primarily for sustaining a farming system. Nonfood legumes play a major role in the Central American humid tropics as weed-suppressing crops in maize-based and plantation systems. Maize is planted into the dense foliage of recently slashed *Mucuna pruriens*, a green manure "slash and mulch" system. This and other promising options for sustainable agriculture are discussed in Chapter 4.

The importance of livestock varies from region to region, and indeed from family to family. Often in dry areas and where rainfall is highly variable, livestock are highly prized and essential to culture and livelihood. Livestock provide a means to concentrate energy and biomass over a large area through grazing and are flexible in the face of periodic or occasional drought. The system of transhumance relies on moving livestock annually to utilize grazing effectively. Chapter 8 discusses in-depth livestock innovations and agricultural development. It is illuminating to consider briefly the role of small ruminants in particular for poverty alleviation. Families that have small ruminants in West Africa were the first to adopt new dual-purpose cowpea. The introduction of a rotational crop of pigeon pea combined with improved, early duration maize varieties and intensified poultry production in Brazil also highlights the role of integrated crop and livestock technologies, where research followed farmer interest in intensified versus extensive production, for different aspects of the farming system.

Researchers have at times prioritized intensification, whether through introducing new crop or livestock varieties that produce more per unit grown or through agricultural input use. We contend here that agricultural system performance and resilience can be enhanced both through extensive and intensive cropping systems but this must be done in consultation with the ultimate end users, the smallholder farmers (see Box 1.4).

Another pressing problem is organic matter depletion under continuous arable cultivation in heavily populated and land-constrained agricultural systems, which has invariably led to decreased land productivity. A great deal of research has been conducted to overcome this problem. Some of the agricultural systems options (discussed in Box 1.5) qualify as "best-bet" natural resource-improving technologies through their potential for adaptability and adoption by the end users.

It is important for agricultural scientists and change agents not to underestimate the substantial biologically based and economic challenges, that act as barriers to farmer adoption of integrated, low-input, and organic matter-based technologies. This is nowhere more evident than in the marginal and risky environments

BOX 1.4 Intensive and Extensive Cropping Systems

Intensification of cropping systems occurs in time and space and includes the following.

1. Intercropping with complementary crop species.
2. Double cropping over time, with two crops a year. One crop may be a soil-building plant species, such as green manure from herbaceous or tree legume species, and the other a nutrient exploitive species that often has high cash value, extracting benefit from the soil-building phase of the speeded up rotation sequence.
3. Intensified plant populations of a monocultural species, often a plant type that has vertically disposed (erect) leaves that can minimize shading while at the same time maximize the interception of photosynthetically active radiation when a very high density of plants is grown in a given space. Although substantial nutrient sources are required, weed control requirements may be minimized, as plant cover is achieved quickly in an ideal situation.

Extensive systems are another pathway and may be pursued if the climate is highly variable, for example, with severely limited rainfall or other critical resources. Livestock are often very important in these environments, and stover may be a primary use, greater than human food value, for many cereal crops.

The tools being used by farmers are not necessarily good indicators of how intensive the management practices are, as, for example, plowing, which may be used in extensive or intensive land use. Plowing can facilitate planting and weeding of an improved fallow or allow a large area to be planted to meet food security requirements, thus reducing pressure to intensify through the use of inputs or related investments.

that many smallholder farmers inhabit. The lack of easy answers has been well documented. Often the areas that are most degraded, such as steep slopes, are those that allow limited plant growth, requiring intensive labor and other investments to overcome a degraded state (Kanyama-Phiri *et al.*, 2000). Emerging technologies, such as improved grain legumes, for example, drought-tolerant cowpea, combine farmer utility as a grain, vegetable, and fodder source, with moderate but consistent soil-improving properties (Fig. 1.7).

Strategic intervention is the key to successful agricultural development programs and is discussed in more depth in relation to different smallholder farming systems throughout this book.

BOX 1.5 Best-Bet Agricultural Systems Options for Improved Soil Fertility

i. Inorganic Fertilizers

Use of nutrients from inorganic sources has the advantage of quick nutrient release and uptake by plants for a consistent yield response. However, the cost of inorganic fertilizers and associated transportation costs has proven to be prohibitive for many limited resource farmers. It has been reported elsewhere (Conway, personal communication) that a nitrogen fertilizer such as urea in Europe costs $70 per metric ton. By the time the fertilizer reaches the coast of Africa, the price will have doubled, to include transport, storage, and handling, and may be much higher if many middlemen are involved in the process of importing the fertilizer and packing it for resale. Eightfold increases in fertilizer costs are not uncommon by the time the fertilizer reaches a farmer located in a central African country, pushing the commodity beyond the reach of most end users. Thus the use of inorganic fertilizers on staple food crops by smallholder farmers requires subsidies, at least in the short term.

ii. Incorporation of Crop Residues and Weeds

Residues from weeds and crop residues have been overlooked at times, as the wide C/N ratio, high lignin content, and low nutrient content generally found in crop residues and weeds limit soil fertility contributions from these organic sources. However, cereal and weed residues build organic matter and improve soil structure for root growth and development. Legume crop residues have higher quality residues and are one of the most economically feasible and consistent sources of nutrients on smallholder farms. Grain legumes such as soybean (*Glycine max* L.), cowpea (*Vigna unguiculata* L.), common bean (*Phaseolus vulgaris* L.), and peanut (*Arachis hypogea* L.) are best-bet options for soil fertility improvement under rotational agricultural systems in sub-Saharan Africa. Countrywide trials in Malawi have documented over a decade that peanuts, soybeans, and pigeon pea consistently and sustainably improve maize yields by 1 ton/ha, from 1.3 ton ha^{-1} (unfertilized continuous maize) to 2.3 ton ha^{-1} (unfertilized maize rotated with a grain legume) (Gilbert *et al.*, 2002; McColl, 1989).

iii. Green Manures from Herbaceous and Shrubby Legumes

A green manure legume is one that is grown specifically for use as an organic manure source. It often maximizes the amount of biologically fixed nitrogen from the *Rhizobium* symbioses that forms nodules in the roots. This fixed nitrogen is available for use by subsequent crops in rotational, relay, or intercropped systems. Green manures also have an added advantage of a narrow

BOX 1.5 (*Continued*)

C/N ratio, which facilitates residue decomposition and release of N to subsequent crops. In southern and eastern Africa, best-bet herbaceous and shrubby legume options for incorporation as green manures have been widely tested. These include *Mucuna pruriens*, sun hemp (*Clotalaria juncea*), *Lab lab* (*purpreus*), pigeon pea intercropped with groundnut, and relay systems with *Tephrosia vogelli* (see Table 4.3 in Chapter 4). Residue management and plant intercrop arrangement are important to consider, along with the species used for a green manure system. Sakala *et al.* (2004) reported higher maize grain yields from early compared to late incorporated green manure from *Mucuna pruriens*. Similarly, on smallholder farmers in the island of Java in Indonesia, threefold increases in maize yields have been reported following incorporation of a 3-month-old stand of mucuna or sun hemp.

IMPACT AT LOCAL AND REGIONAL SCALES

Participatory approaches are being experimented with widely as a means of supporting the generation of local adaptive knowledge and innovation. PAR can have impact at broader scales as well through improving research relevance. This has not been the explicit goal of many PAR projects, but if causal analysis and iterative learning are explicitly included, then research findings can have wider application. For example, participatory, on-farm research on nutrient budgeting has been shown to

FIGURE 1.7 The crop legume "cowpea" (*Vigna unguiculata* L.) is a productive source of high-quality organic matter and multiuse products widely adapted to the semiarid and arid tropics.

be an effective means of improving farmer knowledge of nutrient cycling; however, it has the potential to provide valuable research insights as well. This was shown in Mali, West Africa where participatory nutrient mapping was undertaken to support villagers learning about nutrient loss pathways and integrated nutrient management practices (Defoer *et al.*, 1998). At the same time, Defoer and colleagues (1998) gained knowledge about farm and village level nutrient flows. Some of the information generated will be locally specific, as nutrient losses are conditioned largely by site-specific environmental factors, yet we contend that knowledge generated locally can often be used to improve research priorities and to inform policy.

One of the goals of this book is to support broader learning from the PAR process. Agricultural researchers are charged with a dual mandate: to provide local technical assistance that supports farmer innovation at specific sites, while simultaneously generating knowledge of broader relevance. To work at different scales and meet these dual objectives, careful attention must be paid to choosing sites that are as representative as possible of larger regions. Thus local lessons learned can be synthesized, and disseminated, over time.

Examples are developed in this book of how to "take to scale" participatory natural resource management, crop, and livestock improvement (see Chapter 11). Promising strategies for large-scale impact will vary, depending on objectives. Successful extension examples include FFS and education/communication campaigns that address an information gap and engage rather than preach. Education requires documentation of current knowledge and farmer practice to identify missing information and promote farmers testing science-based recommendations for themselves. This focus on knowledge generation contrasts with promoting proscribed recommendations and is illustrated by a radio-integrated pest management campaign in Vietnam that challenged farmers to test targeted pesticide use for themselves. The campaign resulted in large-scale experimentation among rice farmers and province-wide reductions in pesticide use (Snapp and Heong, 2003). Another innovative example is from Indonesia where participatory research on sweet potato-integrated crop management (ICM) was scaled up through FFS. A unique aspect of this project was that FFS education materials were developed through joint farmer–researcher learning about sweet potato ICM over a number of years. Only after this participatory development of training materials was FFS initiated to communicate with farmers on a range of ICM principles and practices (Van de Fliert, 1998).

Participatory and adaptive research approaches have evolved out of a desire for the most effective, informed farming systems approach possible. Participation helps bridge gaps and enhances communication among researchers, extension advisors, and rural stakeholders. It recognizes the importance of scientific input from both biophysical and socioeconomic enquiry, while at the same time valuing indigenous local knowledge. By so doing it provides a basis for increased understanding and iterative technology development in partnership with stakeholders, especially smallholder farmers.

Agricultural systems science requires attention to synthesis, reflection, and learning cycles. These are key ingredients in maintaining quality and rigor in an applied science that must engage with the complexity of real-world agriculture. Synthesis techniques are emerging that help address these challenges, including statistical multivariate techniques, meta-analysis, and geo-spatial techniques. These are important methods that can help researchers and educators derive knowledge from local experience and to understand underlying principles of change. Elucidating drivers or regulators of change and building in iterative reflective steps are important components of agricultural systems research. Chapter 10 discusses in more depth agricultural systems innovation and approaches to catalyzing change in sustainable directions.

Institutional reform and engagement with policy is an area that agricultural systems science is beginning to move toward, as discussed in Chapter 12. Farming systems may not have sufficiently addressed institutional change, and this newly reborn farming systems movement—called *agricultural systems*—is not only working with farmers and R&E, but also is now taking on and transforming institutions and policy in every part of society.

LOCAL INSTITUTIONS FOR AGRICULTURAL INNOVATION

Linkages between researchers and local users can vary greatly depending on how "ownership" of the research process is distributed between the two. In recent years, most research projects have sought local people's participation, but objectives of such participation are diverse, ranging from legitimizing outsiders' work and making use of local knowledge, to building local capacity for innovation, development, and transformation. This last objective is essential to increase the capacity of marginal groups to articulate and negotiate for their own interests and improve their status and self-esteem (Probst et al., 2003). The type of participation that evolves will define the research process and roles of researchers and farmers in all areas, including planning, monitoring, and evaluating the learning (Box 1.6).

Innovation development may be based on formal research, informal farmer experimentation, or a combination of the two. "Hard" and "soft" systems approaches can be identified (Bawden, 1995). Hard systems approaches attempt to understand entire systems (e.g., farms, groups of farms, or communities) by looking at them from the outside and assuming that system variables are measurable and that relationships between cause and effect are consistent and discoverable by empirical, analytical, and experimental methods. This is the approach taken by farming systems research. Soft systems proponents argue that systems are creations of the mind or theoretical constructs to understand and make sense of the world. Thus, soft systems methods aim at generating knowledge about processes within systems by stimulating self-reflection, discourse, and learning (Hamilton, 1995).

BOX 1.6 Types of Participation in the Agricultural Innovation Process

- *Contractual participation*: the researcher has control over most of the decisions in the research process; the farmer is "contracted" to provide services and support.
- *Consultative participation*: most decisions are made by the researcher, with emphasis on consultation and gathering information from local users.
- *Collaborative participation*: researchers and local users collaborate on an equal footing through exchange of knowledge, different contributions, and sharing of decision-making power throughout the agricultural innovation process.
- *Collegiate participation*: researchers and farmers work together as colleagues or partners. "Ownership" and responsibility are distributed equally among the partners and decisions are made by agreement or consensus among all parties.

Source: Based on Biggs (1989)

We argue that both research methods are needed: "soft" participatory action research on processes of NRM (e.g., organization, collective management of natural resources, competence development, conflict management) and conventional "hard" research that focuses on technological and social issues (e.g., soil conservation, nutrient cycling, agronomic practices, socioeconomic aspects). Successful attainment of goals of increased production and a sustainable environment depends on the meaningful integration of the two (Probst *et al.*, 2003).

Researchers have adopted a range of approaches to agricultural innovation to date, including the following.

1. *Transfer of technology* model—with technology being developed by researchers at the "center," adapted by local researchers, and transferred by extensionists to farmers.
2. *Farmer first* model—where farmers participate in the generation, testing, and evaluation of sustainable agricultural technologies, often based on their own local practices, with researchers documenting rural people's knowledge, providing technology options, and managing the research.
3. *Participatory learning and action research* (PLA) developed through critical reflection and experiential learning through the process of addressing local development challenges. Researchers help different groups develop their knowledge and capacities for self-development, as facilitators, catalysts, and providers of methodological support and opportunities (Roling, 1996).

BOX 1.7 CIFOR's Project on Adaptive Collaborative Management (ACM)

Improving the ability of forest stakeholders to adapt their systems of management and organization to respond more effectively to dynamic complexity is an urgent task in many forest areas. The ACM project addresses a number of research questions.

1. Can collaboration among stakeholders in forest management, enhanced by social learning, lead to both improved human well-being and maintenance of forest cover and diversity?
2. What approaches, centered on social learning and collective action, can be used to encourage sustainable use and management of forest resources?
3. In what ways do ACM processes and outcomes impact on social, economic, political, and ecological functioning?

The project collaborates with many institutions involved in research, implementation, and facilitation of change across a number of countries. Researchers see themselves as part of the system rather than neutral. Because there is no single "objective," the static viewpoint from which forest management dynamics can be observed, forest managers and users are all actively and meaningfully involved in the research. Research outputs are targeted toward different users at local, national, and global levels, including manuals on methods and approaches, toolbox for development practioners, policy briefs, research papers, and software such as simulation models.

Source: www.cifor.cgiar.org

Each approach has its strengths and weaknesses and may be used together to complement each other in different situations. Currently, most research falls under the "transfer of technology" and "farmer first" approaches. The longer term PLA approach requires a reorientation of skills, management, and financing modes (without predefined quantifiable targets) and is only just beginning to be considered by researchers. One example of researchers attempting to put PLA research into practice is the Center for International Forest Research (CIFOR)'s project on adaptive collaborative management (Box 1.7).

Exciting recent developments show the potential of decentralized and client-driven research and extension to be highly relevant to smallholders. Experiments in realigning research and extension are currently underway in a number of developing countries, such as Bolivia and Uganda. The political context is promoting rapid change in agricultural research and extension, with almost universal disinvestment in government agricultural services providing incentives for more client-driven extension and private sector partnerships. In some countries, such as Bolivia, the public sector investment in agricultural research and extension has shrunk to almost nil. Although there are concerns about meeting the needs of the poorest clients under these circumstances, promising shifts toward more responsive and integrated extension have arisen in Bolivia and Uganda. These countries show that it is possible for extension and agricultural advice services to reduce the level of commodity focus and focus more attention on science in the service of client partnerships, catalyzing rural innovation while giving consideration to the whole farming system.

CATALYZING DIRECTIONS OF CHANGE

The rural environment is undergoing rapid change. Globalization, climate change, and epidemics are some of the forces impinging on farming systems. Farmers are developing innovative responses and can be supported to intensify or, in some cases, extend in directions that are sustainable, resilient, and equitable.

The case for working with smallholder farmers as a key engine of development and production is now well known. Not in all cases, but in many situations the ability of smallholders to be highly productive on a per acre basis is outstanding. Farmers will produce given access to resources and incentives, even if their land holding size is small. Examples include China and Russia (and numerous others) who have fed their billions not from the collective farmer movement that built up large farms, but from individual smallholder efforts, where many farmers built very productive small plots (intensive gardens) that fed most of the population. Indigenous knowledge (IK) of smallholders is unsurpassed—they know their resources such as soils and priorities better than anyone!

Researchers have become interested in linking with what they see as a vast untapped resource and many have initiated participatory research projects to try to extract and replicate this knowledge. This can increase the efficiency and effectiveness of development programs, as IK is owned and managed by local people, including the poor. However, the danger is that in joining researcher-driven activities, farmers may abandon some of the very practices that have been built up and continually extend and modify their local knowledge base. A few researchers have taken an alternative "empowering" approach to participatory research, seeking to enhance farmers' capacities to experiment and extend their knowledge. This requires scientists to strive to understand the process of local knowledge generation and then to support and complement it (see Box 1.8).

The goals and interests of smallholders vary widely, but a starting point is identifying where change is occurring and where interest in intensification is high. The challenge is to bring researchers and smallholder farmers together in a productive partnership, based on respect for each others' knowledge, skills, experience, and situation.

A recent development in systems thinking has seen adaptation of the innovation systems approach from industry to agriculture. The agricultural innovation systems approach is based on dynamic multistakeholder partnerships. These can comprise

BOX 1.8 Research Approaches to Indigenous Knowledge

Indigenous knowledge is the basis for local level decision making in food security, human and animal health, education, natural resource management, and other vital economic and social activities. Agricultural and social scientists have been aware of the existence of indigenous knowledge since colonial times, but from the early 1980s, understanding of farmers' practices as rational and valid has rapidly gained ground. Two contrasting interpretations of IK are as follows.

1. Local knowledge is a huge, largely untapped, resource that can be removed from its context and applied and replicated in different places (such as formal science). Proponents of this perspective have scientifically validated IK or sought similarities and complementarities between their knowledge and farmers' knowledge. Farming systems approaches and participatory research and development largely follow this thinking.
2. Indigenous knowledge is based on empirical experience and is embedded in both biophysical and social contexts and cannot easily be removed from them. It follows that the process by which IK is created is as important as the products of this research.

Many participatory research projects have superimposed a Western scientific method of inquiry over local innovators' procedures without first assessing local knowledge and understanding the processes that generate it. This can result in innovative farmers "playing along" or participating but not internalizing or adopting research/extension messages; abandoning their practices and following those brought by outside agents who they see as more knowledgeable (and powerful) than they are; or adopting and adapting elements of Western scientific research modes.

Some participatory research projects, particularly those that aim to empower local people, attempt to support indigenous research as a parallel and complementary system to formal agricultural research. The approach is to enhance farmer capacities to experiment through training in basic scientific and organizational principles. Skills include problem solving and analytical and communication capacity building, particularly in adaptive experimentation and technology dissemination.

Tangible results and impacts on the lives of the poor have been achieved under each approach. However, achieving a lasting impact requires stimulating processes for innovation that are already present in rural communities. Case studies of participatory research empowering IK are given at the accompanying website.

Source: Saad (2002)

farmers, input suppliers, traders, and service providers (researchers and extension staff) who work *together* toward a common objective, such as producing cut flowers for export or increasing the efficiency of cassava production, processing, and distribution for the domestic market. This idea is gaining ground quickly due to sponsorship from some major donors and is discussed further in Chapter 11.

ROAD MAP

In this chapter we have laid out some of the challenges faced by farmers and rural communities, and the dynamic, complex nature of equitable and sustainable development. Chapter 2 presents ways to address the complexity of farmer livelihoods, building on farming systems research, livelihoods analysis experience. Approaches and tools are presented that educators, extension staff, researchers and change agents can use to work with smallholder farmers around the globe. Underlying biological principles to support rural innovation is the focus of Chapter 3, where agroecology theory and practical implications are summarized. Chapter 4 explores design principles for long-term sustainability, and applications to developing resilient, efficient systems.

The next four chapters explore the resources that support rural livelihoods, namely: soil productivity, plant and animal genotypes. Chapter 5 explores lessons from adoption of low input agriculture technology in developing country contexts, summarizing challenges and opportunities for agricultural change at scale. Chapter 6 presents agroecological approaches to nutrient management, including theoretical and practical considerations to improving nutrient efficiency, and enhancing productivity.

Chapters 7 and 8 focus on participatory plant breeding efforts and livestock improvement, including exciting examples of innovation, and genotype improvement that follows farmer priorities. Theory and practice is presented for client-oriented development of technologies, within the complexity of farming systems.

Gender and equitable development is the focus of chapter 9, where the complexity of social relations and access to resources is addressed at the household and community level. The theory of innovation, and approaches to catalyze rural innovation are addressed in chapter 10. Chapter 11 presents new models in agricultural outreach, highlighting extension that is client-oriented, and demand-driven knowledge generation and dissemination. Challenges of the 21st century faced by farmers and rural stakeholders are revisited in Chapter 12, which presents promising pathways and integration across local and global efforts in agricultural systems development.

Beyond these 12 chapters, readers are invited to explore the Web site that accompanies this book, which includes a wide set of case studies and PowerPoint presentations. These are intended as technical references and cutting-edge information to inform on the development of training materials and learning modules. Our intention in the book and Website is to provide powerful examples of research and extension programs that have achieved impact in developing countries context, to inspire and improve understanding of agricultural change.

REFERENCES AND RESOURCES

Bawden, R. (1995). On the systems dimension in FSR. *J. Farm. Systems Res. Extens.* 5(2), 1–18.

Biggs, S. (1989). Resource-poor farmer participation in research: A synthesis of experiences from nine National Agricultural Research Systems. OFCOR comparative study paper. ISNAR, The Hague. .

Defoer, T., De Groote, H., Hilhorst, T., Kante, S., and Budelman, A. (1998). Farmer participatory action research and quantitative analysis: A fruitful marriage? *Agric. Ecosyst. Environ.* **71**, 215–228.

de Jager, A., Onduru, D., and Walaga, C. (2004). Facilitated learning in soil fertility management: Assessing potentials of low-external input technologies in east African farming systems. *Agric. Syst.* **79**, 205–223.

Dixon, J., Gulliver, A., and Gibbon, D. (eds.) (2001). "Farming Systems and Poverty." Food and Agriculture Organization, Rome.

Freebairn, D. K. (1995). Did the green revolution concentrate incomes? A quantitative study of research reports. *World Dev.* **23**, 265–279.

Gilbert, R. A., Komwa, M. K., Benson, T. D., and Sakala, W. D. (2002). "A Comparison of Best-Bet Soil Fertility Technologies for Maize Grown by Malawian Smallholders." A research report on the nationwide on-farm cropping system verification trial by Action Group 1, Maize Productivity Task Force, Malawi Dept of Agriculture and Irrigation, Lilongwe, Malawi: .

Inaizumi, H., Singh, B. B., Sanginga, P. C., Manyong, V. M., Adesina, A. A., and Tarawali, S. (1999). "Adoption and Impact of Dry-Season Dual-Purpose Cowpea in the Semi-Arid Zone of Nigeria." International Institute of Tropical Agriculture (IITA), Ibadan, Nigeria.

Janssen, W., and Goldsworthy, P. (1996). Multidisciplinary research for natural resource management: Conceptual and practical implications. *Agric. Syst.* **51**, 259–279.

Kanyama-Phiri, G. Y., Snapp, S. S., Kamanga, B., and Wellard, K. (2000). "Towards Integrated Soil Fertility Management in Malawi: Incorporating Participatory Approaches in Agricultural Research." Managing Africa's Soils No. 11. IIED, UK. Web site: www.iied.org/drylands

Kristjanson, P., Okike, I., Tarawali, S., Singh, B. B., and Manyong, V. M. (2005). Farmers' perceptions of benefits and factors affecting the adoption of improved dual-purpose cowpea in the dry savannahs of Nigeria. *Agric. Econ.* **32**, 195–210.

MacColl, D. (1989). Studies on maize (*Zea mays* L.) at Bunda College, Malawi. II. Yield in short duration with legumes. *Exp. Agric.* **25**, 367–374.

Nathaniels, N. Q. R. (2005). "Cowpea, Farmer Field Schools and Farmer-to-Farmer Extension: A Benin Case Study." Agricultural Research and Extension Network Paper 148, July 2005. Overseas Development Institute, London.

Norman, D. W. (1980). "The Farming Systems Approach: Relevancy for the Small Farmer." Michigan State University, East Lansing, Michigan.

Pound, B., Snapp, S. S., McDougall, C., and Braun, A. (eds.) (2003). "Uniting Science and Participation: Managing Natural Resources for Sustainable Livelihoods." Earthscan, London.

Probst, K. (2000). Success factors in natural resource management research. Dissection of a complex discourse. *In* "Assessing the Impact of Participatory Research and Gender Analysis" (N. Lilja, J. Ashby, and L. Sperling, eds.). CGIAR Program on Participatory Research and Gender Analysis. CIAT, Cali, Colombia.

Probst, K., Hagmann, J., Fernandez, M., and Ashby, J. (2003). "Understanding Participatory Research in the Context of Natural Resource Management: Paradigms, Approaches and Typologies." Agricultural Research and Extension Network Paper 130, July 2003. Overseas Development Institute, London.

Richards, P. (1986). "Coping with Hunger: Hazard and Experiment in an African Rice Farming System." Allen and Unwin, London.

Roling, N. (1996). Towards an interactive agricultural science. *Journal of Agricultural Education and Extension* **2**, 35–48.

Saad, N. (2002). "Farmer Processes of Experimentation and Innovation: A Review of the Literature." Participatory Research and Gender Analysis Program, CGIAR.

Sakala, W. D., Kumwenda, J. D. T., and Saka, R. R. (2004). The potential of green manures to increase soil fertility and maize yields in Malawi. *In* "Maize Agronomy Research 2000–2003," pp. 14–20. Ministry of Agriculture, Irrigation and Food Security.

Snapp, S. S., and Heong, K. L (2003). Scaling up: Participatory research and extension to reach more farmers. *In* "Uniting Science and Participation: Managing Natural Resources for Sustainable Livelihoods" (B. Pound, S. S. Snapp, C. McDougal, and A. Braun eds.), pp. 67–87. Earthscan, UK and IRDC, Canada.

Van de Fliert, E. (1998). Integrated pest management: Springboard to sustainable agriculture. *In* "Critical Issues in Insect Pest Management" (G. S. Dhaliwal and E. A. Herinrichs eds.), pp. 250–266. Commonwealth Publishers, New Delhi, India.

INTERNET RESOURCES

The International Association of Agricultural Information Specialists has a Web site that provides support for searching different databases on agricultural knowledge and a blog on recent agricultural information-related topics: http://www.iaald.org/index.php?page=infofinder.php

Agricultural knowledge links are available at the Food and Agriculture Organization FAO "Best Practices" Web site: http://www.fao.org/bestpractices/index_en.htm

Knowledge management for development e-journal often has articles of interest to agricultural information specialists: http://www.km4dev.org/journal/index.php/km4dj/issue/view/4

CHAPTER 2

Livelihoods and Rural Innovation

Barry Pound

Summary

This chapter recognizes that farmers face tough, complex decisions every day. They need a framework for deciding what to produce, how to use the products, how to complement farming with other activities, and what livelihood strategy to adopt for the medium to long term. These decisions go beyond technical agriculture and into social, economic, and policy considerations.

We also need a framework so that we can provide balanced advice, develop relevant research programs, and formulate appropriate policies. The sustainable livelihoods framework, combined with the use of carefully selected participatory rural appraisal (PRA) tools, provides a powerful method to analyze the situation of a farming family or community to furnish information needed for the informed design of interventions. Examples from Afghanistan, Tanzania, and elsewhere are used to illustrate the use of this approach.

The results of such in-depth rural analysis reveal the breadth and interconnectivity of the social, economic, technical, and policy issues facing rural people. Conventional linear, separate research and extension systems are not designed to address this complexity. The concept of multistakeholder partnerships within an innovation systems approach is described as a promising alternative, with a number of case studies to convey the diversity of situations to which this approach can be applied.

FARMING-RELATED LIVELIHOODS

Farmers are systems thinkers. They have to decide which crops (if any) to grow and where and how. They have to decide which livestock to keep (if any) and where and how. They have to decide what to sell (when and where) and what to keep for the household. They must balance investment in protecting local natural resources for future generations, immediate requirements to feed and shelter their family, and long-term investments in education, or off-farm employment. Overall, they need to think of today's priorities and tomorrow's sustainability.

They have to decide how to balance their farming activities with opportunities for off-farm and nonfarm income and how much time to devote to social interaction with their neighbors, friends, and relatives. They have to predict the outcome of playing higher levels of productivity against higher levels of risk. Their lives may depend on getting the balance right (Fig. 2.1).

Farmer decisions are based on experience, natural indicators, the information they obtain from other farmers, the radio, the shopkeeper, extension staff, and other service delivery agencies such as nongovernmental organizations (NGOs), private sector vets, and input suppliers. They then make their best guess with the limited information available to them. In particular, it is difficult to predict the future—the climate, market prices, consumer preferences, policy changes, and the security situation. The majority of smallholder farmers live in marginal environments with minimal infrastructure, which sharpens the degree of uncertainty (Fig. 2.2).

The farmer's life and the way in which he responds to his situation also depend on factors outside agricultural science, such as his health and that of his family, his education, and the linkages he has with others in and beyond his community. Also, of course, our farmer may not be "he" at all, but "she," with the additional or different roles, responsibilities, and rights that this implies (Figs. 2.3 and 2.4).

As researchers and agricultural advisers, we have to admire the farmer for being able to think her way through this complexity. We also need a framework that will help *us* think in a logical way about the different factors that influence the decisions that farmers, and those that support the farmer, have to make.

One such framework is the sustainable livelihoods framework (Carney, 1999), which is presented in Fig. 2.5 as a farming-related livelihoods diagram. This people-centered framework starts with the premise that all farmers have some assets, which can be divided among natural assets, physical assets, social assets, human

FIGURE 2.1 Put yourself in his shoes. As an immigrant from the food-deficit high plains of Bolivia, you have traveled to the very different environment of the tropical lowland rainforests and along logging tracks to the end of the road. You now have to make a living from the forest using a machete, a box of matches, and your wits.

FIGURE 2.2 Erosion gulleys and compaction caused by livestock grazing and passage, Nargas valley, Afghanistan. What are the alternatives, apart from opium poppy?

FIGURE 2.3 Women farmers in Uganda sharing written information. Literate members may share information with less literate friends and family.

FIGURE 2.4 A woman farmer in Nepal explains her success in cauliflower production and marketing. Extension services are still predominantly male, and few take the trouble to find out the specific priorities and concerns of women.

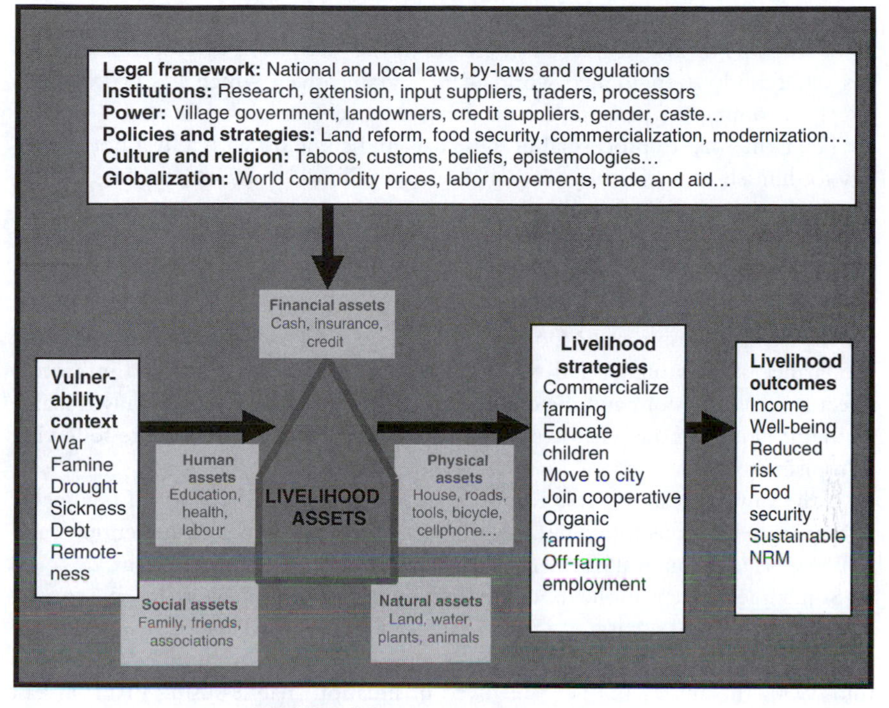

FIGURE 2.5 Farming-related livelihoods.

assets, and financial assets. Some assets might be relatively high and others relatively low at any point in time. The level of assets is dynamic, and a rise in one can mean a fall in another. For instance, one could buy land, thereby trading financial assets for natural assets. In slash and burn agriculture, one might sacrifice trees and shrubs (a natural asset) through burning to provide short-term fertility (a different type of natural asset) to grow a cash crop to provide finance (financial asset) for school fees (education being a human asset). The assets are thus interwoven and, to some degree, interchangeable.

The farming world is one of risk, and Figure 2.5 shows how assets can be affected severely and quickly by war, famine, drought, or falling into debt and how being in a remote place can add to the vulnerability of families and communities through poor access to health services, food aid, or markets.

The ways in which assets are used can be affected by a number of external influences, such as the legal framework (which can include local bylaws[1]), government, NGO, and private institutions that affect the whole value chain from field to consumer, the balance of power within the community and beyond, national policies (and how these are interpreted and implemented locally[2]), and the culture and values of

[1]An example would be "social fencing" in India, where farmers are subject to community fines if their livestock enter forbidden areas.

[2]Government policy on pastoralism in Tanzania is well balanced and appears to favor its continuance, but in practice pastoralists are being squeezed out by local action.

the community. All of us, even in remote parts of the world, are also touched by globalization. For instance, recent increases in the cultivation of crops for biofuels has significantly strengthened world-market grain prices, which has had a ripple effect on commodity prices and behavior right across the globe.

The farmer has to process all of these data and come up with a livelihood strategy for himself and his family that will lead to livelihood outcomes that are beneficial in the short, medium, and longer terms.

It is our responsibility as researchers, advisers, and policy makers to understand the complexity facing farming families. The livelihoods framework helps us to do so by providing a check list of issues. Do we understand the vulnerabilities and shocks facing a community? Do we know the level of each of the five assets that the farmer is operating at? Do we know how external institutions and influences affect farming decisions and the capacity of farmers to take up the options available to them? Do we know what livelihood strategies rural communities are following or aspiring to?

If the answer to all of these is "yes," then we have a good chance of being able to develop relevant technologies, advisory services, development interventions, and policies—always mindful of the fact that situations can change overnight.[3]

Sometimes there is sufficient up-to-date information available to be able to analyze local livelihood situations without the need for fieldwork, but that is rare. More commonly there is little recent, comprehensive information available. Since the mid-1990s, the author has been involved in situation analyses using PRA[4] in east African countries and livelihood analyses in Moldova, Afghanistan, and Yemen. All of these have involved the use of a mixture of PRA-type tools, carefully selected for each individual case. In Afghanistan (Pound, 2004), the objective was to understand the present livelihood situation in villages in Bamyan Province (which had been severely affected by war with the Soviet Union, by conflict with the Taliban, and by drought) so that a large Food and Agriculture Organization (FAO) project could develop relevant farming-related development interventions. Box 2.1 describes the sequence followed and the methods used.

In Moldova (eastern Europe), a similar method for livelihood analysis was used (Pound, 2001), but instead of three wealth rank groups of men and women, 13 social groups were interviewed as these represented the different social constituencies in the communities. It is important to be flexible and inventive in the use of methods so that they achieve the objectives set and conform to the cultural and logistical situation of the study.

[3]For example, the collapse of the Soviet Union also brought about the collapse of the banking system in ex-Soviet states and the overnight loss of the life savings of millions of rural people. This meant that even though land again became a privately owned asset, few had cash to invest in its development.

[4]PRA as a methodology includes a wide range of tools for working in a participatory way with communities to explore their general situation or to deepen knowledge about specific topics.

BOX 2.1 Methods for Sustainable Livelihoods Analysis Applied to the Project "The Development of Sustainable Agricultural Livelihoods in the Eastern Hazarajat, Afghanistan" (FAO Project: GCP/AFG/029/UK)

Objectives

a. Characterize current livelihood systems, including vulnerabilities and constraints
b. Identify development drivers and intervention opportunities
c. Establish a baseline for monitoring the project

Approach

Sustainable livelihoods analysis using PRA tools involving men and women from different wealth groups to gain both qualitative and quantitative information.

Sequence

1. Review of literature
2. Meetings with a range of institutions (government departments, INGOs, local NGOs, UN agencies, research organizations, and donors)
3. Training of four teams of men and women to carry out the field surveys
4. Field surveys of 23 villages over 3 weeks (270 man days)
5. Analysis, documentation, and presentation of results
6. Use of results in developing a program of implementation initiatives

Methods used for field survey

1. *Introductions:* It is always important that village leaders are clear who you are and what you want to do.

2. *Village profile:* This is a PRA technique that is not in the literature. Starting with a clean sheet of flip chart, a circle is drawn to represent the village. A group of villagers describe the buildings, services, trades, and social institutions in the village. A second circle is then drawn and villagers are asked to describe the external influences on their lives (government, links with other villages, market linkages, NGO interventions, etc.), and the vulnerabilities they face.

(*Continues*)

BOX 2.1 *(Continued)*

Village profile developed with fishermen in southern Yemen. Such exercises can be the start of local dialogue, not just a tool to extract information. It is important that those communities that contribute to surveys are also involved in subsequent development initiatives.

3. *Quantitative information on resource distribution and the labor situation:* Village members provide information on the number of households in the village owning large livestock within set numerical ranges (see pie diagram). Similar resource profiles are obtained for land ownership, migration, and labor use. The information helps in understanding the way in which natural resources are distributed among families in the village. The reasons for a skewed distribution can be questioned.

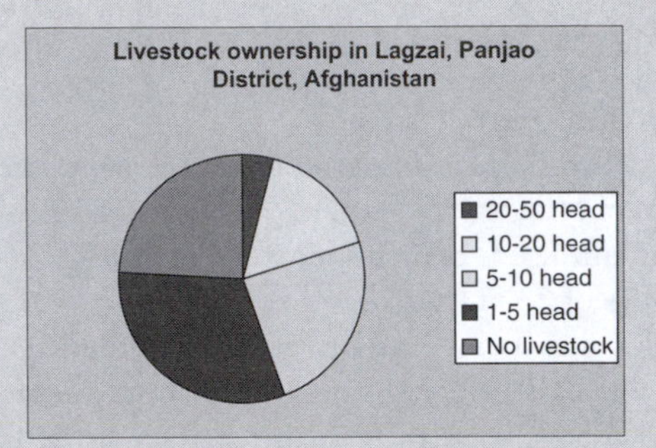

4. *Wealth ranking:* At this point, wealth ranking is carried out to differentiate among the poor, medium, and better-off families in the village. Two or three respected members of the village are selected to write the names of all household heads on separate cards and are then asked to allocate these to three piles, corresponding to poor, medium, and better-off categories.

BOX 2.1 (*Continued*)

Conducting wealth ranking with villagers is a serious business. After separating households into wealth groups, the participants are asked what their criteria for allocation were. What does well-off mean? What makes a family poor?

1. *Activity profiles with women:* Semistructured interviews are then carried out (by women team members) with groups of women from each of the three wealth rank categories to determine the daily and annual activities of women (domestic, farming, nonfarming) by wealth group to identify the main constraints that women of different wealth groups face and to identify their suggestions for improving their livelihoods.
2. *Semistructured interviews with men:* Groups of men from the three wealth categories are also interviewed (separately) to determine household income sources and the debt situation for each wealth category, to identify the main constraints faced by men, and to identify their suggestions for ways to improve their livelihoods.
3. *Farming system diagrams* are developed with a few representative farmers selected by the village. These look at their labor situation, their land (including common property and share cropping), crops and livestock produced, interactions between crops, livestock and common land, sale and household use of farming products, off-farm and nonfarm employment, farming constraints and opportunities, and the sources of farmer knowledge and services (see Fig. 2.6).

Village members in Afghanistan are constructing a farming systems diagram. Using a vantage point like this helps, as one can point to the mountains and ask: Is there seasonal transhumance to the mountains? What about medicinal herbs from the high valleys?

Use of the information

Results were analyzed and development drivers (including nonfarming drivers such as the need for literacy training, nonfarm employment, and microcredit) identified. These were used to develop a set of practical interventions that were then applied successfully.

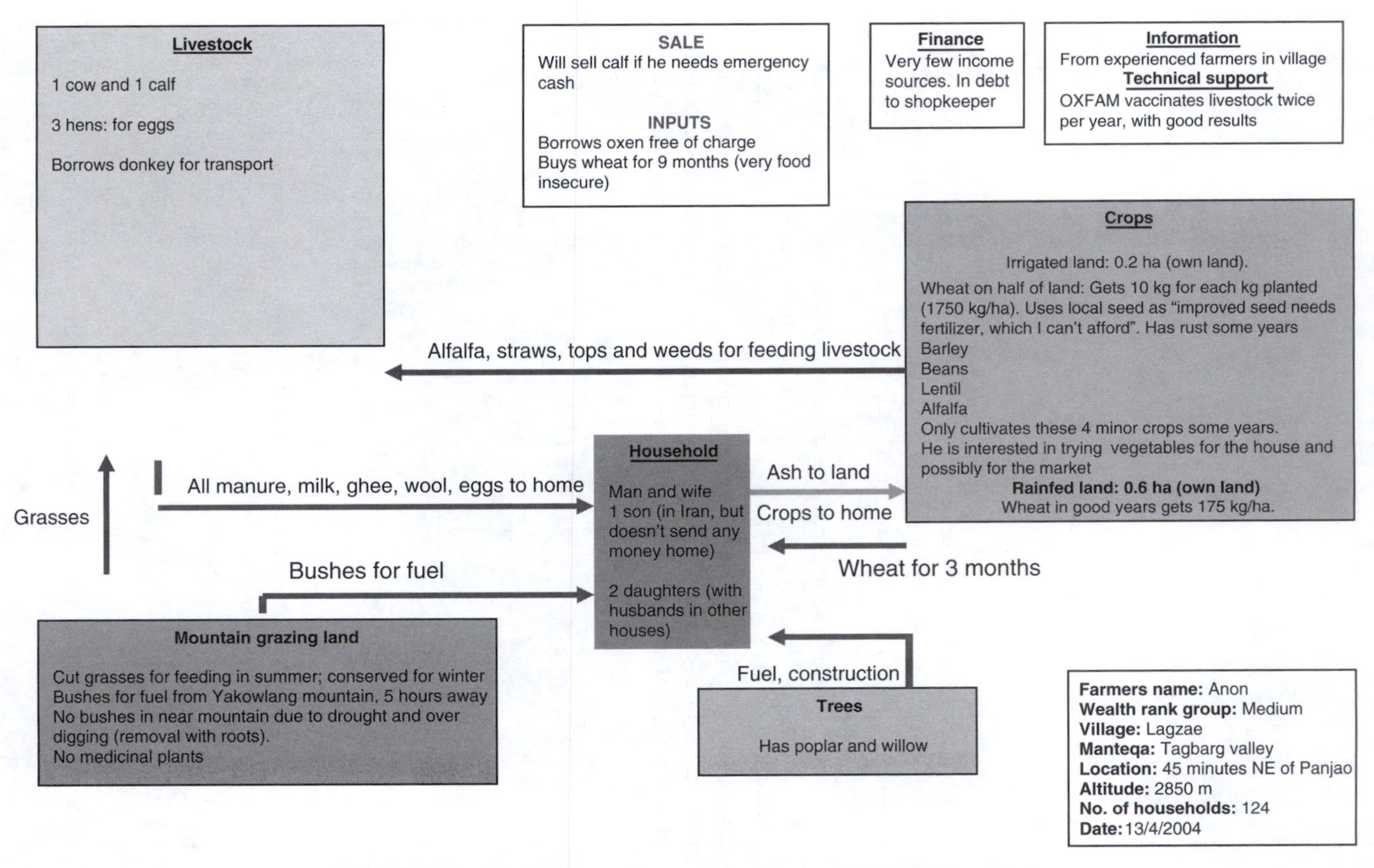

FIGURE 2.6 Farming systems diagram: Lagzae (Panjao District); 2850 m altitude.

Some of the tools described (e.g., the farming systems diagram) take several hours of careful questioning to complete. However, they are worth it as they provide very specific information that can tell a more precise story than a lot of aggregated data. They can pinpoint vulnerabilities and gaps in knowledge, services, or resources that can be addressed usefully by development agencies. Many of the tools described are very visual and are developed with the community members so that they can see exactly what is being produced. It is then easy to "question the diagram" and probe for further in-depth information or suggestions about specific issues (see companion Web site).

Once the field information is available, it is important to work with government officials, research organizations, private sector suppliers, and others to see what interventions are supported politically, appropriate technically, viable economically, and practical logistically.

In a minority of cases, there may be a very good technical case for an intervention, but the policy environment needs to be challenged. This was the case in Tanzania, where the privatization of government veterinary services left most rural livestock keepers without animal health support. The NGO FARM-Africa piloted the use of "community-based animal health workers" to fill the gap in service delivery. These proved very successful in reducing animal mortality and disease, but were initially rejected by the Tanzanian Veterinary Association. It took several years of advocacy and lobbying to change government policy, but community-based animal health workers are now an accepted component of animal health care in the country (Fig. 2.7).

Sometimes, it is difficult to know if an intervention will work without trying it on a pilot basis. One of the findings from the livelihoods analysis in Afghanistan described in Box 2.1 was the very low levels of literacy among rural women. This severely restricted their involvement in any nonfarm employment and their self-esteem. Literacy classes, using agricultural themes such as vegetable growing as the subject matter for the lessons, were tried on a pilot basis and have proved a tremendous success (Fig. 2.8).

FIGURE 2.7 Training community-based animal health workers in Tanzania. Trainees are selected by their communities and are accountable to them.

FIGURE 2.8 Afghan village men look after the children while their wives are at adult literacy classes. It is important that the men are in favor of the lessons, as these seem to be.

While each of the livelihood analyses comes up with specific agronomic or livestock husbandry needs (e.g., in Afghanistan the need for varieties of wheat and barley to replace those susceptible to rust was very clear), most show that poor families (who usually have minimal physical collateral) have little access to credit, which severely limits their enterprise options. Recently there has been an expansion in the use of group microcredit schemes aimed at poor (but economically active) families. These usually require the family to join a group and save into a scheme that then lends from the accumulated capital to its members. Two examples are Self-Help Groups, which are very successful in southern India and now in Afghanistan (Pound, 2006) (Fig. 2.9), and the Savings and Credit Cooperative Societies (SACCOS), which are becoming a feature of NGO projects in eastern Africa (Fig. 2.10).

NEW DIRECTIONS FOR AGRICULTURAL RESEARCH AND DEVELOPMENT

The Need for Radical Reform of Research and Extension

The results of any in-depth rural analysis, such as livelihoods analysis, reveal the breadth and interconnectivity of the social, economic, technical, and policy issues facing rural people. Conventional government agricultural research and extension systems are not designed to deal with this complexity. They are often institutionally separate organizations with little real cohesion. Research is structured most commonly around commodities and focused on technical issues, with a linear delivery process (where technologies are developed by researchers and then passed on to farmers through extension agencies). Government extension typically has been

FIGURE 2.9 Afghan farmer in a Self-Help Group in Badakhshan hands over his monthly savings to the group treasurer. He will have the chance to borrow capital from the group to invest in an agricultural or nonagricultural enterprise.

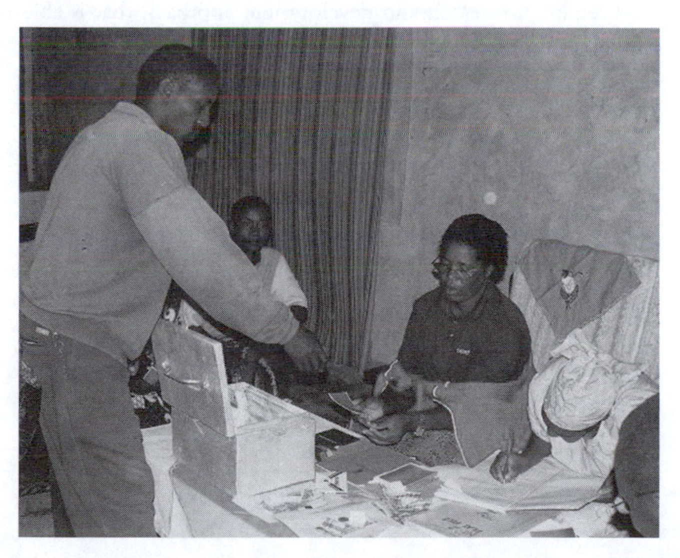

FIGURE 2.10 A SACCOS member hands over his monthly savings to the treasurer of the Cooperative Society in northern Tanzania. Note the careful, transparent registration of this transaction. He will also be able to borrow against the society's capital to invest in a productive enterprise.

top-down, publicly funded and publicly delivered, with inadequate coverage and poor responsiveness to local needs (Fig. 2.11).

The rapidly changing context for agricultural development in sub-Saharan Africa and other developing regions offers new challenges that require the reorientation of research and extension to models that emphasize development outcomes, not academic ones. Some of the changes that particularly affect Africa according to the IAC (2004) are as follows.

- The globalization of agricultural production and trade
- Agricultural development is increasingly being driven by markets
- The private sector is driving innovation, and also increasingly the generation, diffusion, and application of knowledge, information, and technology
- Information and communication technology provides new opportunities
- Biotechnology is making an impact on agricultural production and processing systems
- National development strategies increasingly emphasize support to the commercialization of agriculture (e.g., the Ugandan "Plan for the Modernisation of Agriculture" and the "Horticultural Exports Industry Initiative" in Ghana) and the increasing role of markets and the private sector in agricultural development
- Environmental changes as a consequence of climate change, degradation of ecosystems, genetic erosion, water shortages, social conflicts, and upheavals pose new challenges in agricultural development

The rapid environmental, economic, and social changes occurring at global, national, and local levels require a research and development approach that is able to support

FIGURE 2.11 While many extension services are top-down, there are exciting exceptions, such as the National Agricultural Advisory Services in Uganda, where a privately delivered service responds to priorities decided by farmer groups.

adaptation to those changes, while the complex, risk-prone, and diverse nature of rural life requires a flexible, decentralized type of research that builds on the involvement of farmers and other actors in developing and disseminating technologies and processes.

One approach that is gaining ground with donors and some research and extension agencies is the *agricultural innovation systems* approach. The evolution and theoretical backgrounds to this approach are described in Chapter 8 of this book, while here we look at the characteristics of the approach and some of its applications.

Characteristics of Innovation Systems

1. The most striking characteristic of the innovation systems approach is its emphasis on bringing *multistakeholder partnerships* together to address a need or opportunity. Interaction between stakeholders is essential in order to learn from each other and negotiate the terms of the partnership. Thus the concept embraces not only the science suppliers (traditionally, researchers in national agricultural research institutes and universities) but also the totality of actors involved in production and marketing. Stakeholders that can form part of an innovation partnership include farmers (of different types), market traders, processors, exporters, researchers, extension staff, input suppliers, and others. The mix of stakeholders depends on the problem or opportunity to be addressed, and the composition of the partnership changes over time as the situation changes.

2. Forming, maintaining, and managing partnerships require *facilitation/coordination skills* and resources. Trust is an essential ingredient of success, and any partnership has the potential for asymmetry (power, voice, benefit, etc.) and conflict. Local institutions may need empowerment through building their capacity and the development of links to input supplies, markets, and technical assistance.

3. A major difference with conventional research is that the innovation systems approach recognizes that *innovation can arise anywhere*. It is not the preserve of formal research organizations. Farmers, community-based organizations (CBOs), NGOs, and private enterprises can be the source of the innovation and, in many cases, can develop the innovation independently of formal government structures (see Box 2.2 describing case studies on the development of the cassava-processing industry in Nigeria). In other cases, innovation is stimulated and coordinated by government initiatives that bring the relevant stakeholders together and support them through the innovation process (see Box 2.3 for a case study on cassava processing in Ghana).

4. In other situations the innovation process can be coordinated by an NGO or extension program that provides capacity building and empowerment for farmer groups and *links the farmer groups to local government, research, and private agencies* that can support production and marketing (see Box 2.4 for a case study from Tanzania).

5. Normally, the term "innovation" at farmers' level has been used to refer to farmers' adoption of new technologies coming from outside, rather than the new technologies, management practices, and institutions that farmers and their communities have developed themselves. Many local innovations are not of a technical nature

BOX 2.2 Nigeria: Spontaneous Development of Gari Processing

The southeastern zone of Nigeria accounts for 53% of national cassava production. Processing cassava into gari for food and income is practiced by many Nigerians in rural areas. In order to harness the opportunities offered by the increasing market demand for gari, farmers devised several technical, social, and institutional innovations.

Gari markets began to emerge as middlemen from urban centers besieged rural markets to bulk purchase gari from the farmers. Simultaneously, enterprising farm households discovered that higher cash income could also be earned from the same quantity of gari if they took it directly to urban markets, thus circumventing the middlemen. Gari processors also began to spring up in urban centers, purchasing roots directly from farmers.

Several prevailing factors catalyzed gari marketing:

1. The favorable natural environment
2. Wide sociocultural acceptance of gari in local food systems
3. Income-generating potential of gari
4. Government stimulus of the market by inclusion of gari as one of the essential commodities in the strategic food reserve program
5. Improved processing technology to reduce drudgery

The innovations are of four types:

a. Technological innovations (improved varieties, fuel-saving technologies, adding palm oil to improve color, mechanization of processing, reduction of the fermentation period, adjusting processing for specific market requirements)

b. Social innovations (establishment of cooperative societies, improving access to market, diversifying markets, emergence of private ancillary service providers)

c. Economic innovations (investment in equipment, partnerships to pool finances, use of informal local credit services)

d. Institutional innovations (contractual arrangements between parties, including gari in the strategic food reserve program, government provision of N50 billion loans to farmers).

Adapted from Ekwe and Ike (2006)

(Fig. 2.12), but rather are socioeconomic and institutional innovations, such as new ways of gaining access to resources or new ways of organizing marketing activities (Waters-Bayer *et al.*, 2006; also see http:// www.prolinnova.net and the companion Web site for this book).

6. Research is important, but not always central, and one needs to consider other *bottlenecks to the use of innovations*. We have seen that the livelihoods framework is

BOX 2.3 Ghana: Government-Led Cassava Initiatives

Cassava is a major staple in West Africa and is grown in most agroecological zones of Ghana. Cassava can also be processed into gari and into starch, which is the raw material for other industrial products. The government sought to use the comparative advantage of Ghana as a cassava producer to transform the cassava industry into a major earner of export revenue in industrial starch.

Whereas a presidential special initiative to do this through Corporate Village Enterprises has failed, a much smaller initial initiative, the "Sustainable Uptake of Cassava as an Industrial Commodity," is having much more success. This initiative revolves around the creation of market linkages to provide market access for small- and medium-sized enterprises, new product development, quality assurance, and the management of supplier–buyer business relations.

The critical factors are

- Scientific research institutes: national (Food Research Institute) and international (NRI)
- Policy institutions (especially the Ministry of Food and Agriculture)
- Business promotion organizations (especially the National Board of Small-Scale Industries)
- Producer/processing organizations

A major lesson learned was that market access does not happen by itself, but needs strategic support and deliberate cultivation. Trust also has to be built between suppliers and purchasers. The case study calls for strengthened links not only between research and farmers but also between research and industrialists. Industrialists should challenge scientists to find solutions to the problems they encounter.

Adapted from Essegby (2007)

BOX 2.4 Tanzania: Sustaining Innovation by Farmer Research Groups

In Tanzania, an NGO (FARM-Africa), working closely with district and village governments, has established a number of farmer research groups to stimulate local innovation for improved crop productivity and natural resource management. The groups were successful because of the linkages (facilitated by FARM-Africa) with national and international research organizations, government seed certification and training centers, and private input suppliers.

(*Continues*)

BOX 2.4 (*Continued*)

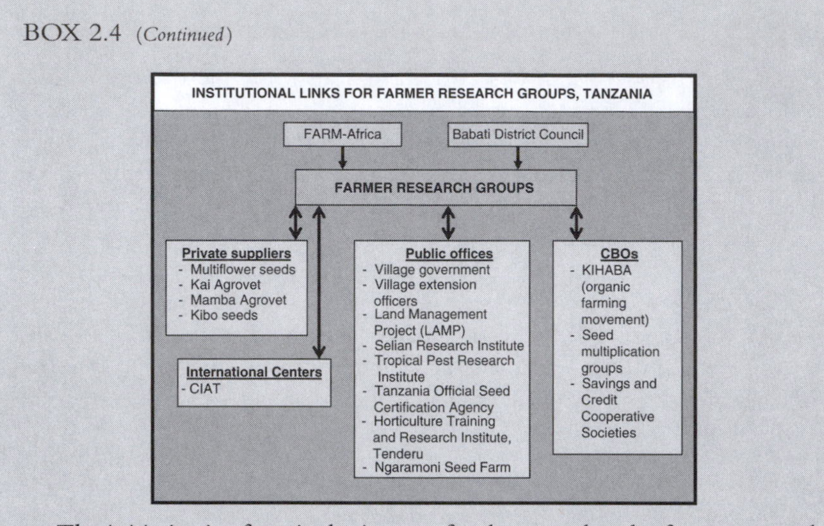

The initiative is of particular interest for the ways that the farmer research groups ensured their long-term access to these linkages and a sustainable capacity to continue to investigate novel technologies of local relevance. They achieved this through

a. The establishment of *community-based seed multiplication schemes*, which provide income that can then be used to finance further experiments.

b. *Community agricultural input supply shops,* established by farmer research groups in response to the need for local access to the technologies identified.

c. *SACCOS*, which enable members to accumulate capital for the purchase of inputs.

Farmer research groups manage input shops in rural areas to ensure that villagers have access to improved seeds (including those multiplied by their

BOX 2.4 (*Continued*)

members), fertilizer, and pesticides. A savings and credit scheme helps those who previously could not afford to purchase improved inputs.

The production increases have resulted in farmers' being able to store grain at harvest and either sell it at a better price later on or use it to reduce food insecurity. They have become financially independent of external donor and government agencies.

Adapted from Pound *et al.* (2007)

FIGURE 2.12 Innovation for agriculture might include disciplines outside the agricultural sciences. Here, low-cost water tanks are developed with farmer groups in Nepal for supplementary irrigation of high-value vegetables. Farmers in this case were supported with advice on production and marketing of the vegetables, as well as training in tank construction and the use of water. The "innovation" covered the whole value chain from production to consumption.

useful in analyzing where these constraints might be through consideration of the social, human, financial, natural, and physical assets available to communities and the legal, institutional, and political influences on them. Regional research organizations such as the Association for Strengthening Agricultural Research in Eastern and Central Africa (ASARECA) have put emphasis on the policy reforms needed to support agricultural innovation.

7. The innovations systems approach is related to some previous approaches, such as the commodity systems approach and the analysis of value chains, both of which, like the agricultural innovation systems approach, consider the whole chain from *producer to consumer* (see Box 2.5 for the case study from Uganda on potato production and marketing).

BOX 2.5 Uganda: "Enabling Rural Innovation"—Nyabyumba United Farmers' Organization

The Nyabyumba Farmers' Group of Kabale district, Uganda, was formed in 1998, with 40 members. The group, supported by the NGO Africare, focused on producing improved potatoes from clean seed provided by the National Agricultural Research Organization (NARO). In 2000, the Nyabyumba group formed a farmer field school to improve their technical skills on potato production and to increase yields. In 2003, equipped with the necessary skills for producing high quality and quantity of potatoes, the group decided to increase their commercial sales and requested support from Africare, NARO, PRAPACE (Regional Potato and Sweet Potato Improvement Network in Eastern and Central Africa), and CIAT.

Through this consortium of partners, Nyabyumba Farmers' Group received training in identifying and analyzing market opportunities and developing a viable business plan for the potato enterprise. From the market study the group identified "*Nandos*," a fast-food restaurant based in Kampala, and the local wholesale markets in Kampala. The group has set up a series of committees to manage, plan, and execute their production and marketing processes. To maintain a constant supply the farmers have set up a staggered planting system to ensure that there are up to 50 tons of potatoes available each month.

To increase the competitiveness of production, the group has conducted research supported by NARO to determine the most suitable nutrient levels of NPK fertilizer and time of dehaulming potato plants that produce big tuber size, with higher organic content, firm skin, and higher yields as required by the buyer. The farmers' group has expanded to a membership of 120 members, 80 of whom are women. They have supplied 190 metric tons of potatoes to *Nandos*, bringing their income to UgSh 60,000,000 or approximately $33,000.

Adapted from Kaaria *et al.* (2006)

8. The agricultural innovation system approach also has much in common with *IAR4D* (Integrated Agricultural Research for Development). The term IAR4D was first used in 2003 by the Forum for Agricultural Research in Africa (FARA, 2004) and follows similar principles and objectives as approaches, such as ICRA's Agricultural Research for Development approach, DFID's Sustainable Livelihoods approach, and the Integrated Natural Resources Management approach developed by the CGIAR.

9. The approach requires that organizations must *act in new ways* (flexibly and in partnership) and with *new skills* (facilitation, conflict management, participation, etc.) (see Table 2.1).

10. The *public sector* has a central role to play through developing legal and regulatory frameworks and providing an enabling policy, trade, infrastructural, and support environment that encourages innovation (Fig. 2.13).

TABLE 2.1 Human Capacity Needs for Implementing Agricultural Innovation Systems Approaches and Some Mechanisms for Developing That Capacity

Human capacity needs for the implementation of innovation systems[a]	Some mechanisms for enhancing human resource capacity[b]
• Management of dynamic partnerships	• Partnerships (e.g., through competitive grant schemes)
• Governance of partnerships	• Exchanges (N–S, S–S) attachments and internships
• Facilitation	• Undergraduate and postgraduate degree studies
• Negotiation and conflict management	• Vocational training
• Communication	• On-the-job learning
• Sourcing, managing, interpretation and "packaging" of information	• Short courses
• Entrepreneurship and business skills	• Distance learning (e.g., professional Ph.D.s)
• Systems thinking	• Conferences and workshops
• Value chain analysis	• Reflection and learning events
• Market evaluation	• Job rotation
• Research methods, including participatory and impact-oriented methods (action research)	• Mentoring and coaching
• Research leadership	• Joint activities (e.g., joint monitoring visits, PRAs, etc.)
• M&E, impact assessment, and learning	• Curriculum reform and the adoption of course delivery methods that stimulate problem-solving abilities
• Mobilization and local organization development	
• Rural finance	
• Demand identification/articulation and priority setting + technical expertise and curriculum reform	

[a]Adapted from Kibwika *et al.* (2007).
[b]Adapted from Pound and Adolph (2005).

11. The innovations systems approach is *not inherently pro poor.* As with other approaches, a real impact on poverty and gender imbalances will only result if special attention is given to meeting those challenges (Fig. 2.14).

12. A change in the mind-set of some researchers can be brought about through *competitive grant research* funds that make it a condition that applicants work together with private and civil society elements to test new technology in the real world of markets and inputs. The Hill Agriculture Research Program in Nepal demonstrated that in its second phase, which emphasized the development of uptake pathways and scaling-up of promising innovations through multistakeholder partnerships, it included traders and input suppliers (Pound and Shakya, 2004).

FIGURE 2.13 Multistakeholder partnerships can improve the community use of natural resources.

FIGURE 2.14 In Bihar, India, there are many landless people with few income options. The East India Rain-fed Farming Project searched for income-generating occupations for the landless that also contributed to the overall well-being of the community. These included blacksmithing, the use of "Sal" (*Shorea robusta*) leaves as plates, the use of grasses to make ropes, and the use of split bamboo to make mats and baskets.

Capacity Development Needs for Agricultural Innovation Systems

While there is no consensus on the precise nature of innovation capacity, its broad features include a combination of

a. Scientific, entrepreneurial, and managerial skills

b. Partnerships, alliances, and networks linking different sources of knowledge and different areas of social, economic, and policy activity
c. Routines, organizational culture, and traditional practices that encourage the propensity to innovate
d. Supportive policies and incentives and governance structures
e. The mechanisms and encouragement to learn continuously and use knowledge more effectively (http://www.innovationstudies.org)

The World Bank (2007) maintains that research capacity should be developed in such a way that, from the beginning, it nurtures interactions among research, private, and civil society organizations An effective agricultural innovation system requires a cadre of professionals with a new skill set and mind-set that encompasses markets, agribusiness, intellectual property law, rural institutions, rural microfinance, facilitation, system analysis, and conflict management.

Capacity Strengthening Implications of Adopting the Innovation Systems Approach

Implications for national research and advisory services include the need to reskill in the areas of facilitation, communication, entrepreneurship, conflict management, value chain analysis, and market research. Reward systems will also need to change to reflect the changed emphasis from academic papers to developmental outcomes. In addition, it also implies a much closer working relationship between research and extension on the one hand and government, civil society, and private sectors on the other. The stakeholders need behavioral change and a reconfiguration of their roles and relationships to ensure a more proactive and interactive engagement for innovation. These sentiments are echoed by World Bank in its book (World Bank, 2007) and in calls for the reform of university curricula to include innovation systems principles and case studies.

Risks and Benefits of Adopting an Innovation Systems Approach

There is a risk that organizations will adopt this approach *in favor* of previous approaches, such as farming systems, sustainable livelihoods, agricultural knowledge and information systems, and participatory approaches. This would be a shame as each of those approaches is still valid, and their concepts should still be brought to bear when considering rural development situations. Rather the innovation systems approach should be *complementary* to these other, still valid, approaches.

A further risk is that the need for technical specialists will be disregarded in favor of those with soft skills. Again, that would be a mistake, as technical specialists are still needed to investigate and provide understanding of complex technical aspects of innovations.

In contrast, benefits include greater efficiency due to the "joined up" thinking in the production to consumption cycle as people work together to address opportunities and the emergence of creative solutions to problems facing rural communities from a range of sources.

Implications for Agricultural Development in Africa

Major donors such as the World Bank and DFID are already advocating the adoption of the approach, and subregional organizations such as ASARECA are incorporating integrated agricultural research for development as the underlying research paradigm for their programs. In addition, there will be a reform of university curricula to include innovation systems approaches, including a shift toward more client-oriented, vocational courses.

Implications of the innovation systems approach for donor support are clear. Formulation of intervention programs should be done in the specific context of the respective countries and localities, and these must emphasize strong linkages among the critical stakeholders and detail the respective roles and expected outcomes. Investments in such programs must be in producing not only tangible outputs, such as improved planting materials, technologies, and products, but also intangibles, such as enhanced skills, knowledge, and mutual trust.

CONCLUSIONS

This chapter identified and encouraged development interventions with rural communities in developing countries. To be sure that our technologies, policies, and advice are relevant, we need to understand the social, human, physical, natural, and financial assets of those communities and the vulnerabilities and shocks faced by them. We need to know how external institutions and influences affect farming decisions and the capacity of farmers to take up the options available to them and what livelihood strategies they are following or aspiring to. Sensitive use of PRA tools within a sustainable livelihoods framework can be a learning experience for communities, as well as for external agencies.

However, gathering information is not enough. Nor is unilateral action by one or other stakeholder in the development spectrum. Instead, dynamic, multistakeholder partnerships are seen as effective ways to bring different interests and perspectives together across the value chain from producer to consumer so that all can benefit. This innovation systems approach is evolving, but there are sufficient case studies available from which to distill the elements of success and good practice.

It is clear from these case studies that organizations (research, extension, NGOs, CBOs, and the private sector) will need to change to enable them to work in a flexible, unselfish way with their development partners. Changes in skills, mind-sets, procedures, reward systems, and institutional relationships are necessary if the approach is to be mainstreamed, requiring political will, institutional commitment, and the reform of university curricula.

Throughout this process, the best of agricultural science must be retained. There is still an acute need for competent agronomists, livestock health specialists, soil scientists, market economists, and social anthropologists. However, this expertise now serves a research-for-development agenda, providing the scientific basis for decisions that affect the appropriate use of resources in the short term, and their sustainability for future generations.

REFERENCES AND RESOURCES

ASARECA IAR4D CB team (2007). Review Paper on Integrated Agricultural Research for Development: Towards Enhancing IAR4D Capacity at ASARECA. KIT, The Hague.

Carney, D. (1999). "Approaches to Sustainable Livelihoods for the Rural Poor." ODI Briefing paper No 2. ODI, London.

DFID (1999). "Sustainable Livelihood Guidance Sheets." DFID, London. Available on www.livelihoods.org/

Ekwe, K.C., and Ike, N. (2006). "Sustaining Gari Marketing Enterprise for Rural Livelihood: Farmers Indigenous Innovations in South Eastern Nigeria." Paper presented at the Innovation Africa Symposium, Kampala, November 21–23, 2006. National Root Crops Institute, Umudike, Nigeria.

FARA (2006). "Agricultural Research Delivery in Africa: An Assessment of the Requirements for Efficient, Effective and Productive National Agricultural Research Systems in Africa," March, 2006. FARA, Accra.

Hall, A. J. (ed.)(2004). "Innovations is innovation." ICRISAT, Patancheru, India.

IAC (2004). "Realizing the Promise and Potential of African Agriculture." Inter Academy Council. (www.interacademycouncil.net).

Kaaria, S., Abenakyo, A., Alum, W., Asiimwe, F., Best, R., Barigye, J., Chisike, C., Delve, R., Gracious, D., Kahiu, I., Kankwatsa, P., Kaganzi, E., Muzira, R., Nalukwago, G., Njuki, J., Sanginga, P., and Sangole, N. (2006). "Enabling Rural Innovation: Empowering Farmers to Take Advantage of Market Opportunities and Improve Livelihoods." Paper presented at the Innovation Africa Symposium, Kampala, November 21–23, 2006. CIAT.

Kibwika, P., Mukiibi, J., and Okori, P. (2007). Draft of the Preliminary Findings of the Capacity Needs Assessment in the ASARECA Region. Report to SCARDA.

NEPAD (2005). "Science, Technology and Innovation for Africa's Development: A Plan for Collective Action." NEPAD Secretariat, South Africa.

Pound, B. (2001). "Report on a Consultancy to Develop and Implement Sustainable Livelihoods Analysis in Two Pilot Villages in the South of Moldova." Natural Resources Institute, Chatham, UK.

Pound, B. (2004). Livelihoods Systems Analysis on behalf of the project "The Development of Sustainable Agricultural Livelihoods in the Eastern Hazarajat, Afghanistan (GCP/AFG/029/UK)." Natural Resources Institute, Chatham, UK.

Pound, B. (2006). Final Technical Report on the project: "Innovative Financial Mechanisms for Improving the Livelihoods of Rural Afghans Currently Economically Dependent on Opium Poppy" to the Research into Livelihoods Fund (DFID). Natural Resources Institute, Chatham, UK.

Pound, B. (2007). "Widening the Definition of Capacity to Respond to the Innovations Systems Approach." Paper presented at the RUFORUM Biennial Conference on "Building Scientific and Technical Capacity through Graduate Training and Agricultural Research in African Universities" held at the Sun 'N' Sand Hotel, Malawi, April 23–27, 2007. Natural Resources Institute, Chatham.

Pound, B., and Adolph, B. (2005). "Developing the Capacity of Research Systems in Developing Countries: Lessons Learnt and Guidelines for Future Initiatives." Paper commissioned by DFID's Central Research Department. Natural Resources Institute, Chatham, UK.

Pound, B., and Essegby, G. (2007). Agricultural Innovation Systems. SCARDA briefing paper. Natural Resources Institute, Chatham, UK.

Pound, B., Massawe, K., and Fazluddin, F. (2007). "Innovation Partnerships for Effective Adaptive Research and Technology Uptake." Paper presented at the workshop "Enhancing Agricultural Innovation" organized by the World Bank and held in Washington, March 22–23, 2007. Natural Resources Institute, Chatham, UK.

Pound, B., and Shakya, P. B. (2004). Final Consultancy Completion Report: Uptake Pathways and Scaling-up. Hill Agriculture Research Programme (DFID-Nepal). Chatham, UK: Natural Resources Institute.

Waters-Bayer, A., van Veldhuizen, L., Wongtschowski, M., and Wettasinha, C. (2006). "Recognizing and Enhancing Local Innovation Processes." Paper presented at the Innovation Africa Symposium, Kampala, November 21–23, 2006. PROLINNOVA.

World Bank (2007). "Enhancing Agricultural Innovation: How to Go Beyond the Strengthening of Research Systems." World Bank, Washington.

livelihoods-connect@ids.ac.uk

www.prolinnova.net

www.innovationafrica.net

CHAPTER 3

Agroecology: Principles and Practice

Sieglinde Snapp

Summary

Ecological principles provide a foundation for resilient and sustainable agriculture that supports rural livelihoods. Climate and soil resources are key drivers of the constraints and opportunities that smallholders can exploit, and the dynamic nature of the agroecological response is discussed in this chapter. The concepts of organism niche and functional groups are presented as a means to group plant and animal types and to combine organisms for high-performing assemblages. Design principles explored here include complementarity, redundancy, and mosaics. Agricultural systems that are biologically consistent with these principles are presented, including indigenous and novel technologies. Constraints and opportunities associated with humid and arid environments are discussed as examples of agroecological principles and practice.

Agricultural Systems: Agroecology and Rural Innovation for Development

INTRODUCTION

Agriculture is an example of a managed ecosystem. The principles that are useful in understanding natural ecosystems apply to agriculture. This chapter provides a brief introduction of ecosystem concepts and how these can be applied to improve management of agricultural systems, with particular reference to tropical agroecologies.

Ecology is concerned with different scales over time and space, from the individual to the community, from populations to ecosystems, and the evolution of species (Fig. 3.1). The socioeconomic context is particularly important to organization of agroecosystems and is the focus of other chapters in this book, including Chapter 2 on livelihoods and Chapter 9 on equity and social dynamics.

Gradients of resources define the environmental context within which plants, animals, and soil biota interact and profoundly influence the complex communities of macro-, meso-, and microorganisms that evolve. Temperature and moisture are the primary gradients that delineate ecological zones. Soil properties are another primary resource gradient, one that influences ecosystem development, which in turn is influenced by ecosystem biota. Soil parental materials interact with moisture and temperature gradients, and living organisms, over long time periods as soils evolve.

FIGURE 3.1 Levels of ecosystem organization with agroecology examples of the individual organism, community, and watershed. (**A**) Individual organism: a bean plant.

FIGURE 3.1 Cont'd (**B**) A community: potato plants in the foreground are growing in a field with an erosion prevention strip planting of vetiver grass intercropped with a mango tree. (**C**) A highland watershed in southern Africa. Note the shelter for field watchers (protecting crops from foragers such as monkeys and stray livestock), a maturing maize crop, vegetable gardens near water sources, and a mosaic of woodlots, grain fields, and pasture land throughout the watershed, reaching to the forest edge on steep mountain slopes.

ECOSYSTEM DRIVERS

In addition to resource gradients, major disturbances and fluctuations in resource availability are key regulators of productivity in ecosystems. Fire, flooding, soil disturbance, and herbivore damage from insects or livestock are the most common irregular disturbances in agroecosystems. Planned disturbances as farmers perform soil tillage, weeding, and harvest operations are also important regulators of system performance and resilience in agroecosystems. Turning over the soil and burning residues just before crops are planted are the primary means of enhancing nutrient availability in synchrony with crop demand, as well as providing weed suppression (Fig. 3.2).

Fire is an effective means to break pest and disease cycles, as well as to reduce the potential for residue decomposition bottlenecks in areas where low-quality residues are incorporated or where a long dry season is a constraint to rapid decomposition (Fig. 3.3). It may be one of the only practical means of reducing weed presence in subhumid to humid farming, replacing large inputs of labor for weeding. However, frequent fires will alter species succession and soil resource quality. They can have unfortunate side effects, such as favoring the invasion of aggressive grass species and reducing nitrogen (N) inputs into a system. Upon burning, the majority of N in the residues will be lost through volatilization; how much will depend on the heat of the fire as N losses increase at high temperatures.

Farmers have developed methods of smoldering organic materials, which tend to minimize N losses. This may involve slow burning of trash heaps, which can serve as an informal method of compost preparation (Fig. 3.4). More research is

FIGURE 3.2 Oxen tillage prepares the land to plant cotton in southern Mali.

FIGURE 3.3 Burning residues in Zambia, a common field preparation practice.

FIGURE 3.4 Smoldering compost heap on a smallholder farm in northern Malawi.

needed to compare controlled, slow burning of waste materials to more traditional compost processes. This has particular relevance where rainfall is unreliable and where water limits decomposition in compost heaps.

Flooding is another type of disturbance that can greatly alter weed populations and change soil conditions. Farmers use strategic flooding as an important management tool for crops tolerant of water-saturated conditions. The world's most important grain crop, rice, is well adapted to growth in a flooded environment. This has contributed to the popularity of rice, as timed flooding can be used to suppress the vast majority of weeds.

Tillage is the most important tool farmers have to prepare the seed bed for crop plants, while burying weed seed and enhancing nutrient availability. On smallholder farms, "tillage" may vary from a planting stick, used to disturb a localized area around the seed, to the greater disturbance possible with hand hoes and oxen-pulled implements, such as moldboard plows. A central challenge to long-term sustainability of farming systems is that disturbance to enhance nutrient mineralization also enhances mineralization of soil organic carbon (C). Productivity is mediated in almost all soil types by C status, as this largely determines soil water-holding capacity and nutrient buffering and supply, as well as soil structure and aeration. Thus agroecological management requires attention to replenishing organic materials through manure and residue additions (Fig. 3.5) and mitigating loss pathways to enhance soil C assimilation. Tillage depletes soil C through oxidation and by breaking up aggregates that protect soil C. Soil aggregation is supported preferentially by root biomass, thus growing long duration plants that enhance the presence of active roots over the year is an efficient means to build soil C.

FIGURE 3.5 Pigeon pea residues at harvest. The leaves of this multipurpose plant senesce at different times, which enhances the amount of residues that accumulate. Crop residues are an important source of organic matter and can be managed to protect soil from erosion.

FIGURE 3.6 A living cover intercrop system of sweet potato grown between maize plants.

Reduced tillage systems have been the subject of farmer experimentation in many locations around the world, from Brazil to Zambia. To develop a practical and successful reduced tillage system requires a major shift not only in the planting method but also in weed and nutrient management strategies. Where it has been adapted successfully to local environments, reduced tillage systems have been shown—in some but not all cases—to improve soil organic matter and water-holding capacity through enhanced residue retention on the soil surface and through reduced oxidation of soil C. Weed management can be a tremendous challenge, particularly in humid environments, and the vast majority of reduced tillage systems rely on herbicide inputs. Widespread adoption has occurred in some countries, where economics and access support use of herbicides. Living cover systems and layered mulch are also being experimented with as means to suppress weeds through careful introduction and management of smother crops (Fig. 3.6).

AGROECOZONES

Agroecologies vary from the subartic to temperate, with pronounced warm and cold seasons, to subtropical and tropical, with a wide range of moisture regimes. Tropical regions are defined by warm temperatures with no frost and a varying intensity of precipitation from dry to wet. These include the arid to semiarid tropics, subhumid tropics, and humid tropics. The constant warmth and moisture availability in the humid tropics can support high levels of productivity; however, soil resources are often limiting in this environment. Deeply weathered soils predominate, where nutrients have leached from excess moisture in relationship to evapotranspiration demand and chemical weathering has reduced nutrient supplies. Pests, as well as agriculturally desirable plants and livestock, thrive in the subhumid and humid tropics, which often limit productivity compared to temperate and drier zones where cold periods or aridity provides breaks in the growth of pests as well as economically important crops. Of considerable importance in tropical agriculture are the highland ecologies of hill and mountain agriculture that are relatively cool in temperature. These often vary markedly in precipitation over short distances depending on topography and heterogeneous microclimates.

An agroecology is defined by more than the long-term average temperature or moisture pattern. The distribution and variability of resource distribution are critically important. A detailed zonation of agroecosystems often takes into account information on temperature and precipitation variation, particularly in relationship to plant growth requirements. Where information is available about large-scale climatic patterns, this can be of considerable value in understanding and predicting agricultural performance (Patt *et al.*, 2005). In Zimbabwe, a concerted effort to evaluate rainfall patterns in the unimodal row crop agricultural zone was successful in documenting a constituent pattern for the timing of the rainy season cessation; in 8 years out of 10 years, rains stopped within a 2-week time period. This was regardless of timing for the onset of the rainy season and resulted in high probability of low precipitation in years when the rainy season started late. Shifts to lower fertilization rates and drought-tolerant varieties are thus recommended upon late rainy seasons in specific agroecosystems of Zimbabwe (Piha, 1993). Recent changes in global climate and increased variability in climatic patterns need to be considered carefully where long-term precipitation averages or timing of rainy season onset may not be a reliable guide.

COMMUNITY STRUCTURE

Communities of species have coevolved within ecological zones. The presence and diversity of plant communities and associated animals, soil organisms, and symbionts can be categorized in relationship to gradients of moisture and temperature. Because isolation enhances speciation, diverse ecologies have evolved where barriers to species movement are present, such as oceans that surround islands and the effective "island" ecology of an inaccessible mountainous area. A tremendous diversity of species has been documented for island nations and highland countries around the world. This diversity may account in part for the large number of crops that originated from mountainous areas, including some of the most important food crops, such as maize (*Zea mays* L.), potato (*Solanum tuberosum* L.), barley (*Hordeum vulgare* L.), mustard (*Brassica juncea* L.), common bean (*Phaseolus vulgaris* L.), and many fruit trees.

Crops grown within three relatively isolated regions are shown in Table 3.1. These crops of smallholder farms in the Andes, Ethiopia, and Madagascar illustrate both the striking extent to which a handful of crop species have penetrated farms across the globe and the diversity of local species unique to particular regions.

TABLE 3.1 Crops Grown in Isolated Ecologies from the Mountain Highlands of Ethiopia and the Andes to the Island Nation of Madagascar

Ethiopia	The Andes	Madagascar
Staples	Staples	Staples
Tef (*Eragrostis tef*)	Quinoa (*Chenopodium quinoa*)	Rice
Ensete	Kiwicha (*Amaranthus caudatus*)	Plus cassava, maize, yams, indigenous fruits
Plus wheat, maize, sorghum, millet, barley	Oca (*Oxalis tuberosa*)	
	Ulluco (*Ullucus tuberosus*)	
	Maca (*Lepidium meyenii*)	
	Mashua (*Tropaeolum tuberosum*)	
	Plus potato, maize, rice, barley	
Oilseeds	Oilseeds	Oilseeds
Niger seed	Quito palm (*Parajubaea cocoides*)	Peanut
Flax		
Rapeseed		
Castor bean		
Plus peanut, sunflower, safflower, sesame, soybean		
Pulses	Pulses	Pulses
Fava bean (*Vicia faba*)	Lupin (*Lupin mutabilis*)	Rice beans (*Vigna umbellate*)
Chickpea	Lima bean	Bean (*Phaseolus lunatus*)
Vetch (*Vicia dasycarpa*),	Fava bean (*V. faba*)	Cowpea (*Vigna unguiculata*)
Plus common bean, lentils, peas, pigeon pea	Dry pea	Pigeon pea
	Nunas or popping bean (*Phaseolus vulgaris*)	
	Basul (*Erythrina edulis*)	

Adaptation of a crop genotype to a specific ecozone has been used by some scientists as part of the criteria for choosing genotypes—species or varieties—to test, as they are likely to succeed in an area with similar environmental traits. A successful example of this targeted deployment approach is the adoption of new potato varieties in Rwanda, East Africa, where genetic materials were chosen for testing from a similar climatic zone in South America (Sperling and King, 1990). Highland initiatives have begun to catalyze discussions through the Internet, providing an opportunity to exchange knowledge and genetic materials among those who work in similar mountainous zones from around the globe (Tapia, 2000).

A key criterion for adoption is meeting local quality trait requirements, as well as adaptation of varieties or species to climatic conditions. Success in meeting local demands has been achieved recently through participatory plant breeding methods, as discussed in Chapter 7. Although some genotypes are particularly well adapted to microclimates, there are also examples of species that show tremendous adaptability. Many of these species have spread around the globe. Genotypes that are adopted rapidly by smallholder farmers often have traits in common, such as vigorous growth, weed suppressive plant architecture, and robust, pest-resistance traits of such successful crops as maize and soybean (*Glycine max* L. Merr.).

Species diversity is an important characteristic of a community and is defined—at the simplest level—as the number of species present. Organisms that are straightforward to count are often used to describe community diversity, which leads to an emphasis on macroorganisms in soil ecology, where it is challenging to enumerate or even define the concept of species for microorganisms. Organisms that have the greatest impact on the ecosystem, and in many cases the most visible organisms, are sometimes referred to as the dominant species. Ecosystems vary markedly in diversity over time and space and in the presence of dominant species. The scale at which measurement is undertaken will influence which organisms are perceived as dominant and the diversity characteristics of the ecosystem. Many saltwater marshes, for example, are dominated by a few plant and bird species, compared to the impressive diversity of species present in freshwater ponds; however, microorganism diversity is tremendous in both environments.

Species diversity in agroecosystems is determined in large part by human intervention. Management practices favor specific species by planting propagules, such as seeds, clonal materials, and seedlings. The disturbance regime imposed by management also greatly influences the species present in agroecosystems. Crop species and common weed species tend to be adapted to the highly disturbed environment of conventional agriculture, with frequent tillage to release nutrients. In developing lower disturbance systems, such as reduced tillage row crop production, it was necessary to select new crop genotypes with the appropriate adapted traits for this different environment.

ORGANISM NICHE

An organism can best be understood in relationship to its environment: the multidimensional habitat or niche that it exists within. Organisms acquire different resources through space and time, often minimizing competition through exploiting a specific niche. The concept of niche evolved, in part, out of interest in understanding how species are able to share habitats.

The fundamental resources being competed for include light, water, and nutrients. As shown in Fig. 3.7, crop species are adapted to different niches across the farmscape. Consider, for example, the moisture gradient from a flooded paddy where rice (*Oryza sativa* L.) thrives, using the unique morphological biological traits that allow rice to grow in a water-saturated environment, to the highly

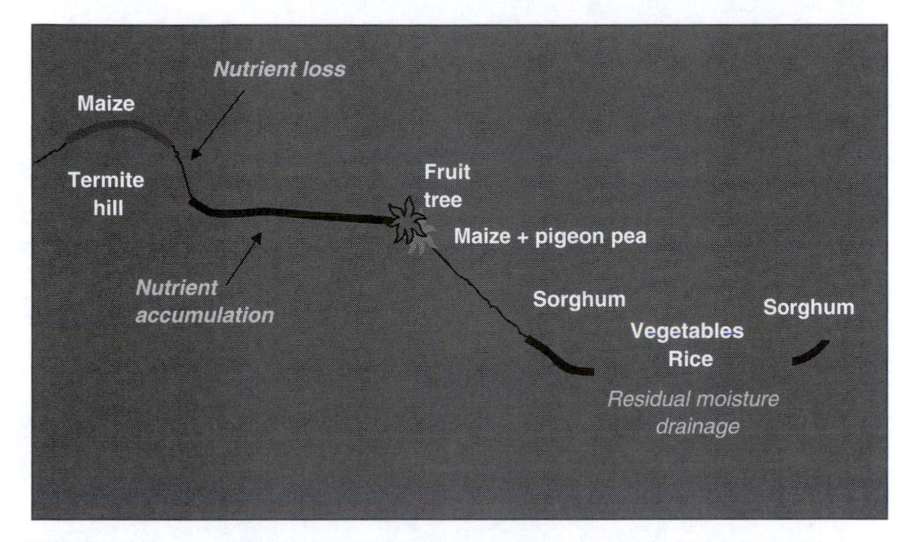

FIGURE 3.7 Crop placement across a farmscape. Farmers often locate high value and nutrient-demanding crops at low spots in the topography, where moisture and soil fertility are least limiting.

fluctuating environment at the paddy edge, which requires stress-tolerant crops such as sorghum (*Sorghum bicolor* L. Moenche), which can survive wide swings in moisture status. A bank next to the paddy provides a niche for a deeply tap-rooted species such as pigeon pea (*Cajanus cajan* L. Millsp.), which requires a well-drained environment. Maize is planted on fertile microsites across the farm. Variable maturity genotypes of maize are often planted at different times in the growing season to take advantage of shifting temporal niches; this is an important means to manage climatic risk.

A classic combination of plant species that illustrates the concept of sharing a niche is the maize–bean–squash (*Curcurbita pepo* L.) triculture grown by farmers from the semiarid southern United States to the subhumid tropics of eastern Africa. Maize is a fast-growing grass, which photosynthesizes through the C4 pathway, which allows it to thrive in hot weather and to photosynthesize at times of the day and season when C3 pathway plants such as bean and squash are not growing fast. Through its symbiosis, the bean plant fixes N, thus minimizing competition for N with the other two species, while the sprawling growth habit of the squash tends to suppress invasive weed species and use a different portion of the canopy than the upright maize and bean complex (as the viney bean grows on the maize stalk).

Temporal separation of niches is common, as shown in Fig. 3.8A, where pumpkin flourish after maize is harvested. A spatial niche is illustrated in Fig. 3.8B, where soybeans are growing on the tops of ridges while a pigeon pea crop is growing in the furrow between ridges. This is expected to reduce competition, as root systems have minimal interaction when spatially separated in this way; this facilitates sharing a habitat.

FIGURE 3.8 (**A**) Pumpkins grown as a relay intercrop in maize, producing a crop after maize is harvested. (**B**) Soybean growing on ridges and pigeon pea growing in the furrows between the ridges.

COMMUNITY EVOLUTION

Community composition is influenced by the resources available and, in turn, may influence resource quality and the trajectory of the evolving community. That is, plants, associated symbiotic organisms, and animals that enrich soil nutrient status alter the habitat and change species composition over time. One of the most ecologically important examples of plant species interaction with the soil resource is the interaction of N-fixing associations, such as the legume symbiosis with *Rhizobium* bacteria, which enhance soil N status through biological N fixation. The presence of N-fixing symbioses also alters soil pH through associated acidification. Soil that is N enriched may favor the dominance of grass species adapted to N acquisition early in the growing season, at the expense of legume species. Slowly, species composition will shift from legume-dominated communities to grass-dominated communities.

Understanding species impact on agroecological resources is complex, as species composition is driven by human intervention, both through propagation of desired plants and through suppression of undesired plants. Management practices in many ways speed up the process of succession observed in natural communities. Consider the following successional systems.

1. A crop rotation, where a sequence is followed, such as a 3-year rotation of maize–soybean–cotton
2. A pasture, where a mixture of grasses and legumes is planted and the pasture community evolves over time as livestock graze
3. A tree plantation, where trees are intercropped initially with an annual crop, such as common bean, and then as the trees mature they are intercropped with a living cover crop, such as mucuna, to suppress weeds and renew fertility

Farm managers should take into consideration, and manipulate, the ecological processes that govern succession in each of these agricultural systems. A central question under active investigation is the ideal ratio of the legume component to the other components in the agricultural systems and how this ratio is expected to change over time (Drinkwater and Snapp, 2007). Initially, the intensity of legume presence is expected to be a substantial component, particularly for a nonfertilized system. As the soil N status increases and managers become concerned about the cost of expensive legume seed in a cropping system or the bloat potential of legume tissues in a pasture system, a decreased legume presence will be appropriate. A pertinent example is pasture mixture recommendations, which often involve a shift from about 40 to 25% legume component, after a pasture is well established. Farm priorities, animal requirements, and environmental conditions will all influence the ideal percentage of legume.

Intensified production and removal of harvestable products in agricultural systems often acerbate the N-limited status of soil type, as grain and leaf products used for human consumption are N enriched. However, N status is dynamic, and at many points in time, or at specific locations in the rural landscape, phosphorus (P) limitation may be a greater determinant of productivity than N (Fig. 3.9). In slash and burn forms of agriculture, P may be initially limiting as a flush of N is released from the ash of accumulated organic materials and from the disturbed soil of a newly cleared field. The phosphorus status will depend greatly on soil type, as well as on management in the recent past and on the species grown at the site.

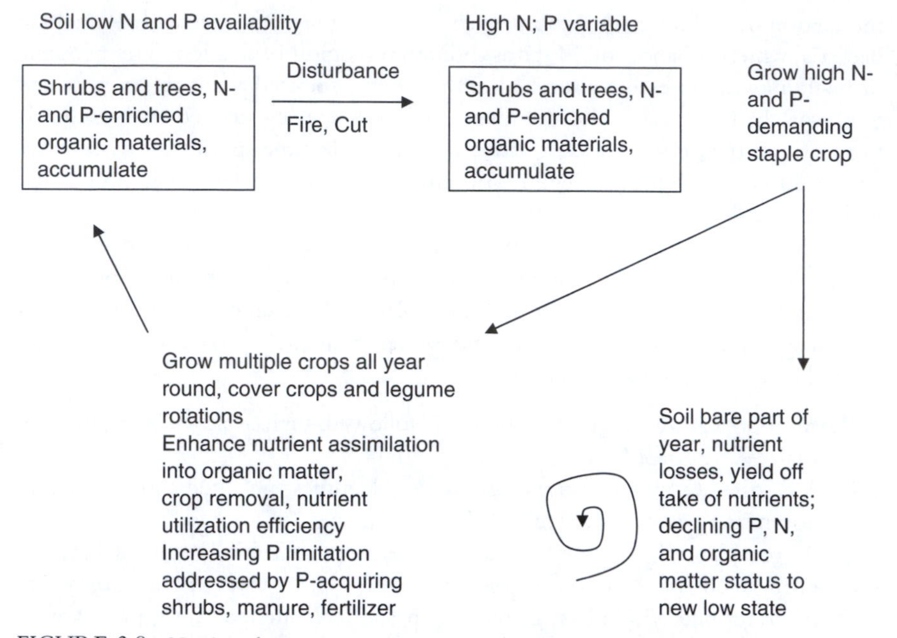

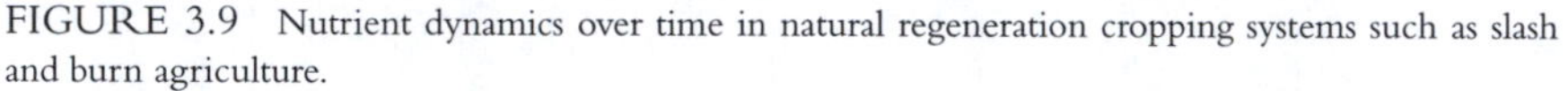

FIGURE 3.9 Nutrient dynamics over time in natural regeneration cropping systems such as slash and burn agriculture.

Research has shown that the P availability varies markedly from field to field in smallholder agriculture (Tittonell *et al.*, 2005). Future investigations are required to determine the effect of plant species on plant-available P and on the efficiency of P fertilizer. Preliminary evidence is consistent with legumes playing a significant role in enhancing P availability, as shown by long-term studies of tropical pastures (Oberson *et al.*, 1999) and crop rotations (Gallaher, 2007).

Promoting the growth of deep-rooted, mycorrhizal, and semiperennial species as intercrops or improved fallows will imitate a natural system, such as a rapid succession that occurs on streambeds or other disturbed environments. The principle of integrating plants with extensive rooting structures has been promoted as a means to enhance nutrient status. Nutrients are recycled from deep in the profile, and soil quality is built up through the maintenance of continuous soil cover. Paradoxically, perennial-dominated cropping systems are highly competitive with, and will often suppress, annuals. During the early growing season, which is critical for establishment and yield potential of annual crops, the presence of actively growing perennials will be highly competitive for water and light. In natural systems, specialization of species allows cohabitation, as different photosynthetic pathways and morphological traits favor C3 grasses to grow most actively and photosynthesize during cooler times of the year, whereas C4 grasses tend to dominate during hotter times of the year. In agricultural systems, farmers can manage to promote preferred species by frequent cutting of perennial branches or by limiting competition by growing species in different locations and separating plants in time.

PLANT GROWTH TYPES

The ability of plants to cohabit in an ecosystem relates to many factors. Plant life strategies include growth types, resource acquisition mechanisms, and other key competitive and survival traits. Plant traits often vary in a coordinated manner, forming plant functional types. In agroecology, an emerging principle is to design combinations of functional types with complementary traits that form a community with resilient properties.

A useful typology for describing functional types is to use life strategies, where species are grouped into stress tolerators, ruderals, and competitors (Table 3.2; Grime, 2001). There is extensive plant ecology literature based on three strategy groups, which will be drawn upon in this chapter, although classification into two functional strategies—r and K selected species—is also common.[1] Disturbance from fire or herbivory (insect or livestock damage) and stress from limiting resources (insufficient water, nutrients, or light) have been shown to be key factors determining plant success. They combine to describe three environments where organisms can survive: low disturbance, high stress (e.g., stress tolerators, such as perennials adapted to arid zones); low disturbance, low stress (competitors, including many rapid growing species with high plasticity); and high disturbance, low stress (ruderals, such as rapid maturing annuals).

Annual crops and successful weed species are often ruderal plants. They are adapted to highly disturbed and high nutrient environments, with rapid growth that maximizes the ability to acquire resources and utilize nutrients and fixed carbon for reproductive purposes. The leaf area index of ruderal crop plants is generally

TABLE 3.2 Survival of Organisms is Divided into Three Strategies, Adapted to Ecosystems That Vary in Stress and Disturbance[a]

	Low disturbance	High disturbance
Low stress	Competitors	Ruderals
High stress	Stress tolerators	Mortality (no viable strategy)

[a]A survival type adapted for each habitat is described, except for the combination of high disturbance and high stress, which induces mortality.

[1]r strategists are opportunists with a high intrinsic growth rate and preferential allocation to reproduction, usually producing many offspring; K strategists tolerate or avoid interference, allocate to vegetative and other nonreproductive activities, and produce a few large seeds, or care for their young among animals.

high, which is consistent with what might be termed a competitive ruderal, and is important to the ability of crop plants to compete with weeds, as discussed later. Ruderal plant traits include minimal investment in stress toleration traits, such as tissue defense compounds against herbivores. The consequence is that insects and mammals prefer to consume plant tissues from ruderals that have few of the recalcitrant plant chemical compounds (e.g., phenols and lignins) that make tissues unpalatable and indigestible. There are trade-offs for farmers between growing crops that are highly insect resistant, which grow slowly and produce defensive compounds, and growing crops that produce desirable food products rapidly, which are susceptible to herbivore damage.

Among crop plants, all three growth types are represented; for example, common bean is classified as a ruderal given its rapid growth habit, early reproductive phase, edibility, and highly pest-susceptible tissues. In contrast, soybean is also a grain legume, yet has many traits that are associated more often with the stress-tolerator group. This includes a range of plant growth types and defense traits, such as pubescence that discourages insect herbivory and biochemical compounds that require processing for humans to obtain nutritional value from soybean grain. Maize has competitor traits, including highly effective nutrition acquisition mechanisms, such as N mineralization-inducing root exudates, high N uptake activity, and a large nutrient demand (sink capacity). The design of cropping systems should take into consideration what are complementary combinations of functional traits, such as a highly competitive crop mixed with a resource-sparing ruderal type.

Weed species have life cycle characteristics that mimic or are closely aligned with the life strategy of the infested crop species. Diversification with different life forms in rotational and intercrop systems will reduce the success of a mimic strategy. That is, weed species with similar life cycles to crop species will be disturbed and suppressed to the extent that an agricultural system incorporates different life cycles. Examples include use of a pasture or forage rotated with crops to suppress persistent weeds or relay planting of a long-growth duration, viney legume crop into a short season grass crop to suppress annual grass weeds.

Identifying organisms that contribute to sustainability while simultaneously producing harvestable products is central to the design of ecologically based farming. However, the coordinated evolution of plant traits and biological constraints has led to close linkages among characteristics and to many traits that are not compatible with agricultural production. For example, it has been the goal of researchers for decades to develop stress-tolerant plants that can grow in saline soil or with minimal water. However, stress-tolerant traits include a slow growth rate, a perennial habit, and production of defense compounds. To adapt farming to a highly stressed environment, a shift may be required from the focus of current research, which emphasizes annual production and a large "harvest index" (yield as a percentage of aboveground biomass), to consider instead perennial plants with indeterminant growth habits. Perennial food crops can be used as models, including oil crops, for example, avocado (*Persea Americana* Mill.) and the west African shea nut (*Vitellaria paradoxa* L.), and carbohydrate crops, for example, banana (*Musa acuminata* Colla), and the Ethiopian enset (*Ensete ventricosum* Welw. Cheesman).

Developing perennial crops from annual grains is a far-reaching goal, yet scientists have initiated investigations into domesticating perennial relatives of annual crops and breeding perennial traits into annuals. There has been progress in selecting for perennial grain sources among wheat breeders, where perennial wheat selections produce about 40% of the yield of annual varieties. As expected, trade-offs are initially severe between perenniality and annual grain yield, as photosynthate investments in stress tolerance structures, such as a cold tolerance mechanisms and deep rooting systems to improve survival through a dry season, will reduce photosynthate available for the reproductive grain. In general, perennials are associated with much lower human "off take," around 10% of aboveground plant productivity compared to 50% or more for annuals (Cox *et al.*, 2006) (Box 3.1).

Rural inhabitants in many environments use indigenous knowledge of plant products and have found slow-growing, stress-tolerator perennials to be important food and medicine sources, particularly in drought years (Fig. 3.11). The use of Basul (*Erythrina edulis*) or "tree bean" of the Andes, which is a shrub that is grown often along property lines or in gardens where it provides important risk mitigation in drought years, as its dried seeds are an important nutritional safety net, is an example of this system. It is a relatively rapid-growing pioneer species so is a ruderal among shrubs, but is a stress tolerator compared to short-duration annual grain legumes.

Ecologists widely debate which traits to consider and which categorization systems to use to evaluate species and predict performance. Grouping of species into functional groups can be conducted using many different traits, from life forms to growth habit. Useful criteria for agroecology applications include grouping species by length of growing season, determinancy and indeterminacy, and ability to grow in a compensatory manner. These traits influence which species are adapted to different locales and timing sequences within a farming system. Photosynthetic pathways, and thus growth response to hot and dry environments, are also useful characteristics to consider as functional categories. For example, C3 pathway species such as wheat are adapted to cool conditions, whereas C4 pathway species such as maize thrive in warm seasons and locations. Combinations of C3 and C4 grasses are the basis for many successful pasture systems.

An approach that has begun to be explored for designing cropping systems is to consider combining plants from a continuum of symbioses. Trisymbioses are represented by the vast majority of legumes, which are associated with both rhizobia and mycorrhizal symbioses. Most crop plants have mycorrhizal symbionts only. Some crops are highly mycorrhizal dependent, such as cassava (*Manihot esculaenta* Crantz)

BOX 3.1 Perennial Crops

In Kansas, the Land Institute has pioneered efforts to develop perennial grain crops, including genotypes related to sorghum, sunflower (*Helianthus annuus* L.), and wheat (*Triticum aestivum* L.). At Washington State University, plant breeders have worked for over a decade to produce a variety of perennial wheat that can be planted about once every 5 years and grain harvested each year (Fig. 3.10). This conserves soil, reduces input costs, and generally diversifies farmer options for environmentally friendly farming. The winter hardiness, yield potential, and quality traits of perennial wheat varieties are highly variable, and genotypes will require further development. Farmers are interested in perennial grains, particularly those that grow crops on steep slopes, or are looking for means to alter weed population dynamics radically while reducing tillage. A much longer development process is required to develop a perennial grain legume crop for temperate or subtropical regions. There is a semiperennial tropical grain legume available, if one considers that pigeon pea is often "ratooned" or cut back after harvest and a second or third harvest obtained.

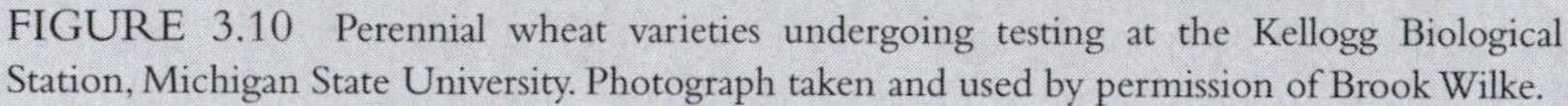

FIGURE 3.10 Perennial wheat varieties undergoing testing at the Kellogg Biological Station, Michigan State University. Photograph taken and used by permission of Brook Wilke.

and onions (*Allium* species). Brassica species and a few other crops are interesting exceptions as they are not mycorrhizal and are often associated with suppression of soil fungi. This has implications for using brassica species as a "break crop" to include in a rotation sequence when the goal is to alter soil community structure markedly.

FIGURE 3.11 A wide range of products are produced by perennial legumes.

ECOLOGICAL PRINCIPLES APPLIED TO AGRICULTURE

Natural plant communities illustrate principles that can be used to improve the resilience of cropping systems and to enhance efficiency (Knops *et al.*, 1999). Agricultural systems that cycle energy and nutrients in an efficient manner can be termed semiclosed and will have low requirements for external inputs. There is a continuum from natural ecosystems where no products are removed, which can be termed closed systems, to semiclosed systems with foraging (e.g., natural ecosystems with limited removal of products such as reproductive parts of trees and fungal organisms), to semiclosed systems such as those discussed in this book (e.g., those that rely on biology to reduce farming systems losses and replenish resources), to an open, and highly productive, conventional agricultural system. Open systems have high resource requirements and the potential for significant leakiness, particularly if they are productive.

Agroecology aims to improve ecological understanding and use ecological principles to design semiclosed and resilient farming systems with high environmental services. The principles of agroecology must also meet "relevance criteria," such as reasonable yield potential. The off take from agricultural systems must meet farmer goals and have feasible requirements for labor and other investments. The principles for sustainable, long-term agricultural practices are developed in depth in the next chapter. This section discusses what have been termed the ecological pillars of tropical agroecology: complementarity, redundancy, and mosaics (Ewel, 1986). The concepts of complementarity and mosaics, in relationship to diversity, successional patchiness, and landscape ecology, are described in depth by Wojtkowski (2003).

Complementarity

The theory of niche differentiation is a useful concept to optimize the design of plant combinations for complementarity of species across time and space. To maximize productivity in agroecology requires careful consideration of Gause's (1934) theory that two species cannot occupy the same ecological niche at the same time. This is the origin of our understanding of "competitive exclusion"; if two organisms have similar niches, one will generally exclude the other over time. Overlapping patterns of land use, such as relay cropping and taungya, semisequential tree and crop systems, must minimize competitive exclusion while maximizing resource capture through complementarity. Resources such as light and nutrients may be underutilized in cropping systems that are based on monocultures of annuals, particularly during the period at the end and beginning of the planting cycle when resources tend to be in excess of plant demand. For a detailed discussion of the theory and practice, see Gliessman (2007).

Species used in mixed cropping systems ideally have complementary traits, such as short- and long-duration growth habits, which combined will ensure the capture of sunlight and recycling of water and nutrients throughout the growing season. Characterizing species based on such traits as growth habit, pest tolerance, maturity date, and nutrient acquisition strategies are useful first steps in considering which crops combine well. These factors are the basis of complementary intercrops that are widely grown, such as mixtures of cereal and legume species, for example, maize–bean and sorghum–cowpea (*Vigna unguiculata* L. Walp.). Not only is resource utilization enhanced through the complementarity of these species combinations, but fostering a diversity of species in a farming system will enhance pest resistance through promoting beneficial insects and suppressing pest outbreaks (Knops *et al.*, 1999).

There is a diversity of traits present even in closely related organisms. This can be illustrated for domesticated legume species (Fig. 3.12). Many have contrasting growth habits, which have consequences for the harvest index and the nutrient status of residues. If a plant has a short maturation period and a determinant growth habit, then it will usually have a high harvest index. There has been a considerable plant breeding effort over decades to select for greater determinacy, resulting in a wide range of yield potential and plant growth types. An example of the range possible is provided by pigeon pea, a species that is a short-lived perennial managed as an annual or biannual, where indigenous varieties may take more than 300 days to mature. Plant-breeding efforts have developed very early maturing pigeon pea

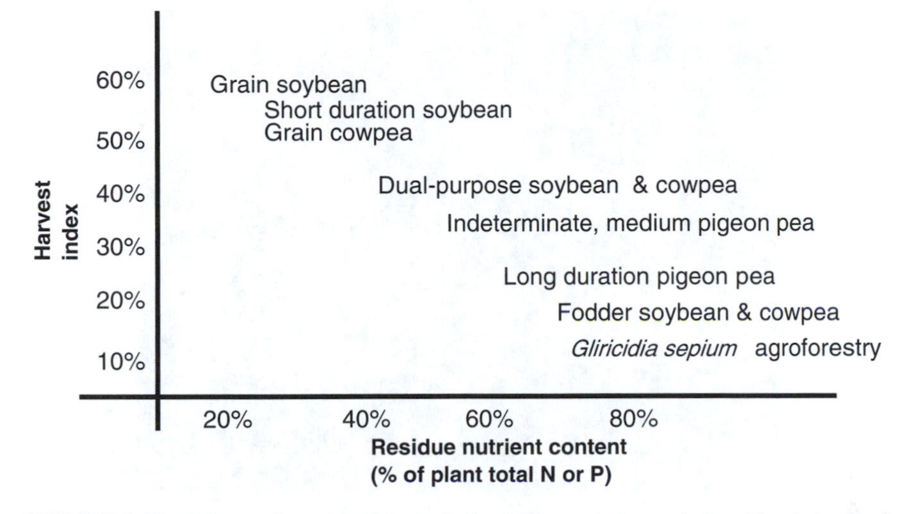

FIGURE 3.12 Legumes domesticated for agricultural use vary in harvest index; this trait is related inversely to nutrient content of plant residues as nutrient off take is high as harvest index increases in high-yield potential crops.

growth types that can mature in less than 80 days. This provides a tremendous diversity of plant growth types for farmers to experiment with and to integrate into a farmscape.

There is generally a trade-off between residue nutrient content in determinant and indeterminant growth habits (see Fig. 3.12). That is, determinant plants have limited amounts of low nutrient content residues, as nutrients have been remobilized to reproductive tissues and removed as harvestable products. In contrast, plants with a long maturation period and indeterminant growth habit often have nutrient-enriched residues. Although yield potential may be limited, multiple benefits are associated with plant types that combine a modest amount of food production with nutrient-enriched vegetation that can be used as a vegetable, as livestock fodder, or to build soils. An example is provided by long season, climbing bean genotypes, which fix more N and acquire more P than determinant bean types.

Indeterminant, long-duration plants are generally successful candidates grown as multipurpose crops on field margins and around field perimeters. It is important to take into account the range and type of products that indeterminant plants produce. Multiple harvests of leaves for vegetable use may occur along with grain yield at the end of the growing season, and yield potential may be substantial from indeterminant plants. Farmers often value these secondary products highly, but they are difficult to measure as they require labor-intensive, multiple harvests and attention to complex quality traits associated with high moisture vegetables or medicinal products. For all of these reasons, secondary products may be undervalued by researchers. Many legumes produce, in addition to grain, vegetable products, such as pods and leaves. Small amounts of fuel wood, construction materials, and leaves

FIGURE 3.13 Mali farmers value sorghum and millet stover for many uses, including livestock fodder and construction materials.

are produced by indeterminant crops, such as sorghum (Fig. 3.13). Stover from dry land crops are essential livestock fodder sources. Perennial agroforestry species, such as *Gliricidia sepium*, are shown on the lower right corner of Fig. 3.12, as these plants have few or no immediate food products (harvest index approaching zero), but have many secondary products, including soil fertility enhancement, forage, fuel wood, and construction materials.

The design of complementary species mixtures must take competitive interactions into full account. Ideally, plants with complementary root and shoot systems can be grown together. As shown in Fig. 3.14, pigeon pea has a deep tap root compared to an intercrop determinant species, thus minimizing competition with a shallow-rooted crop, such as peanut (also termed groundnut; *Arachis hypogaea* L). Pigeon pea has a very slow growth rate initially, which facilitates temporal compatibility with most crops. However, it is important to consider the tremendous plasticity of root systems. If the topsoil has substantially higher nutrient and water content than the subsoil, then the vast majority of perennials will explore the topsoil, as well as foraging deep, and thus will directly compete with shallowly rooted annual crops.

Active management in agroforestry systems is critical to successfully suppressing tree root activity in favor of crop roots. Management practices include frequent shoot pruning, which enhances root die off, and partial burning of tree branches or aboveground portions of a tree on an occasional basis. Careful placement of trees is one of the most effective means to separate tree and crop roots. To this end, trees are often grown along terrace bunds, in furrows between ridges, along perimeter mounds, or separated in time, as illustrated by improved fallow systems (see Chapter 4).

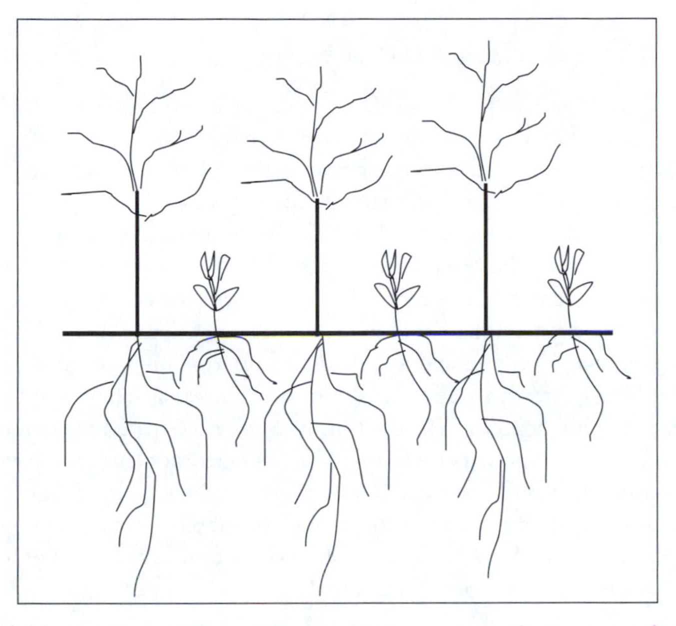

FIGURE 3.14 Spatially and temporally compatible legume intensification system: long-duration pigeon pea intercropped with medium-duration soybean or peanut.

In the subhumid tropics, N is the nutrient that limits productivity by a substantial margin (phosphorus and micronutrients rarely limit growth, except in the more extreme environments of specific soil types and in very arid or humid climates). Thus, minimizing competition for N is important, which is often addressed by using legume intercrop species. However, the presence of a legume does not guarantee an effective symbiosis. The consequences of intercropping maize with the nonfixing legume *Senna spectabilis* is presented as a cautionary tale (Box 3.2). There are inherent time demands associated with managing multiple species and inherent biological variability, which can be significant barriers to adoption of agroforestry systems (Sirrine *et al.*, 2008).

Sequential rotation systems are one of the most successful means of integrating complementary species over time, where sufficient land is available to follow this practice. In the Andes, a traditional system is the 7-year potato crop rotation that incorporates (a) a nutrient-accumulating phase (several years of grazed pasture with manure additions, 1 year of grain lupin [*Lupin mutabilis* L.], which is a N-fixing and P-solublizing legume followed by a small grain such as barley), which builds soil aggregation and organic matter, and (b) a high nutrient-demanding tuber crop, grown every 7th year, such as potato (or at higher altitudes indigenous tubers such as oca, *Oxalis tuberosa*), which provides cash income and a staple food product. The 7-year cycle may seem long, but it disrupts and suppresses a major nematode pest with a 6-year life cycle. The diversity of crops (six crop species) over time and space

BOX 3.2 Competition and Hedgerow Intercrops

Senna spectabilis is a tropical legume that has been promoted as a hedgerow intercrop species for agroforestry systems. It has N-enriched leaves (≈3.2% N) and is an easy-to-establish species, one that produces large amounts of N-enriched biomass faster than many N-fixing species (Table 3.3). However, the discovery that *S. spectabilis* did not fix N symbiotically raised an urgent question: where was the N-enriched status derived from? Plant traits that support high leaf N content in *S. spectabilis* were shown to include a highly extensive, plastic root system that could branch rapidly in the presence of inorganic soil N and rapid N uptake and assimilation capacity. This agroforestry species acts as a highly effective weed species and has traits that maximize its ability to compete with cash crops. Promotion of *S. spectabilis* based on tremendous biomass production potential was an insufficient criterion and may have been based on performance-on-research station trials, where soils of high organic matter may not have been representative of smallholder farm environments. Biological review of *S. spectabilis* and testing on farm revealed the highly competitive nature of this species and its unsuitability as an intercrop species.

TABLE 3.3 Information on Tropical Legume Biochemical Composition Based on the Tropical Residue Quality Database Developed by Palm *et al.* (2001)

Species (Latin name)	Leaf biochemical composition[a]	Growth duration	Uses
Mucuna	3.8% N 4.0% sPoly 6.6% Lignin	Annual indeterminant viney bush (long duration annual)	Soil fertility enhancement; grain (requires processing)
Pigeon pea (*Cajanus cajan*)	3.5% N 3.0% sPoly 10% Lignin	Short-lived perennial bush; annual varieties (termites and nematodes reduce life expectancy)	Grain; vegetable pods; fuel wood; forage; soil fertility enhancement; medicinal
Crotelaria species	4.1% N 2.6% sPoly 4.0% Lignin	Short-lived perennial bush; annual varieties	Soil fertility enhancement; forage
Tephrosia vogelli	3.0%N 5.9% sPoly 8.0% Lignin	Short-lived perennial bush (termites reduce life expectancy)	Soil fertility enhancement; fuel wood; improved fallow

TABLE 3.3 (*Continued*)

Species (Latin name)	Leaf biochemical composition[a]	Growth duration	Uses
Sesbania sesban	3.4% N 3.8% sPoly 6.7% Lignin	Short-lived perennial tree	Soil fertility; poles; improved fallow
Senna spectabilis	3.1% N 3.4% sPoly 15% Lignin	Perennial nonfixing legume tree	Soil fertility enhancement as a hedgerow
Gliricidia sepium	3.5% N 3.8% sPoly 15.5 % Lignin	Perennial, managed as a bush	Soil fertility enhancement as a hedgerow; fuel wood
Leucaena	3.0% N 8.8% sPoly 16.7% Lignin	Perennial, managed as a bush	Soil fertility enhancement as a hedgerow; fuel wood

[a]N, nitrogen; sPoly, total soluble polyphenols as a percentage of leaf weight; lignin, complex polymer that acts as a binder in cell walls.

provides pest cycle disruption, a habitat for beneficial insects, and a wide range of food species that buffer against weather risk. Traditionally, these complex rotation systems were systematized by dividing fields into seven or more parcels and rotating crops through the designated areas (Fig. 3.15).

FIGURE 3.15 An Andean 7-year rotation cropping system involves partitioned fields and planned sequences of crops to suppress pest populations.

An interesting multiple species system used in the subtropical and subhumid tropics of Australia and in the Americas revolves around rotational pasture grazing. A mixture of grasses and forage legumes is rotationally grazed by cattle. This system optimizes plant growth and quality by carefully controlled and intermittent grazing. Controlled grazing stimulates forage regrowth, maximizing the production of highly palatable vegetative tissues. In a multispecies version of this system, poultry are either included with cattle or are sequenced immediately after cattle graze using movable poultry pens. This is a labor-intensive, but ecologically sensible, design. Poultry are one of the most energy-efficient forms of livestock and are highly effective at consuming grubs, including cattle pests that thrive in manure. Thus system performance and resilience to pests are both enhanced by including organisms that fit different components of a complex food web (Fig. 3.16).

Animal–crop interactions involve careful consideration of species composition, weather, and timing. Climate variability can be addressed through attention to complementarity in species choice and utilizing the mobility inherent in livestock that can be used to reduce, or intensify, grazing as required. Combinations of species can reduce risk and improve resource utilization. If one plant or animal species does not thrive at a specific locale in a given year because of a precipitation or temperature regime that year, then another species has the potential to compensate. To optimize the productivity of farming systems, attention to plant associations is essential, as plants are the primary producers on the farm, supporting livestock, the soil food web, and, ultimately, the human consumers (see Fig. 3.16). Agroecological principles of complementarity provide a base for the design of plant associations over time and space, where animal interactions are beginning to be elucidated as well.

Redundancy

A high degree of redundancy is often found in natural ecosystems, in addition to complementarity. Redundancy has the potential to suppress productivity, as competitive interactions are high when similar organisms are grown together. However,

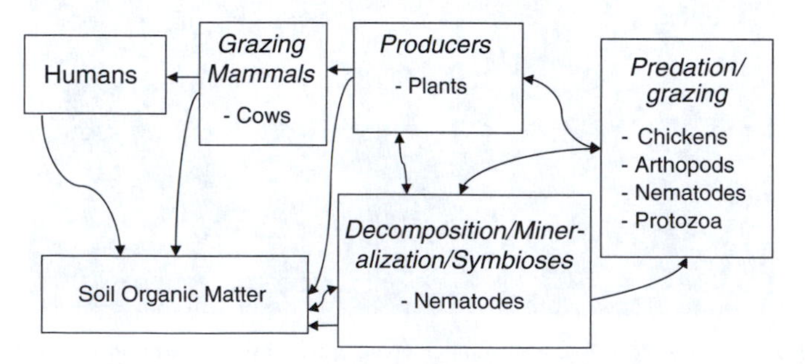

FIGURE 3.16 Soil food web associated with a pasture farming system.

this problem has been minimized in cropping systems as generations of breeders have selected for ideotypes that have high tolerance to dense populations. Examples include wheat and maize; these crops illustrate how an upright plant type and erect leaves can minimize intraspecies competition for light. Advantages of managing for a high density of plants with redundant features can include reduced pest problems and an enhanced ability to exhibit compensatory growth (Ewel, 1986).

The environment plays a large factor in the relevance of redundancy features to agroecosystem design. If water is not scarce, this strategy has great potential for success in tropical environments, as light is rarely limiting in the tropics and competition for water is the major factor that reduces productivity in close plant associations. Conversely, there are many advantages to "buildin" redundancy under high moisture environments, as multiple layers of leaves will reduce rainfall impact and protect the soil resource from erosion. In addition, phosphorus nutrition and availability to crops depend on the organic fraction of soils in many humid tropical soils. Highly leached soil chemistry has a high P fixation capacity that can be circumvented through maximizing P uptake in plant residues and cycling P through the organic pool, which requires large amounts of vegetative growth.

Redundancy is a feature of cropping systems based on genotypes that exhibit diversity within, as well as between, species. It is often possible to find species that have overlapping sets of traits, with stress responses that are related, but distinct. Enhanced pathogen resistance can be achieved, for example, through combining isogenic or closely related varieties that have slight differences in plant resistance genes. This has been termed a "multiline" approach and has been adopted to control disease outbreaks in Chinese rice production. The deliberate combination of varieties is a useful strategy, as complementarity between specific genes is united with the redundancy inherent in combining closely related genotypes.

An intercrop system that has both complementary and redundancy features is a legume–legume intercrop that combines early and late maturing species (see Fig. 3.14). An example is a pigeon pea–peanut intercrop, which is grown by smallholder farmers in India. A variation of this system is the soybean–pigeon pea intercrop, which is being experimented with in northern Malawi (Snapp *et al.*, 2002; see Fig. 3.8B). This intercrop has the redundancy of including two legumes. The N fixation pattern of the two species mixture is expected to be of longer duration then the N fixation of either crop grown alone. Furthermore, species vary in tolerance to various stresses, and the combination provides a buffered response to a stressful environment. If we consider the earlier example, pigeon pea is sensitive to flooding events, but relatively tolerant of soil acidity. Thus the combination of pigeon pea and soybean will be able to respond to either stress through the redundant presence of two N-fixing symbioses.

The intermediate growth habit of a short-lived, N-fixing shrub has many redundant features. The moderate stature and indeterminant growth habitat are generally adapted to browsing by mammals and insect herbivory, with rapid regrowth capacity and unique suites of plant biochemical compounds. The tissues of shrubby species often have defense compounds such as polyphenols that provide intermediate effectiveness against herbivores, but not the highly lignified or waxy tissues of long-lived trees or the high palatability of vegetative tissues of annuals (see Table 3.3). Investment in relatively "low-cost" defense compounds such as polyphenolics is evolutionarily sensible for tissues of an intermediate life span. Examples of common tropical legumes, including leaf composition and plant growth type, are shown in Table 3.3. Tissue biochemistry varies markedly in these species and is influenced by soil nutrient status, tissue type, and age of the plant organ. The consequences of plant biochemistry are only beginning to be understood, including the impact of residues on soil organic matter, on N mineralization, and on the diet of insects and mammals. Legumes play unique, multipurpose roles in farming systems, but require careful testing to determine the long-term influence of above- and belowground residues.

Mosaics

Landscape ecology provides many insights into the impact of land use structure and the function of agricultural systems (Wojtkowski, 2003). For example, a mosaic pattern of growth has been shown to be common in natural ecosystems. Localized disturbances tend to bring about a mosaic of diverse age structure in a community, as long-lived plants are blown over or uprooted by storms and young plants colonize the location where light and nutrients are suddenly available. In a similar manner, perennial and annual plantings of different age structures in an agricultural landscape can be planned as strategies both to reduce risk and enhance returns. A mosaic pattern is often associated with more stable productivity over time than more uniform land management. This is in part because of climatic variability, as extremes in weather will often be tolerated by some of the diverse species present in mixtures spread across a landscape, while a monoculture of any given species could be devastated.

High productivity is common at the boundaries of different land uses, such as the interface of perennial trees and annual fields. An edge effect is often observed on the perimeters of experimental plots as well, which is why yield measurements are conducted toward the center of plots, away from a potentially distorting edge. This has led to the contention by some that mosaic land use is inherently more productive than monoculture agriculture. Indeed, light availability and altered wind patterns are some of the processes at work that influence yield potential at edges and interfaces between land uses. However, productivity will vary greatly over a mosaic of widely varying plant type and mixtures. The presence of perennials displaces some annuals, which leads to reduced production potential for staple foods across the entire land area, despite local improvements in yield potential. If land access is limited, then the success of a mosaic may depend on the presence of perennials that produce relatively high value crops. For example, if perennials of different age and size categories produce marketable products such as coffee or nuts, then mixtures of perennials may be economically feasible as well as environmentally sensible.

A mosaic land use pattern may be particularly valued by managers facing diverse landscapes and changeable weather. Farmers contending with a highly variable climate will find a diversity of plant types and age classes important as means to buffer weather extremes and erosive forces. Mosaics are most suited to farmers engaged in marketing a range of products, in high rainfall areas, with access to sufficient land. In contrast, high population density areas with small farm sizes and relatively uniform conditions will tend to prioritize productive cereals such as maize or rice.

Microclimate variation is high in mountainous regions, and small differences in elevation can be utilized in areas that have topography. The Andes has been an ideal location to develop highly complex cropping systems that utilize diverse altitude niches (Fig. 3.17). This spreading of crops across the landscape and at different elevations prevents catastrophic crop failure from localized weather or insect pests. Producing seed crops is often carried out at higher altitudes to take advantage of low insect and virus loads in colder zones. Instead of using expensive insecticides, the use of high-altitude ecozones is a prevention strategy that takes advantage of the isolation and minimal insect pressure at high-altitude locations, ideal conditions

FIGURE 3.17 Mountainous terrain and alluvial valleys provide a wide range of niches for growing crops at different altitudes and in a range of environments, a common risk avoidance strategy in the Andes.

for producing high-quality seed. This strategy is particularly important for clonal propagation materials, such as potato tubers, for seed.

The interplay of socioeconomic and biophysical complexity is nowhere more dramatic than at the interface of land and water. This is often the most contentious of land use areas, with very high production potential, and a mosaic of diverse and often conflicting management objectives. Farmer management is often informal but intensive, as it is beginning to be appreciated for drainage zones, wetlands, and riverine environments throughout the developing world. The Amazon flood plain, for example, was once thought to be a natural area exploited by farmers through recessional planting after flood waters recede. However, recent findings illustrate that human interventions along the Amazon include centuries of channel building, mound erection, and soil dredging to replenish soil fertility and intensify crop production. In China, the recycling of nutrients through dredging and hauling soil from waterways to fields was historically a primary nutrient cycling pathway, one that required considerable labor but was highly effective.

Land use is a dynamic process. Extensive use, such as livestock grazing, is replacing crop production along many Latin America waterways, whereas the opposite trend of intensification is occurring along drainage zones in southern Africa. Promoting farmer experimentation and diversification of land management requires close attention to indigenous knowledge and current land use patterns, as highly valued crops and nutrient responsive crops are often located very precisely within intricately managed landscapes (see Fig. 3.7). Alterations in land use have long-term ramifications that are difficult to predict. In sum, management of mosaics is a complex undertaking that requires attention to climate, topography, and stakeholder objectives, as well as a long-term planning horizon.

FARMING SYSTEM BY AGROECOLOGY

Cropping systems challenges are specific to climatic zones. Agroecology design requires comprehensive understanding of variability in moisture and temperature gradients and in soil resources. For example, the length of the growing season will determine the intensification options that can be pursued. If cold or dry conditions limit the growing season to a few months, there will be few viable plant-based technologies that restore soil fertility, which increases the requirement for external inputs. Extensive options, such as silvopastoral systems, are also well suited to short season climates. An agroecological perspective is presented here on the specific challenges faced in dry versus humid environments.

Arid environments are marginal for many farming system endeavors, as plant productivity is limited by insufficient and erratic rainfall. Return to inputs is often limited and highly variable, increasing risk markedly. It is difficult to predict when or where to apply external inputs and labor. Livestock play a unique role in dryland farming, being able to move in response to climatic variability and concentrate nutrients in a low productivity environment. Grazing animals convert low-quality plant residues into valuable products, such as meat, milk, and manure. Livestock integration with crop production helps reduce risk through transfers between the two systems. Application of manure builds soil organic matter and water-holding capacity for improved crop production, whereas crop residues are crucial components of livestock feed. Box 3.3 illustrates how climate and farmer investment in fodder and animals interact to determine resource use efficiency in crop–livestock systems.

Extensive versus intensive use of land and labor is a challenging question in dry environments. Many traditional systems rely on reducing risk by minimizing investment to seeds. Planting seeds may be required several times over, if the start of the growing season involves sporadic rains, as is common in West Africa. Weed management is often not a major investment in this dry environment, as only a few plants can survive and weeds are removed by farmers for use as fodder or food. If they are supported by access to market opportunities, many farmers are genuinely interested in making larger investments in crop production. A successful example of a productivity-enhancing investment that utilizes soil biota is the Zai hole, a technology developed in northern Burkina Faso and being experimented with widely in West Africa (Kaboré and Reij, 2004). Small basins are dug in land to be rehabilitated and handfuls of manure are added, which attracts termites that dig channels and improve water infiltration. The basins capture wind-blown residues, and

BOX 3.3 Livestock and Crop Integration

The ratio of livestock to crops varies depending on aridity, the extent of grazing area and cropland, and socioeconomic context. An animal unit of one cow, or two small ruminants, can produce 1 to 2 tons of manure annually and requires about a hectare of grazing land. However, the quality of plants grown will markedly alter the area of required grazing, from 15 ha of dry savannah per cow in southern Zimbabwe to 0.2 ha of planted legume–grass fodder per animal in Kenyan stall-fed dairy systems. The amount of manure required to support cropland will also vary, depending on feed quality and animal species. Approximately 5 tons/ha of manure will appreciatively improve cereal grain yields, based on recent findings from subhumid and semiarid on-farm trials carried out in east and southern Africa (Ncube *et al.*, 2007). Recommended rates of manure application tend to be higher, as much as 10 to 40 tons/ha for maize production in Zimbabwe. Overall, crop–livestock system ratios of about 5:1 appear to be sustainable: five animals on 5 ha of grazing land for every hectare of cropland. Access to land is often insufficient to support this ratio, and innovations in fodder and livestock systems are required, as discussed in depth in Chapter 8.

the site-specific concentration of nutrients and water supports plant growth in the Zai hole.

Targeted input use can markedly enhance crop tolerance to environmental stress. Irrigation is one of the most widely used technologies to enhance return to other investments, such as fertilizer and high-quality seed. However, irrigation is an expensive technology that is only applicable where water is available and economic returns are sufficient. A novel technology for dry areas is the use of "microdosing," where small doses of fertilizer (5 to 10 kg nutrients/ha) are point applied directly to the base of a plant. This supports the growth of healthy plants, with large root systems, and has been shown to markedly improve drought tolerance. Microdosing of sorghum with phosphorus fertilizer in West Africa has been shown to markedly improve water use efficiency and yield potential for environments ranging from arid to semiarid (Buerkerht *et al.*, 2001). The economic risk of fertilizer use is considerable in a dry environment, and technologies must be tested thoroughly over long time horizons to assess climatic risk fully. A successful agricultural development strategy in West Africa has been the combination of biological risk mitigation through microdosing and inventory credit systems to reduce economic risk (see "smallholder productivity" PowerPoint presentation on the companion Web site).

To summarize, the dry environment strategies discussed here involve either (1) flexible and extensive approaches, such as concentrating nutrients and energy through grazing and manure transfers to cropland, or (2) intensification, targeted to specific locations. Examples include watered niches in dry environment and targeted microdoses of fertilizers or organic amendments. Understanding trajectories of intensification in marginal environments is challenging and requires multidisciplinary teamwork between social and biological scientists.

Humid environments face quite different challenges from arid ones. Biological productivity is high and growth-limiting factors tend to be competition from weeds and herbivory by insects. Disease is another growth-limiting factor for both plants and animals. Understanding processes and timing of interventions are critical in this rapid growth environment. Biologically sensitive management in the humid tropics, for example, requires knowledge of pest and predator growth dynamics. Not only is this essential to integrated pest management, but also to belowground pest management and organic matter decomposition. Interestingly, research indicates that timing is critical to the health of crops planted into soil amended with organic residues (e.g., manure and leaf litter). Widespread seedling damage from grubs, termites, and soil-borne root rot organisms will result if crops are planted and insufficient time is allowed for decomposition, as facultative organisms on decaying residues transfer to young, vulnerable seedlings. However, if sufficient time between organic amendment and planting is allowed, then a diverse soil community asserts itself; there are many examples of specific suppression of soil disease and parasitic organisms.

A key challenge in the humid and subhumid tropics is competition from weeds. Weed management efforts take up the majority of labor inputs in many cropping systems, as farmers are caught in a cycle of weeding and generation of weeds.

Commonly, weed management relies on shallow tillage with an oxen plow or hand hoe, which leads to soil disturbance that enhances weed germination. Managing these weeds requires further disturbance, which promotes yet more weeds. This is a difficult cycle to break, but novel technologies are used to alter the seedbed environment, such as dust mulchers, which cut rather than turn over weeds and thus reduce surface disturbance, and the exposure of weed seed to germination conditions (Renner *et al.*, 2006). Use of a "stale seedbed" relies on a regime of high initial disturbance to kill several generations of weeds before planting a crop with minimum disturbance into this prepared bed.

Ecologically based weed management relies on principles such as asymmetry competition, where early crop growth is enhanced relative to weed growth. If a high-density planted crop, or augmented crop (e.g., transplants), can outcompete weeds initially, this will allow canopy closure and weed suffocation through denying sunlight, water, and nutrients. Cereals are ideal candidates for this approach, as they have an erect shoot growth habit that facilitates a high density of plants within a row and rapid achievement of height, shading weeds. Limiting resource availability between rows, through point-applied water and nutrients, will support an asymmetrical cropping system design. An intercrop system can also be designed to maximize rapid canopy closure and expansion of the leaf area index, as shown for the triculture discussed earlier of maize, bean, and squash. In this widely grown intercrop, maize provides a fast-growing element that initially shades weeds, the bean climbs the vertical maize and increases leaf area index, and squash forms a ground layer that shades weeds attempting to germinate.

Overall, management of moist environments requires understanding of processes that control trade-offs in investments. Enhancing fertility will have little or no effect on productivity without augmenting the control of weeds and other pests. Smallholder farmers have limited resources and need advice about less than ideal systems, such as combinations of modest investments versus large investments in either nutrient augmentation or weed control (Snapp *et al.*, 2003).

APPLICATION OF ECOLOGICAL PRINCIPLES TO A CHANGING WORLD

The ecological literature has developed the useful concept of a dynamic equilibrium (Botkin, 1990), replacing historical views of ecosystem evolution moving toward a climax system. The theory is that ecosystems evolve with feedback between the environment and organisms, each influencing each other so that a permanent steady state does not occur but rather equilibrium states are achieved that offer a temporary optimum balance. For example, infertile soils on a sand dune site condition the type of plants that will thrive at that site. Over time the pioneering plants and their symbionts enrich the dune soil through N fixation and mineralization, which sets up a new state whereby a different set of plants is favored. Eventually—as longer lived plants such as trees become established and residues shift toward acidifying and recalcitrant tissues—the soil environment is slowly altered again, and the dynamic equilibrium continues. Plant and symbiont evolution within agricultural systems may be following similar dynamic equilibrium patterns, although more research is required. For example, mycorrhizal fungi may be parasitic within fertilized systems compared to low-input systems (Kiers *et al.*, 2002). Agroecological management may involve selecting plants and associated organisms to perform well within a changing, and minimally resourced, environment.

A cropping system rotation can be viewed as a rapid form of plant succession. Many cropping systems involve deliberate alternation of complementary plant types, including following nutrient-enriching plant sequences with nutrient-demanding plants. For example, a rotation might start with a legume forage that builds soil fertility (e.g., alfalfa, *Medicago sativa* L.), followed by a nutrient-demanding and vigorous C4 grass (e.g., maize), then a moderately nutrient-enriching crop such as a legume grain crop (e.g., soybean), and finally a C3 grass (e.g., wheat) that provides effective soil cover and high surface rooting density to regenerate compacted soil. In some Mediterranean systems, wheat is intercropped or rotated with a very deep tap-rooted brassica (e.g., rapeseed), which enhances nutrient recycling and provides "break crop" properties by altering soil biology drastically.

In summary, agroecology requires flexibility: design for a dynamic rather than a steady-state system. It requires understanding of spatial heterogeneity over time and space, and a planning horizon of many years. Outdated concepts in sustainable agriculture are based on the transfer of technologies and set recommendations. In contrast, agroecology is based on knowledge, participatory action research, and education. Farmers and other rural stakeholders have local knowledge and access to the raw ingredients of plants, animals, and natural resource; agroecology has the mandate to broaden the range of organisms and technologies available and to provide support for the development of knowledge and adaptive responses to rapid change.

REFERENCES AND RESOURCES

Bezner Kerr, R., Snapp, S., Chirwa, M., Shumba, L., and Msachi, R. (2007). Participatory research on legume diversification with Malawian smallholder farmers for improved human nutrition and soil fertility. *Exp. Agric.* **43,** 437–453.

Botkin, D. (1990). "Discordant Harmonies: A New Ecology for the Twenty-first Century." Oxford University Press, New York.

Buerkert, A., Bationo, A., and Piepho, H. P. (2001). Efficient phosphorus application strategies for increased crop production in sub-Saharan West Africa. *Field Crops Res.* **72,** 1–15.

Chapin, F. S., III, Autumn, K., and Pugnaire, F. (1993). Evolution of suites of traits in response to environmental stress. *Am. Natur.* **142,** S78–S92.

Cox, T. S., Glover, J. D., Van Tassel, D. L., Cox, C. M., and DeHaan, L. R. (2006). Prospects for developing perennial grains. *BioScience* **56**(8), 649–659.

Drinkwater, L. E., and Snapp, S. S. (2007). Nutrients in agroecosystems: Re-thinking the management paradigm. *Adv. Agron.* **92,** 163–186.

Ewel, J. J. (1986). Designing agricultural ecosystems for the humid tropics. *Ann. Rev. Ecol. Syst.* **17,** 245–271.

Gallaher, C. (2007). "Phosphorus Availability in Annual and Perennial Cereal Legume Systems." M.S. thesis Michigan State University, East Lansing, MI.

Gause, G. F. (1934). "The Struggle for Existence." Williams and Wilkins, Baltimore, MD.

Gliessman, S. R. (2007). "Agroecology: The Ecology of Sustainable Food Systems," 2nd Ed. CRC Press, Boca Raton, FL.

Grime, J. P. (2001). "Plant Strategies, Vegetative Processes, and Ecosystem Properties," 2nd Ed. Wiley, New York.

Jackson, L. L. (2002). Restoring prairie processes to farmlands. *In* "The Farm as Natural Habitat: Reconnecting Food Systems with Ecosystems" (D. L. Jackson, and L. L. Jackson, eds.). Island Press, Washington, DC.

Kaboré, D., and Reij, C. (2004). "The Emergence and Spreading of an Improved Traditional Soil and Water Conservation Practice in Burkina Faso." EPTD Discussion Paper No. 114. [E. a. P. T. D. I. F. P. R.] Institute, Washington, International Food Policy Research Institute: 28.

Kiers, E. T., West, S. A., and Denison, R. F. (2002). Mediating mutualisms: Farm management practices and evolutionary changes in symbiont co-operation. *J. Appl. Ecol.* **39**(5), 745–754.

Knops, J. M. H., Tilman, D., Haddad, N. M., Naeem, S., Haarstad, J., Ritchie, M. E., Howe, K. M., Reich, P. B., Siemann, E., and Groth, J. (1999). Effects of plant species richness on invasion dynamics, disease outbreaks, insect abundances and diversity. *Ecol. Lett.* **2,** 286–293.

Ncube, B., Dimes, J. P., Twomlow, S. J., Mupangwa, W., and Giller, K. E. (2007). Raising the productivity of smallholder farms under semi-arid conditions by use of small doses of manure and nitrogen: A case of participatory research. *Nutr. Cycl. Agroecosyst.* **77,** 53–67.

Oberson, A., Friesen, D. K., Tiessen, H., Morel, C., and Stahel, W. (1999). Phosphorus status and cycling activity in native savanna and improved pastures on an acid low-P Colombian Oxisol. *Nutr. Cycl. Agroecosyst.* **55,** 77–88.

Palm, C. A., Gachengo, C. N., Delve, R. J., Cadisch, G., and Giller, K. E. (2001). Organic inputs for soil fertility management in tropical agroecosystems: Application of an organic resource database. *Agric. Ecosyst. Environ.* **83,** 27–42.

Patt, A., Suarez, P., and Gwata, C. (2005). Effects of seasonal climate forecasts and participatory workshops among subsistence farmers in Zimbabwe. *Proc. Natl. Acad. Sci. USA* **102,** 12623–12628.

Piha, M. (1993). Optimizing fertilizer use and practical rainfall capture in a semi-arid environment with variable rainfall. *Exp. Agric.* **29,** 405–415.

Renner, K., Sprague, C., and Mutch, D. (2006). "Integrated Weed Management: One Year's Seeding," p. 130. Michigan State University Extension Bulletin E2931, Michigan State University, East Lansing, MI.

Sirrine, D., Shennan, C., Kanayama-Phiri, G., Kamanga, B., and Snapp, S. S. (2008). Agroforestry, risk and vulnerability in southern Malawi: Improving recommendations resulting from on-farm research. *Agric. Ecosyst. Environ.* (in press).

Snapp, S. S., Blackie, M. J., and Donovan, C. (2003). Realigning research and extension services: Experiences from southern Africa. *Food Policy* **28,** 349–363.

Snapp, S. S., Kanyama-Phiri, G., Kamanga, B., Gilbert, R., and Wellard, K. (2002). Farmer and researcher partnerships in Malawi: Developing soil fertility technologies for the near-term and far-term. *Exp. Agric.* **38,** 411–431.

Sperling, C. R., and King, S. R. (1990). Andean tuber crops: Worldwide potential. *In* "Advances in New Crops" (J. Janick and J. E. Simon, eds.), pp. 428–435. Timber Press, Portland, OR.

Tapia, M. E. (2000). Mountain agro-biodiversity in Peru. *Mount. Res. Dev.* **20,** 220–225.

Tittonell, P., Vanlauwe, B., Leffelaar, P. A., Shepherd, K. D., and Giller, K. E. (2005). Exploring diversity in soil fertility management of smallholder farms in western Kenya. II. Within-farm variability in resource allocation, nutrient flows and soil fertility status. *Agric. Ecosyst. Environ.* **110,** 166–184.

Wojtkowski, P. A. (2003). "Landscape Agroecology." Haworth Press, New York.

INTERNET RESOURCES

http://www.agroeco.org/index.html. Agroecology in Action. Web-based courses and case studies of applying ecology to agricultural systems are presented at this site, which has examples from Latin America and beyond.

http://www.agroecology.org/case.html. This Web site presents international case studies—California and Latin America are well represented—that illustrate agroecological principles and practice.

http://www.icimod.org/home/. Web site of the International Centre for Integrated Mountain Development, Kathmandu, Nepal, with links to mountain agroecological efforts around the globe.

http://ipmnet.org/cicp. The Database of IPM Resources: A compendium of customized directories of worldwide IPM information resources accessible through the Internet.

http://www.iwmi.cgiar.org/africa/West/projects/Adoption%20Technology/Technology_Adoption.htm. This Web site describes a wide range of resource conservation technologies, and the extent of adoption in sub-Saharan Africa.

www.landinstitute.org. The Land Institute Web site is hosted by this pioneering organization and links to efforts to develop perennial grain crops.

www.nwaeg.org. New World Agriculture and Ecology Group (NWAEG) is an international organization that analyzes the problems of contemporary agriculture and ecology in order to support the development of alternatives.

CHAPTER 4

Designing for the Long Term: Sustainable Agriculture

Sieglinde Snapp

Summary

Farming systems must simultaneously address production goals and environmental sustainability in a constantly changing world. A first step is to understand the sophisticated and complex systems farmers have devised. Sustainable technologies must be relevant to stakeholders, which require multifunctionality, flexible options, and risk mitigation. Biological principles for sustainable systems are explored here, including biodiversity, resource efficiency, economic production, and resilience. Diversity involves many dimensions and is discussed for individual plant and animal combinations, as well for the landscape. Processes and case studies are presented to highlight efficient nutrient cycling, resource conservation, and cropping system resilience. Finally, sustainable agriculture is discussed in terms of equitable development goals.

Agricultural Systems: Agroecology and Rural Innovation for Development

INTRODUCTION

In a world where change is occurring at an increasingly rapid pace, developing sustainable and adaptable farming practices is central to rural livelihoods. In locations where farmers are at the edge of survival, it is possible to see widespread cutting of trees for fuel wood or charcoal production, and tillage of steep slopes (Fig. 4.1). At the same time, farmers everywhere attempt to protect their children's

FIGURE 4.1 (**A**) Charcoal is often produced by those with few livelihood options, shown here being transported to urban markets in southern Malawi. (**B**) Production on steep slopes is also common in southern Malawi, where population densities are high and land is used intensively.

heritage, soil productivity, natural resources, and water quality. Technical advice for sustainable management should focus on improving adaptability and resiliency in the face of uncertainty.

There is no defined set of practices that comprise sustainable farm design because agroecosystem management takes place within an inherently dynamic (nonstatic) context. Increasingly, as the world economy continues to globalize and with projections of severe impacts due to climate change, farmers face an unpredictable environment (Fig. 4.2). Farmers are responding to these pressures in many ways. Some farmers are increasing intensified agricultural production, producing higher value crops, and using water control methods such as drainage and informal irrigation systems. Intensification is both an opportunity and a challenge: it can provide greater food security and the potential for income generation, but there are concurrent environmental and financial risks. Intensification involves increased labor

FIGURE 4.2 (**A**) Floods are becoming more frequent, devastating smallholder farms in southern Africa. (**B** and **C**) Increasing population pressures, globalization of markets, and drought are factors increasing the intensification of agriculture in many parts of Africa and enhancing reliance on irrigation systems such as treadle pumps.

and inputs, which requires farmer investment of scarce resources and often involves increased exploitation and tillage of areas that may be environmentally sensitive.

Farm families require technologies that help reduce risk and enhance productivity, but at the same time avoid biodiversity reduction and resource degradation. Conservation and productivity enhancement are not necessarily competing agendas. However, they may be points of contention and community tension. Agriculture can have a profound influence on water quality and supplies, and regenerative capacity of an ecosystem. Intensification of production in wetland and riverine areas, for example, is a highly efficient strategy for optimizing return to scarce resources and enhancing food supply around the year, but it can profoundly alter this environmentally sensitive land and water interface.

INDIGENOUS SYSTEM MODELS

Smallholders have devised sophisticated and complex farming systems that utilize the variability associated with specific microsites and climates. Along the slopes of Mt. Kenya, for example, farmers use their knowledge of soil type, climate, and market demand to determine which mixture of annual crops and perennials grow at different locations (Fig. 4.3). Because some crops are widely adapted and produce staple grains, they appear frequently in the landscape, such as rice and corn. Many

FIGURE 4.3 (**A**) A range of crops grown along mountainsides in eastern Africa.

FIGURE 4.3 Cont'd (**B**) Banana and mucuna are commonly grown at low spots in the landscape. (**C**) Maize and cassava are often grown on well-drained, steep slopes.

crops have specific climatic requirements, such as the adaptation of cotton to hot, dry conditions and clay soil types, or potato to cooler climates and well-drained soils. Box 4.1 illustrates the complex landscape management and farming methods used by smallholders in northern Thailand to produce hundreds of crops.

Agricultural development addresses the complex, heterogeneous environment of the smallholder through different approaches, including improved genetics, knowledge-based interventions, and livelihood strategies (Table 4.1). Variability in soils, climate, and water availability are often buffered, through inputs. New plants and animals are then

BOX 4.1 Innovations in Thai Farming Systems

Near Chaing Mai, in northern Thailand, mountain farming systems involve a tremendous diversity of vegetables, fruits, and grains for market and home consumption (Fig. 4.4). Thai mountain farming is evolving rapidly through farmer innovations. For example, integrated pest management and organic production have been improved using a technique called "netted farming."

FIGURE 4.4 Diverse vegetables (**A**) and fruits (**B**) sold in northern Thailand markets.

BOX 4.1 (*Continued*)

As shown in Fig. 4.5, crops are grown under tunnels of netted materials. This helps control—without pesticides—a range of tropical pests and produce unblemished vegetables for a market that demands high quality. In addition to controlling foliar pests, farmers are developing novel methods to control soil-borne pests. Compost preparations and biofumigants are being experimented with to prepare healthy soil in the planting beds and support vigorous crop growth.

FIGURE 4.5 Netted tunnels used for organic production of vegetables and flowers in northern Thailand.

introduced to utilize this high potential environment. Productivity gains have been tremendous from this "green revolution" approach to plant breeding. Local resource cycling efficiency is often enhanced as well, as plant genotypes are adopted that utilize nutrients and other inputs. Over the long term, the sustainability of this approach will depend on the level of inputs required, the resource use efficiency, and the scale of the system under consideration (see Table 4.1). The resource efficiency of a green revolution approach at a regional scale is going to be different from a local scale, as cropping systems with high yield potential varieties rely on mining for nutrient procurement as well as energy-intensive processing and transport.

There are complex ramifications of productivity gains, which do not lead directly to food security gains. That said, it is important to note that crop improvement has made recent progress, moving beyond irrigated systems in developing countries. In Latin America and Southeast Asia, yield gains of over 800 kg/ha have been documented over the last decade, improving average grain crop yield to 2700 kg/ha. In sub-Saharan Africa (SSA), however, the yield of staple grain crops—maize, wheat, and

TABLE 4.1 Sustainability and Agriculture Development Approaches within the Heterogeneous and Constantly Changing Environment Faced by Smallholders

Agricultural development	Genetics	Agroecology	Livelihoods
Historical approaches	Breed high-yield varieties for irrigated and fertilized environments Continuous breeding for resistance	Characterize environmental variation Design farming systems on agroecological principles	Characterize farmer livelihood strategies Visioning to develop new opportunities Catalyze development of value chains
New approaches	Breed varieties adapted to stressed environments Participatory plant breeding	Participatory action research to improve knowledge and innovation capacity Local adaptation of ecologically sensible options Web-linked WikiAg[a] information	Support effective, demand-driven extension through farmer organizations Education for nutrition and market opportunities
Socioeconomic constraints to sustainability	Ensuring smallholder access to new genotypes Seed systems	Inadequate education Conflicting interests Organizational inadequacies	Stable and transparent government Market and policy limitations
Biological constraints to sustainability	Challenges to reproduction of organisms Epidemics	Environmental constraints Erosion of resource base Reliance on external inputs	Insufficient options that are ecologically and economically sound

[a]Wikipedia approach to community refereeing of agricultural and technical information generation. The following Web site is a portal to agricultural information: http://www.iaald.org/index.php?page=infofinder.php.

rice—has remained at the same level for several decades, around 1000 kg/ha. This may in part be because of the risk mitigation strategies of SSA farmers.

Overall, genetic improvements related to stress tolerance have proved elusive to develop. The complexity of yield improvement in a heterogeneous, poorly resourced environment requires a multidisciplinary approach and a tremendous research effort. There have been promising developments in maize breeding targeted to smallholder farms in southern and eastern Africa based on multienvironment trials and participatory breeding. The maize genotypes are being bred with attention to locally acceptable quality traits, as well as improved tolerance to drought and, in a few cases, tolerance to low nitrogen (N) fertility (Bänziger *et al.*, 2000). Similar efforts are underway to improve stress tolerance and farmer quality traits into improved varieties of sorghum in West Africa and upland rice in Southeast Asia (see Chapter 6). As more genotypes adapted to smallholder environments become available, this will expand options and help build more sustainable farming systems.

Moderate-scale producers with limited resources tend to rely on strategies that take advantage of resource heterogeneity rather than focus on ameliorating variability through inputs. Agroecology-based approaches to development have focused on documenting this environmental complexity (see Table 4.1). Targeted planting of crops in different niches across a farmscape are risk avoidance strategies, as performance will vary with climatic conditions. Rather than optimizing yields, many smallholder farmers attempt to ensure sufficient food supply through multiple plantings and complex management practices. Indeed, farmers rely on multifunction systems that encompass not just fields, but "common use" areas of fallow, unimproved forage, savannah, semiwild borders, and woodland. The poorest among the community, in particular, often rely on marginal lands and small gardens for survival. Semidomesticated foraging areas often do not produce large amounts of staple food, but they do ensure the provision of multiple services, from protecting soil and water quality to maintaining cultural integrity (Bennet and Balvanera, 2007).

It is important to be aware that modernizing production by ameliorating a heterogeneous and semidomesticated area can have negative impacts on the environment and on disadvantaged members of the community. Rural people are keenly aware of the wide range of ecosystem services conditioned by land management choices: water quality regeneration, climate regulation, erosion and water flow control, disease regulation, and cultural services. These are integral to sustainable practice and have begun to be considered by economists as valued products in addition to the conventionally valued production of food, fiber, timber, and biofuels (Fig. 4.6).

BIOCOMPLEXITY OF SUSTAINABLE DEVELOPMENT

Agricultural development requires close attention to understanding processes involved in complex systems, as introducing new practices or technologies can have unpredictable consequences. Ways forward for researchers and extension educators are to work closely with farmers, pay attention to indigenous knowledge and local practice, and consider carefully different scenarios to evaluate the potential for negative, unintended consequences, as well as unexpected positive outcomes (see Table 4.1).

It is not just ecological consequences that researchers need to be aware of, it is important to consider power dynamics within communities as well. Complexities of social change, equity, and livelihoods are discussed in more depth in Chapters 8 and 9, but a cautionary example is provided here. Irrigation can greatly enhance

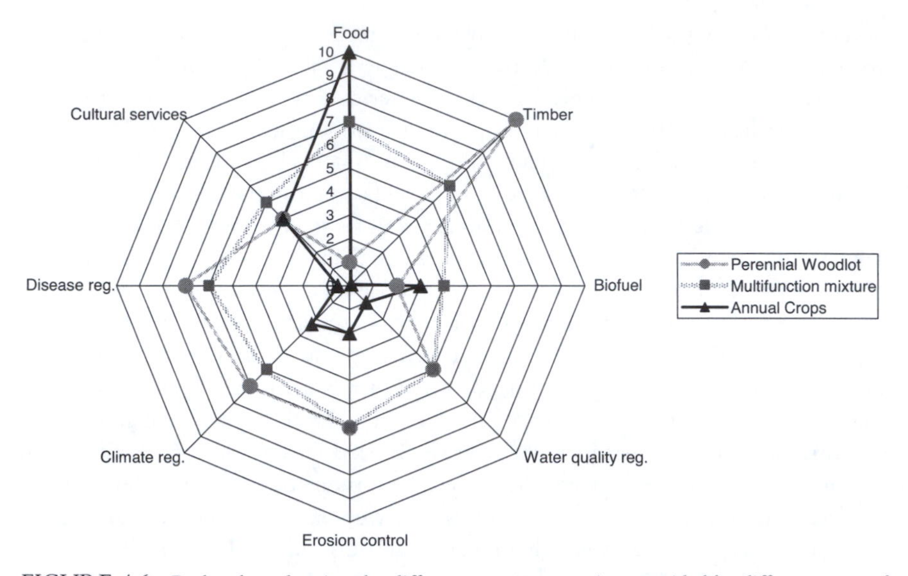

FIGURE 4.6 Radar chart showing the different ecosystem services provided by different types of land use systems, where annual crops are the most effective at producing food and woodlots at producing wood, whereas a mixed farm system can incorporate a wider range of services (adapted from Bennet and Balvanera, 2007).

production potential, but it is a particularly problematic development intervention. Research has shown that large-scale and mechanized irrigation schemes have the potential to degrade soil if not implemented properly through local salt buildups and soil quality decline. In addition, irrigation projects often alter community power relationships. Enhanced income and water access are captured primarily by male-headed and well-off families. This is most notable if customary land access pathways are disrupted through the implementation of formalized land titles and irrigation committee membership, which are often conferred on only one household member (e.g., the male head of household). Households using less formal irrigation methods, such as recessional flood agriculture, shallow well, or treadle pump irrigation, may lose out in the development process as the implementation of new irrigation schemes diverts water and alters land access. Irrigation projects are not implemented in a vacuum. This is illustrated by photographs from Malawi showing the intensity of informal irrigation and high value gardens displaced by a formal irrigation scheme (Fig. 4.7).

A partnership approach is suggested here as the foundation of sustainability. It values local, indigenous, and science-based knowledge. Enhancing the local knowledge base and innovation capacity requires a base of trust and quality relationships. This requires attention to bridging worldviews and working across different vocabularies. Thus it is important not only to document indigenous knowledge, but also to spend time on translating, both across languages and across terms used. The theory and practice of a participatory action approach to agricultural research are presented in Pound *et al.* (2003). Case studies have been developed that provide guidance on "good practice" and techniques that promote colearning, relevance, and scaling up in participatory natural resource management (NRM).

The bottom line is that iterative cycles of learning are at the core of sustainable agriculture. Some components of a learning cycle are outlined in Fig. 4.8 to illustrate how researchers and farmers can systematize attention to reflection and reviewing priorities over time, to continually learn together and correct the course of development as it unfolds.

Gauging the effectiveness of sustainable development requires benchmarks or success criteria. Conventionally, the value of a sustainable practice, such as compost making,

FIGURE 4.7 (**A** and **B**) Malawi cultivation of riverine habitat through informal irrigation.

(*Continued*)

FIGURE 4.7 Cont'd (**C**) Formal irrigation scheme replacing informal irrigation systems in Malawi.

has often been determined by measurements of crop yield or biomass productivity over time. This chapter suggests that the assessment of sustainable systems should consider a range of indices, such as impact on ecosystem services, environmental goods, economic returns, and variability of returns. Farmers and rural stakeholders have their own criteria, which can be documented (Vernooy and McDougall, 2003). Multidisciplinary teams of economists, social scientists, ecologists, agronomists, farmers, and community members can work together to develop criteria. This requires commitment to a colearning process and communication to bridge different conceptions of sustainability.

SUSTAINABLE AGRICULTURE DESIGN

There is growing evidence that some key, ecologically based concepts inform sustainable agriculture design. Building on the agroecological principles discussed in Chapter 3, core practices chosen to expand on here are as follows.

- Biodiversity
- Resource efficiency
- Productivity and economics
- Farm system resilience

BIODIVERSITY

Central to sustainable farm design is the complex challenge of maintaining biodiversity. There is debate among agroecologists regarding how much biodiversity is enough to promote system stability, and the scale at which biodiversity should be

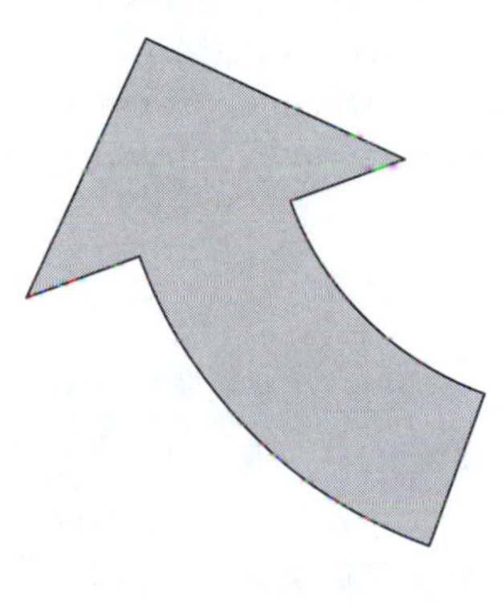

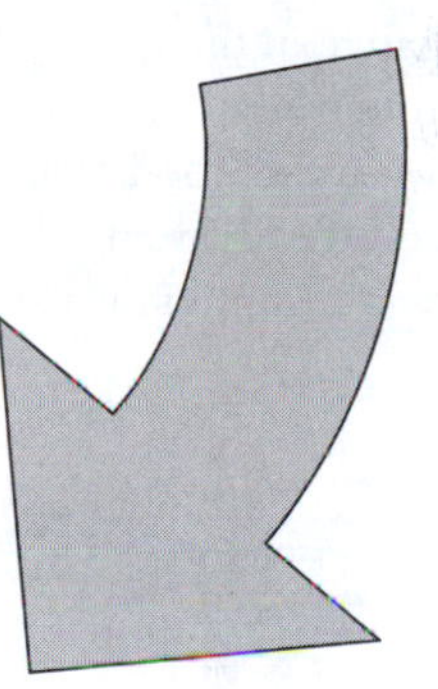

FIGURE 4.8 Iterative learning cycles to support development of sustainable agriculture.

promoted. This section considers the individual species level and diversification at the community level. The presence of different plant types and growth habits, such as mixtures of annual and perennial species, can provide a range of ecosystem services (see Fig. 4.6). These include the closing of nutrient cycles, protection of soil and water quality, and regulation of pest populations. Chapter 3 describes in more depth the biological theory behind combining growth forms in systematic ways to develop community "assemblages" of complementary species.

Diversifying Plant Species

Highly simplified cropping systems have come to dominate much of the world's agriculture, including continuous rice, corn–soybean and wheat–rice rotations. These are highly productive systems in terms of calorie returns to investment and in terms of economic returns within current policy environments. Many of these monoculture and biculture systems are, however, dependent on substantial inputs in the form of externally purchased nutrients and pesticides. The wheat–rice system, for example, is fertilized as much as seven times per year (Fig. 4.9). There are considerable ecological gains to be made from increasing diversity, even modestly, in these highly simplified systems. This is shown by the successful control of a rice pathogen in China by growing a mixture of rice varieties using what has been termed a multiline approach (Zhu *et al.*, 2000).

Resource efficiency and resilience can be enhanced markedly by diversification, as will be explored further in this chapter; however, it is important to keep in mind that carbohydrate production will often decline on an area basis when a cereal is replaced with another crop. An example of this trade-off is the replacement of wheat with the grain legume chickpea in the wheat–rice double crop system; this reduced N fertilizer requirements by 60%. However, there was a concomitant change in the quality and quantity of grain produced, as chickpea produces a high protein grain at a lower yield potential than wheat. For farmers that have some measure of food security, it may be possible to reduce the total amount of calories produced and enhance the sustainability of the cropping system by integrating a grain legume in place of a cereal.

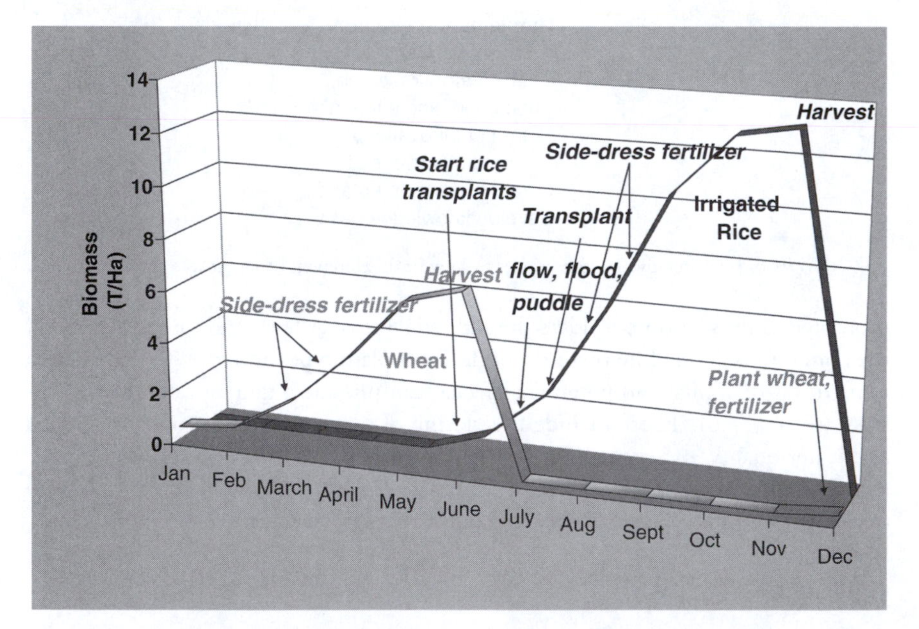

FIGURE 4.9 Wheat–rice double-cropping system, where wheat operations are shown in gray and irrigated rice operations in black. The biomass productivity of this system is very high, as are labor, fertilizer, water, and seed requirements.

Enhancing diversity through a focus on plant species is an important strategy in agroecology because plants are the primary producers that support herbivores and the entire soil food web of micro- and macroorganisms. This is termed "bottom up" management in ecological terms: the plants provide the biochemical quality, and quantity, of root and shoot residues that regulate food and habitat availability. Plants also act as "top-down" regulators through supporting the presence of natural enemies that suppress pests. Indeed, plants are the key providers of shelter, nectarines, and other alternate food sources for beneficial insects. In a few cases, plants act as repellants, producing volatile biochemical compounds such as odors that discourage insects from landing or feeding nearby. Figure 4.10 illustrates the use of basil as an intercrop with eggplant by an organic farmer. Basil is a strongly aromatic plant that provides protection against flea beetles and other pests.

Introduction of new species will enhance diversity to the extent that they have different physiology, morphology, growth habits, and diverse reproductive strategies. These traits impart functional diversity, which is a central principle in the design of buffered, resilient, and sustainable systems. Integrating plant diversity can be accomplished over space and time using "accessory crops" (Drinkwater and Snapp, 2007). Accessory plants are not grown primarily for economic sale or consumption; they are instead planted for a wide range of ecological purposes and often have multiple functions, producing minor crops of medicinal or cultural significance. Examples of accessory plants include green manure crops grown to enhance soil organic matter and nutrient availability to subsequent cash crops. Many agroforestry species are accessory crops and provide multipurpose functions such as soil improvement, fuel wood, fodder, or fruit production (Table 4.2).

Over many generations, farmers have developed highly successful combinations of plants. These include the sorghum–pigeon pea intercrop in semiarid India

FIGURE 4.10 Basil intercropped with eggplants on an organic farm in Michigan.

TABLE 4.2 Polyculture Farming Systems from around the Globe

System name and description	Species components	Sites
Improved fallow Perennials grown to improve soil, followed by a cash crop	*Gliricidia sepium* and *Erythrina peoppigiana* trees for stakes and soil improvement, followed by a climbing bean cash crop	Costa Rica, East Africa highlands, Malawi
	Sesbania sesban, Sesbania rostrata (stem-nodulating legume), or *Crotalaria ochroleuca* 1 to 2 years, followed by rice	East Africa, Tanzania
	Tephrosia vogelii bush 2 years (first year relay intercropped with maize), followed by maize	Malawi
	Sesbania sesban trees 2–3 years (first year intercropped with maize), followed by 2 years maize	Zambia
Green manure/mulch Vegetative and indeterminant annual or biannual (often a legume), followed by a cash crop	*Mucuna pruriens* green manure system or slashed mulch, followed by maize or maize–bean intercrop	Central America, humid to subhumid tropics
	Tephrosia vogelii, Canavalia ensiformis, Crotalaria ochroleuca, and *Lablab purpureus* relay intercropped with maize, followed by maize	Kenya, Malawi
Agroforestry intercrops Perennial intercrop with a cash crop	Coffee and cacao	Central America
	Coffee and *Eucalyptus deglupta*	Costa Rica
	Coffee and *Erythrina poeppigiana*	Central America
	Maize intercropped with pigeon pea (*Cajanus cajan*)	Eastern and southern Africa
	Maize intercropped with *Gliricidia sepium*	Malawi
	Leucaena leucocephala hedgerow intercrops in annual crop fields, terrace edges, rice paddy bunds	Southeast Asia (promoted elsewhere; rarely successful)
	Low-density mango tree intercrop with the maize–bean–squash complex	Subhumid to humid tropics
	Faidherbia albida tree (sheds leaves in growing season) intercropped with a cereal	Semiarid tropics
Silvopastoral Animal grazing or forage production under trees	*Pinus ponderosa* trees intercropped with *Stipa speciosa* native grass	Argentina
	Populus species, trees intercropped with *Bromus uniolides, Trifolium repens*, and other forage species	Central and South America
	Alnus acuminate trees intercropped with *Pennisetum clandestinum* grass	Central America
	Faidherbia albida tree grazed with native species	West Africa; semiarid to arid tropics

and eastern Africa and the complex, multistory intercrop of maize–bean–cucurbit (pumpkin or squash) and low density of mango trees that is popular among smallholder farmers throughout the subhumid tropics from Mexico to Malawi. Agroforestry systems are among the most complex, as they include plants with different life spans, from annuals to short-lived perennials (often a shrub), and longer-lived perennials, such as trees that may require decades to mature and produce fruit (see Table 4.2).

Polycultures provide multiple ecosystem services and efficient use of resources, but they are knowledge-intensive systems that require considerable experience to adjust timing and spacing and manage the different growth habits. Researchers have often tested species for integration into a cropping system as single plant introductions, whereas smallholder agriculturists have developed dynamic and complex cropping systems that demand study in their own right (Alteri, 1999). There is considerable value in experimenting with more than one new species at once through the combination of likely candidates based on an ecological assessment. A review of indigenous systems is also an important place to begin. Research on improved fallows has begun that will enable farmers to test mixtures of short- and long-duration species that combine stress tolerance features (Arim *et al.*, 2006). This is in contrast to conventional attempts to identify the single best species for a system.

Candidate species and mixtures for diversification purposes should be evaluated carefully based on the best information available, using different sources of information whenever possible. Some species are "championed" by organizations without sufficient attention to an ecological zone of adaptation or potential problem traits. Awareness of invasive species is beginning to grow among development educators, and all new species should be assessed carefully for invasive potential before introduction. Web sites listed at the end of the chapter provide a starting place for obtaining a balanced view of how species perform.

Promising legume species that have been characterized as "best bets" for diversifying East African farming systems are presented in Table 4.3. These are legumes that can be used as green manures and grown in rotation with a staple crop or as an intercrop. Some of the legumes are multipurpose, providing benefits beyond soil amelioration, such as livestock fodder or food products (in some cases after processing the seed). A general observation is that a trade-off exists between insect damage and edible product. Legumes that produce high-quality grain and leaves that can be consumed directly tend to be susceptible to insect herbivores, whereas legumes with tissues that contain toxic products are less desirable to insects. In the learning modules of the accompanying DVD there is a presentation on legume best bets, or promising options.

Two particularly promising legume species for diversification of staple crops are pigeon pea and *Mucuna pruriens* L., also known by the common name of velvet bean. Research from Kenya and Malawi shows the high production potential (biomass and grain) and widespread adaptation of *M. pruriens* to smallholder farms in the subhumid tropics (see Table 4.3). The weed-smothering features and

TABLE 4.3 Green Manure and Multipurpose Legume "Best-Bet" Species from Testing Conducted On-Farm across Kenya and Malawi[a]

Legume species	Time to maturity	Uses	Biomass yield[b]	Comments
Large seeded				
Mucuna pruriens	4–12 months	Weed suppression Soil fertility enhancement Food (after processing grain) and fodder	5–9 T/ha	Toxin is L-Dopa Highly effective weed suppressor Medicinal uses
Canavalia ensiformis	6–12 months	Moderate weed suppression Soil fertility enhancement	4–6 T/ha	Toxin is concanavalin Not edible Fodder of dried plants
Lablab purpureus	3–10 months	Weed suppression Food (seeds, pods, leaves) and fodder Moderate soil fertility	5–8 T/ha (fast growing)	Drought tolerance (varies) Edible Insect susceptible
Cajanus cajan	4–24 months	Food (seeds, pods, and leaves) and fodder Moderate soil fertility (higher in year 2) Secondary uses: fuel wood, medicinal	2–5 T/ha (ratooned/ cut year 1, for more biomass)	Edible P- and N-enhanced residues Insect susceptible
Medium seeded				
Vicia benghalensis	4–5 months	Soil fertility enhancement Fodder	5–8 T/ha	High-altitude crop >1800 m altitude above sea level Not edible Moderately antinutritional
Small seeded				
Crotalaria ochroleuca	3–4 months	Soil fertility enhancement, rapid Some food (vegetable) and fodder	6–8 T/ha (fast growing)	Edible vegetable and fodder Insect susceptible Moderately antinutritional effects on animals
Desmodium uncinatum	8–12 months	Soil fertility enhancement Fodder	3–9 T/ha	Insect susceptible Not edible *Striga* suppressor

[a]Adapted from the Legume Research Network Project, http://www.ppath.cornell.edu/mba_project/CIEPCA/exmats/LRNPbroc.pdf and from Soil Fertility Management for Smallholder Farmers, Malawi Ministry of Agriculture and Irrigation and the International Crops Research Institute for the Semi-Arid Tropics (2000).

[b]Biomass yield presented as dry matter, T/ha where T = 1000 kg.

cereal-yield enhancement associated with a *Mucuna* rotation have led to greater than 70% adoption within hillside maize production in some regions of Central America (Fig. 4.11; Buckles *et al.*, 1998). Pigeon pea has important multipurpose features as well, including producing nutritious vegetable pods and grain, with growth traits uniquely suited to intercrop systems, including a slow early growth pattern that minimizes competition and the production of N- and phosphorus (P)-enriched residues for soil improvement.

It is recommended that researchers and educators consider farmer indigenous knowledge and local shrubs. Legume species from genera *Acacia, Calliandra, Centrosema, Crotalaria, Desmodium, Gliricidia, Leucaena, Mimosa, Sesbania*, and others are widely found in the tropics and subtropics. As shown in Figure 4.12, Sesbania was introduced in Southern Yemen to stabilize sandy soils, yet supplementary fodder production has enabled a substantial increase in land-carrying capacity for camels. It is worthwhile investigating the wild species growing in an area to see which may be adapted for use as an accessory plant.

Community Diversity

Plant mixtures require management that minimizes competition with the main crops. Combinations of short- and tall-statured species or plant canopies that are complementary in the structuring of branches and leaves and deep versus shallow root systems are all means to reduce competition. However, even with complementary architectural traits, plants will compete to some degree for limited resources. Separation of plants in time or space will reduce competition. Crop management practices to partition plants include placement, where one crop is located on the

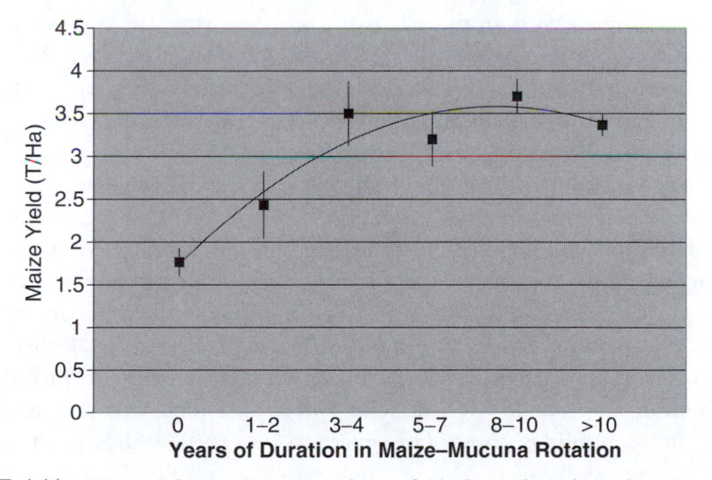

FIGURE 4.11 Transect of Central American farmer fields shows the enhanced maize grain yield associated with increasing years of duration within a maize–mucuna rotation system (adapted from Buckles *et al.*, 1998).

FIGURE 4.12 Introduced leguminous trees such as the Sesbania shown here provide crucial forage for camel during the long dry season in Yemen.

top of a ridge, for example, and the other crop in the furrow between ridges in a ridge–furrow cropping system (see Chapter 3, Fig. 3.8). Another means relies on separation in time. An example is a relay intercrop, where plants are seeded into the understory of a main crop, often after a main crop is weeded (see Table 4.2).

Hedgerow intercrop systems are a form of agroforestry designed to reduce competition through frequent pruning of the hedge and placement of hedge species on ridges at some distance from crop rows. This system was promoted widely in the tropics, where legume trees such as *Leucaena leucocephala* were planted as hedgerow species intercropped with crops such as maize or cassava. When managed intensively, crop competition was shown to be limited to nil and there were considerable soil-building benefits; however, large-scale extension efforts did not lead to farmer adoption of hedgerow intercrops. As illustrated in Box 4.2, some of the challenges were biological, whereas others were socioeconomic in nature.

The hedgerow intercrop system illustrates well a central challenge of ecological management: how to promote diversity of species, soil-building organic inputs, and weed suppression, without excess labor demands or unacceptable competition with the main crop. Short-lived cover crops from genera with rapid growth cycles, such as *Brassica* or *Crotelaria*, are biological technologies that have a "built-in" off switch. They are programmed to produce biomass and then die quickly. Other ruderal[1] type plants may be usefully developed by agricultural scientists to provide farmers with means to produce layers of living and dead residues that provide complementary and continuous cover. This smothers weeds, as well as reducing water,

[1]Ruderal plants are usually annuals, fast growing, and adapted to disturbed environments; see discussion of plant growth strategies in Chapter 3.

BOX 4.2 Hedgerow Intercrop Agroforestry

In a hedgerow intercrop system, trees are pruned intensively to develop hedges spaced 1 to 5 m apart, while crops are grown in the alley or space between the hedges. Nutrient recycling is accomplished through the deep rooting of the tree component, which ideally forms a "safety net" below the crop root zone. The presence of hedges reduces light penetration; combined with mulch (derived from hedge cuttings), this substantially suppresses weed growth between crop cycles, in addition to limiting wind damage. *L. leucocephala* and *Cassia spectabilis* have been identified as promising hedgerow species, being easy to establish, with high growth rates. Hedgerow intercropping was promoted widely as a sustainable alternative to bush fallows (Young, 1989). Frequent pruning was recommended to reduce tree transpiration demand and to enhance root turnover. Over time it became clear that hedgerow species often competed with crops, particularly under on-farm conditions where pruning was more sporadic. Hedgerow species *C. spectabilis* was found to be a non-N-fixing legume and highly competitive for water and nutrients.

Overall, agroforestry performance was variable and appeared to be best adapted to sites with minimal moisture competition, such as high organic matter soils in humid environments or sites overlaying shallow groundwater. Farmers have, on occasion, incorporated novel hedgerow species into boundary plantings and gardens, but have not adopted hedgerow intercrops on a large scale. Further research is needed to reduce labor requirements, minimize competition, and enhance the consistency of returns (Snapp *et al.*, 2002).

nutrient, and soil losses from a system. Herbicide management or vigorous mowing can also be used to suppress growth at critical time points, leaving behind a mulch of residues that can be used for no-till planting of crops.

A diversity of plant species will generally support healthy crop plants and suppress pests. Rotational cropping systems are often associated with ≈15% higher yields compared to monocultures; this has been termed the "rotation effect." Continuous wheat is one of the few exceptions, as it appears to be a long-term viable monocultural cropping system. The soil conservation properties of this long-duration crop and its large, finely branched root system may explain the apparent sustainability of long-term wheat production. The observation of rotational yield responses in most cropping systems could be because of many reasons. A few of the better documented processes are listed here: suppression of soil diseases and organisms, improvements in nutrient synchrony from diverse residues, and crop–health-promoting mycorrhizal interactions.

TABLE 4.4 Influence of Biodiversity on Sustainability and Pest Management at Different Scales

Scale of diversity	Components of sustainable management of pests
Plant	Defense compounds that resist insect herbivores and aromatic biochemistry that drives away pests
Community	"Bottom-up" control from diversified resources and habitat that confuse pests
	"Top-down" control through fostering beneficial insects and dampening of predator and prey dynamics
	Layers of dead and living plant tissues that suppress weed germination and foster soil health
	Rotate crops with biofumigant plants to alter soil flora and fauna and control soil-borne pests
Landscape	Insert species at critical foci in the landscape to regulate pest management, e.g., tall species to filter wind-blown pests
	Structure diverse plantings of multiple age groups to reduce vulnerability to pest outbreaks

Diversifying at Larger Scales

Across rural landscapes there are tremendous opportunities for diversification at different scales (Table 4.4). The pattern of managed and wild in a landscape is a determinant of air flow, nutrients, water, soil, and biological elements. Dispersed or aggregated combinations of domesticated and natural areas will determine the extent of interface area. Corridors can be developed to link wild areas, facilitate conservation of wildlife, and enhance the extent of interface areas, which are often highly productive and biodiverse. This illustrates that there are foci in the landscape where inserting species can have a regulator influence, such as planting perennial species strategically to maintain wildlife corridors. Perennial strips can act as buffers in a landscape, including plantings along riverine areas to protect water quality by acting as living filters to take up excess nutrients or silt. Landscape ecology provides important insights into the influence of landscape structure and land use patterns on resource flow and insect dynamics (Landis *et al.*, 2007).

RESOURCE EFFICIENCY

Resource concentration and utilization are central to agricultural management practices. Many farmers use mined fertilizers to augment nutrients in cropping systems. Many also rely on practices that concentrate organic sources of nutrients and then grow successively less nutrient-demanding crops until the nutrient supply is spent. Traditional slash and burn techniques such as bush fallows utilize this approach and are sustainable if farmers have a sufficient land base to support very long (decade or more) rotations. In a bush or natural fallow system, a regenerated area is cleared after 10 or more years of growth, plant materials are piled and burned, and then nutrient-demanding cereal crops, such as millet or maize, are planted, followed in a rotational sequence by a crop with tolerance to low fertility, such as cassava.

This system is called *chitemene* in Zambia and is highly suited to acid, infertile, and leached soils. Nutrients are less susceptible to leaching losses when applied in an organic residue form compared to a fertilizer source. A widely used variation throughout southern Africa involves grass residues that are not burned, but instead piled in the center of a mound of soil that is then planted to crops with low nitrogen requirements, such as beans, sweet potato, or cassava (Fig. 4.13).

Recycling of nutrient resources and efficient use of energy are widely recognized hallmarks of sustainability. In addition to reducing dependence on nonrenewable resources, minimizing the use of expensive inputs can reduce farming costs, which is vitally important to limited resource farmers. The bush–fallow cropping system described earlier requires a long time horizon and sufficient land, but is an effective means to recycle nutrients. Soil type will often determine which management strategies are sustainable. The maintenance of soil organic matter, nutrient pools, and yield potential is challenging in arable sandy soils unless large doses of organic inputs are applied in the form of manure or rotational soil-improving crops. Heavier soils have sufficient buffering capacity to be able to maintain productivity with inputs of mineral fertilizers alone or with moderate doses of manure combined with fertilizer.

FIGURE 4.13 Grass mounds from northern Malawi planted with cassava.

The stable production of moderate yields at levels that provide acceptable rates of return, rather than attempting to optimize production of high yields, is an overlooked goal of many smallholders. Efficiency of returns from small doses of fertilizer or pest control measures can be quite high compared to the incrementally smaller returns from inputs at the high end of the yield–response curve.

Long-term experimental trials carried out in SSA have provided some of the best evidence that integrated nutrient management strategies, for example, modest doses of fertilizer, less than 50 kg nutrients per hectare, combined with a modest rate of manure 2 to 4 tons/ha, can support stable production in legume–cereal rotation. This is illustrated by the maintenance of crop yields in a peanut–cereal rotation over several decades on a sandy alfisol (Pieri, 1992). However, it is notable that integrated nutrient management (INM) has only been adopted sporadically in SSA, in specific localities. Small-scale farmers instead tend to rely on the relatively low labor input, fertility-enhancing systems of bush fallows, or, in some cases, fertilizer. In other regions of the world, integrated use of organic and inorganic fertility sources is common, and evidence shows that experimentation with INM is increasing in SSA, along with increased market access and education opportunities.

Examples of intensified soil management and organic amendment use by smallholders tend to be associated with land being in short supply. Consider, for example, the tremendous diversity of intercrops, compost systems, and complex land and water management technologies historically documented among Chinese peasant farmers and among Central American farmers (Netting, 1993). An African example is from the island of Ukara, in Tanzania, where indigenous farming systems capitalized on the soil-replenishing nature of the leguminous green manure *Crotalaria striata*, grown as a relay intercrop in bulrush millet (*Pennisetum typhoides*) (Ludwig, 1968). More recently, in Zambia, the use of organic inputs from improved fallows and animal manure is being combined successfully with small doses of fertilizer in maize-based cropping. Suppression of the parasitic weed "striga" has driven much of the interest in INM in Zambia. Pest management issues are interrelated with soil fertility as land use intensity increases, as serious infestations of parasitic weeds are often associated with continuous cropping of staple cereals.

The nutrient balance required for a sustainable cropping system can be approximated by estimating nutrient inputs added to and removed from the farm system and adjusting the addition of nutrients accordingly. This has been advocated as a means to assess long-term sustainability of a farming system. It does provide an indication of the trajectory for nutrient sustainability. However, it is important to remember that available nutrient pools are different than total nutrient stocks, especially in the case of phosphorus where availability to crops of P often depends on the crop species present and long-term management practices rather than on recent inputs. Thus budget approaches often overestimate nutrient loss pathways and do not take into full account the effect of enhanced efficiency or small changes in nutrient availability; for example, shifts of P from inaccessible to accessible pools can alter P availability to plants more than the P input rate. The scale of analysis is also essential to consider, as erosion losses from one part of the watershed may lead to deposition in other areas of the watershed. Further discussion of the complexity of integrated nutrient management can be found in Chapter 5.

Farmers often focus resources on landscape positions that provide the highest rates of return, for example, low-lying areas. The ingredients for high returns to investment are in place, as water is available, soil fertility is potentially high, and market demand is high for out-of-season produce. Intensification is occurring throughout southern Africa and other regions undergoing rapid rural development, particularly at wetland and riverine sites where high efficiency and productivity can be achieved. The AIDS epidemic is another driver of changing land use, as high returns to labor are critical to household survival in southern Africa.

Conservation, productivity, and equity issues are all raised at this crucial interface where land meets water. The ecosystem function of wetlands and riverine areas are profound, protecting fresh water quality and quantity for communities all along the watershed. Riverbank cultivation is prohibited in many countries precisely to protect this at-risk ecosystem. However, complete prohibition of use is not compatible with local demands and priorities, nor does it engage local communities in potentially sustainable land use or community protection activities.

Intensified and sustainable use of wetlands is feasible, as illustrated in a case study of bean production along a Chingale watershed in southern Malawi (see the DVD that accompanies this text). Agricultural development near the edge of the water occurred through the actions of individuals supported by nongovernmental organizations and technical advisors (extension and researchers); however, this development was fully integrated with local communities and soil conservation practices were followed. What was not under the control of traditional authorities and community norms was the edges of the watershed, where resources were degrading rapidly under informal charcoal production and cultivation along steep slopes. Officially, these areas were under government management, yet control mechanisms had broken down. Participatory action research documented that community leaders in the watershed were aware of the destruction and required support to engage policy makers rather than technical education on conservation. This is an example of resource conservation at the water–land interface being maintained and even regenerated through agriculture. Effective support in the intensification process included educational efforts and locally controlled management. However, a comprehensive watershed level approach was needed to address other significant environmental pressures.

PRODUCTIVITY AND ECONOMICS

Productivity within a sustainable farming context should not be defined simply through measurement of yield. Economic return is one important means of assessing performance within or across farming systems. Economic assessment through net benefit or gross margin analyses is one of the only means to compare technologies that are very different in nature. Organic and inorganic sources of nutrients, for example, have quite different costs, rates of return, and opportunity costs. Economic tools provide insights into comparing such contrasting technologies. Figure 4.14 provides an example of net benefit returns from soil fertility-enhancing technologies as a means to evaluate performance within maize-based cropping systems in Malawi. There are inherent challenges in economic evaluations, as product and input prices vary markedly over time and location, particularly where storage facilities are not well developed and for minor crops such as legumes. Labor is also difficult to value appropriately within the context of a smallholder farm family. Given all these challenges, it is still markedly insightful to evaluate technologies on the basis of net returns, and often insights emerge regarding adoption potential and the relevance of technical recommendations.

Development of systems that prioritize stability of production over high yields is particularly important to risk-adverse farmers, many of whom live in highly variable environments. The timing of production, when harvestable products are ready over the year, is critically important to farmers who face food insecurity on a recurrent basis. Developing varieties and planting arrangements that provide harvestable yield early as well as late in the season has particular significance to farm-

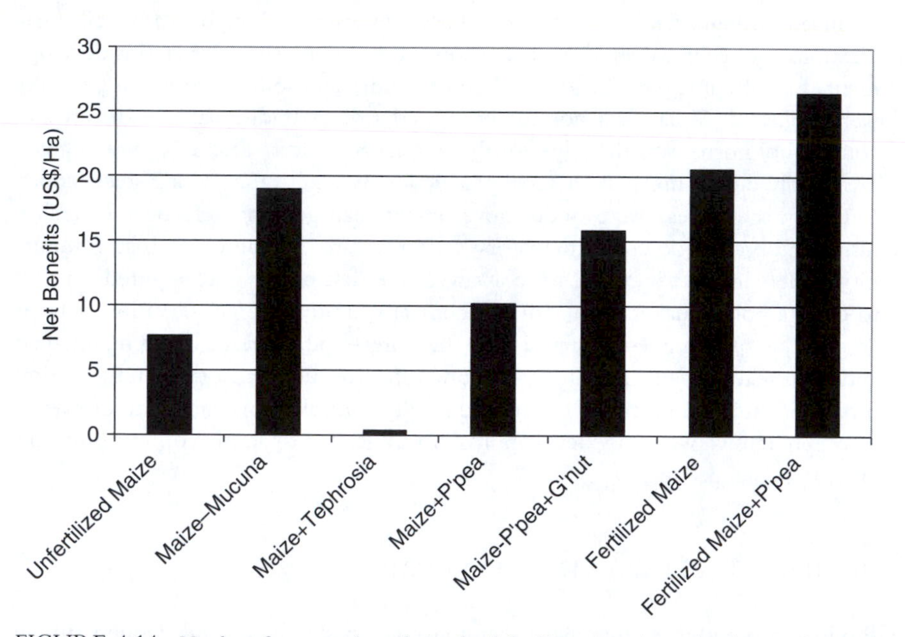

FIGURE 4.14 Net benefit economic analysis of soil fertility-enhancing technology performance based on maize yield value and input costs associated with seeds and fertilizer from on-farm trials conducted in central Malawi from 1997 to 1999 (Snapp, unpublished data).

ers who suffer through a hungry season before crops mature. Postharvest storage is an ongoing challenge and a source of tremendous losses among poor farmers. Production goals must include diversification of harvest times, to the extent feasible, given environmental constraints of rainfall and temperature.

Direct comparison of products from different species or plant parts, such as evaluating cereal yield in relationship to tuber yield or animal products, is not meaningful, as fresh weight and biochemical constituents vary tremendously. An ecological concept that can be used here is "net primary productivity," a term used commonly to assess species' performance within ecosystems. Measurements of biomass, calories, lipids, or proteins provide diverse means to assess farming system productivity. The production of high calories in cereal grains may be somewhat offset by the production of nutrient-enriched, more valuable legume grains. Nutrient-enriched foods should be rewarded on the marketplace with high prices. However, markets are not always responsive to nutrition, and calories remain an essential "coinage" for comparing the productivity of systems, as they are closely related to fulfilling family food requirements.

An on-farm study in Malawi compared calories produced by unfertilized maize to calories produced by legume–maize rotation systems (Fig. 4.15). Legume products have nutritional benefits that complement cereals and support children's health

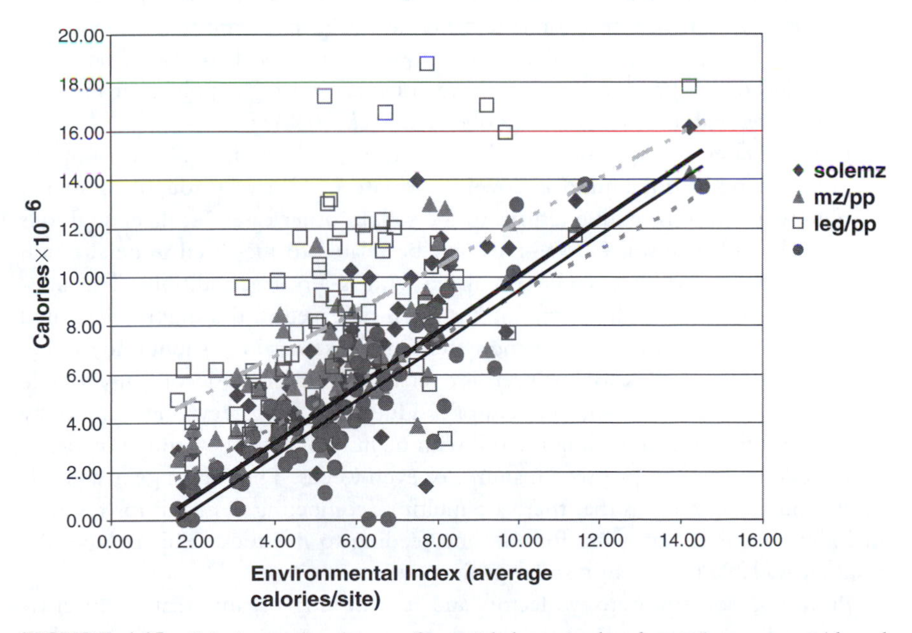

FIGURE 4.15 Calories produced in on-farm trials by maize-based cropping systems with and without legume diversification (Snapp, unpublished data).

and growth, and educational efforts on human nutrition have been shown to encourage farmer adoption of legume crops (see Chapter 9). However, farmers will rightly remain skeptical of production systems that have substantially lower calorie returns than systems currently in use.

Sustainability requires attention to the entire process of developing, testing, and supporting adoption of a new practice. It must not only perform well over time but also meet local criteria for performance and be supported by innovations in local regeneration of the technology. How the technology will be maintained is a key sustainability question. Can farmers obtain access to a new variety or technology over the long term; is there an effective seed system and local technical expertise? What are the economics of reproduction of plant and animals, and manufacture and repair of equipment? These issues are an integral part of the regeneration of a technology and its long-term prospects for adoption and performance.

RESILIENCE

Sustainable resource management is essential to the resilience of rural livelihoods. Building soil quality is at the foundation of a farming system that can respond to disturbance or disaster. Long-term experiments have been carried out around the globe to test the sustainability of farming practices by examining changes in soil organic matter and yield potential over time. These trials consistently show an initial, rapid decline in topsoil organic matter as tillage and grain removal enhance carbon (C) loss. Over time, soil C loss is slowed or reversed in cases where tillage is reduced or where organic inputs are enhanced through applying manure or growing perennials or cover crops (Robertson *et al.*, 2000).

In some cases, soluble nutrient additions from fertilizer have positive impacts on soil C status (maintaining or slowing the rate of C loss) through improving crop growth and thus residue returns to the soil. In other cases, fertilizer additions can actually enhance soil C loss. This may be related to increased mineralization of soil organic matter that can be stimulated with N fertilizer additions. This phenomenon, sometimes called "priming," is often observed in the highly N-limited environment of cereal-based cropping systems in the semiarid to subhumid tropics. The key to sustainable use of fertilizers is to ensure that residues accompany soluble fertilizer use. There are many sources of residues, including recycling residues by livestock feeding and amending the soil with manure or directly through incorporating residues from crops, green manures, or even weeds. The challenge in a developing country context is that there are multiple, competing demands for residues, including burning, removing for sale, and feeding to livestock. This reduces the residues available to replenish soil organic matter.

Protecting soil from erosive factors and replenishing organic matter are critical to building system resilience; thus, it is crucial to promote long-duration plant cover wherever feasible in a given farming system. However, it must be acknowledged that it is challenging to find plant species that are high performers in environments that are marginal and climatically unreliable. In the experience of this author, plants and animals that have high tolerance to stress are some of the only technologies within the reach of the majority of smallholders, including many who cannot afford fertilizer or labor-intensive technologies. However, improving organism tolerance to climatic stress and poor soils is not an easy target, and gains may be incremental compared to readily observable increases in yield from high-yield potential varieties grown under optimum conditions.

Understanding and enhancing tolerance traits deserve higher profiles in agricultural science; farmers urgently need access to species that perform well in marginal environments. Scientists in Australia and Kenya are assessing drought tolerance in legumes such as lablab (*Lablab purpureus*), a promising forage and green manure species (see Table 4.3; Pengelly and Maass, 2001). Availability of novel species types is a key ingredient to supporting innovation and sustainable farming. To this end, participatory approaches to plant and animal system improvement are discussed in Chapters 6 and 7.

Resilient Plant Communities

Sustainable management relies on knowledge-intensive practices and ecological manipulation, replacing the use of purchased inputs wherever possible. Biology is manipulated to improve plant access to nutrients, to build soil capacity, and to reduce susceptibility of crops to pests. Assembling plant communities based on ecological understanding of functional traits was described in some detail in Chapter 3, which presented information on complementary and redundant combinations of species. The three plant group types discussed here are presented in Fig. 4.16 and represent a continuum from brassicas with no symbiosis, to cereals with one symbiotic fungal partner, vesicular–arbuscular mycorrhizal fungus (VAM), and legumes with two symbiotic partners, VAM and rhizobia.

Enhancing farming system reliance on biologically fixed nitrogen (BNF), substituting for purchased nitrogen in the form of fertilizer, is an instructive example of the ecological role that N-fixing symbiotic associations play. Legumes and N-fixing bacteria *Rhizobium* spp. and *Azolla* spp. in rice systems are primary examples of BNF-reliant farming systems. Nitrogen fixation capacity has been improved in Brazilian cropping systems, as explored in Box 4.3.

A sustainable property of systems that rely on biological N fixation is the feedback mechanism within the BNF process: this feedback reduces the rate of N fixation in the presence of soluble N. This reduces the potential for N losses within farming systems that rely on BNF and enhances N cycling efficiency. However, these same feedback mechanisms and the high energy-demanding requirements of BNF symbioses tend to limit the yield potential of BNF crops. Further, cropping systems that rely on BNF require investment of labor and often land and relatively expensive seed as well.

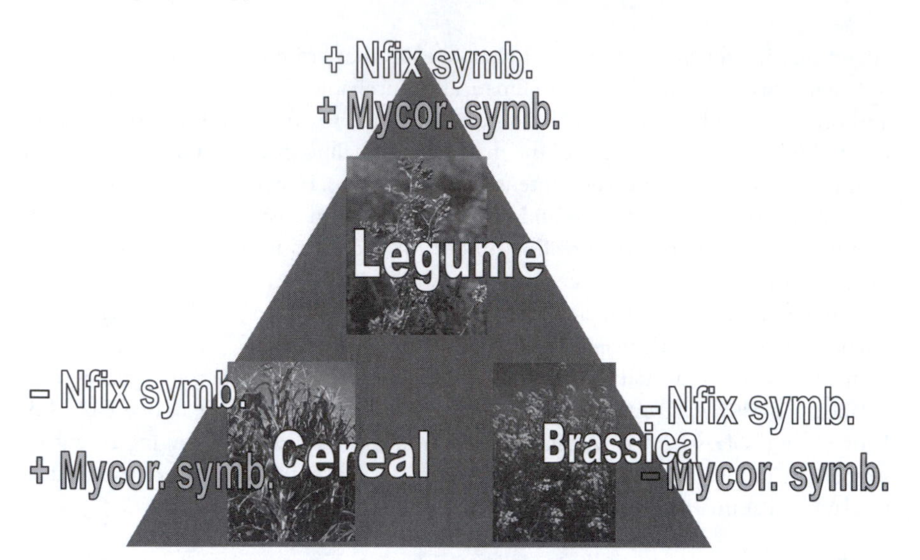

FIGURE 4.16 Functional traits that vary in plant groups; three types of symbiotic groups, from zero symbioses (brassica) to two symbioses (legumes), are shown.

Break Crops for Healthy, Resilient Crops

A radical diversification strategy could have mixed consequences, as illustrated by considering the inclusion of a "no symbiosis" plant type, the brassicas shown in Fig. 4.15. There are conflicting research findings on the consequences of rotating with brassica species, but under specific circumstances, substantial soil and root health benefits have been seen. It may be that brassica species enhance the impact of a crop rotational strategy through altering soil flora and fauna communities. Brassica species, and some other crops, such as sorghum–Sudan grass, have been shown to be effective biofumigation agents. In the soil, plant tissue biochemical

BOX 4.3 Biological Nitrogen Fixation (BNF) in Cropping Systems

There are examples of cropping systems that combine high productive capacity and reliance on BNF. This is illustrated by an example from Brazil. A multidisciplinary, concerted effort has markedly improved soybean BNF rates in commercial cultivars, while at the same time enhancing yield potential by 40% (Alves *et al.*, 2003). The team combined expertise in soil microbiology, agronomy, and plant breeding and worked together over 2 decades to achieve this remarkable improvement in plant-symbiont performance. A negative example is the inadvertent selection against BNF activity within Asian rice cropping systems linked to increasing reliance on N fertilizer.

compounds break down and produce isothiocynates (ITC), which kill or suppress growth of many soil-borne diseases (e.g., *Rhizoctonia pythium*), plant parasitic nematodes, and even weed seed. It is important to consider potential negative impacts as well. For example, some crops, such as onions and cassava, are highly dependent on mycorrhizal infection for normal growth, and a brassica crop might have an inhibitory effect. In contrast, crops such as potatoes are highly susceptible to soil-borne diseases, and rotation with a brassica "biofumigant" crop may enhance tuber and root health significantly (Snapp *et al.*, 2007).

Results are not always dramatic, as biofumigation is a complex process. The release of ITCs depends on many factors, from plant genetics and tissue biochemistry to soil environmental conditions and farm practices used to mow and incorporate residues. In some regions and soil types, notably coarse soils where few brassica crops are currently being grown, substantial benefits have been observed. Farmers have adopted the use of biosuppressive cover crops in southern Italy and the Pacific Northwest of the United States over large acreages of high-value crops such as potatoes and vegetables (Box 4.4). More research is needed to define where brassicas are beneficial. Preliminary evidence points to benefits being highest in systems that are dominated by few crop species.

Brassicas are adapted to a cool environment. There is a need to identify biofumigant or similar "break crops" for tropical environments. One species that has shown potential in Kenya is the green manure crop *Canavalia ensiformis* (Jack bean).

BOX 4.4 Brassica Cover Crops in Potato Production

Potato production in the Pacific Northwest and the upper Midwest regions of the United States is a case study of where brassica cover crops are being adopted rapidly. Research had shown the potential impact of an oriental mustard cover crop on crop health, where plant tissues were as effective as chemical fumigants in producing healthy potato tubers (Snapp *et al.*, 2007; Fig. 4.17). Farmer experimentation was critical to developing practical and effective means to manage cover crops for biofumigant activity. This included choosing the correct genetics, varieties that had biofumigant compounds at high levels (which turned out to be the ones with leaves "hot" to the taste), and determining the window within crop rotation sequences where mustards could be grown without interfering with cash crops. Planting in late summer for fall growth, with supplemental irrigation in dry weather, was found to support high biomass production, and a flail-mowing operation was adapted to incorporate macerated, green tissues in the soil for maximum biofumigation. This intensively managed cover crop has been shown to enhance the health and yield of subsequent crops, particularly disease-susceptible crops such as potatoes.

Plant parasitic nematodes such as *Pratylenchus* species cause widespread and often underappreciated damage to root health and crop yields for staple crops such as maize and potatoes. An increased maize yield of 20 to 35% and a concurrent suppression of *Pratylenchus zeae* infection were shown in maize grown after a green manure of *C. ensiformis* (Arim *et al.*, 2006). *M. pruriens* was found to be less effective at nematode suppression in this study, but other research has documented its potential as a break crop that can alter soil-borne pest populations. *M. pruriens* is a widely adapted green manure crop, as shown by farmer uptake in the subhumid tropics, from Honduras to Benin.

It is important to consider the overall impact of a green manure crop: does it enhance the health of main crops in the farming system? Some green manure species can have negative effects through enhancing plant parasitic nematodes or soil-borne pathogens. Although nematicidal and other biocides are produced by many legume species, particularly *Crotelaria* species, *M. pruriens*, and *C. ensiformis*, other legumes such as *Tephrosia* species and pigeon pea (*Cajanus cajan*) have been shown to enhance pathogenic species of nematodes. All experimentation with new species should be carried out carefully to test for potentially negative impacts as well as positive impacts.

A central feature of resilient systems is diversity. Providing a diverse habitat and food sources supports biota above and below ground, which will tend to damp the cycling of pest and prey to more steady cycles (see Table 4.4). A habitat for beneficial insects is one basis for a reduced occurrence of pest outbreaks in diversified

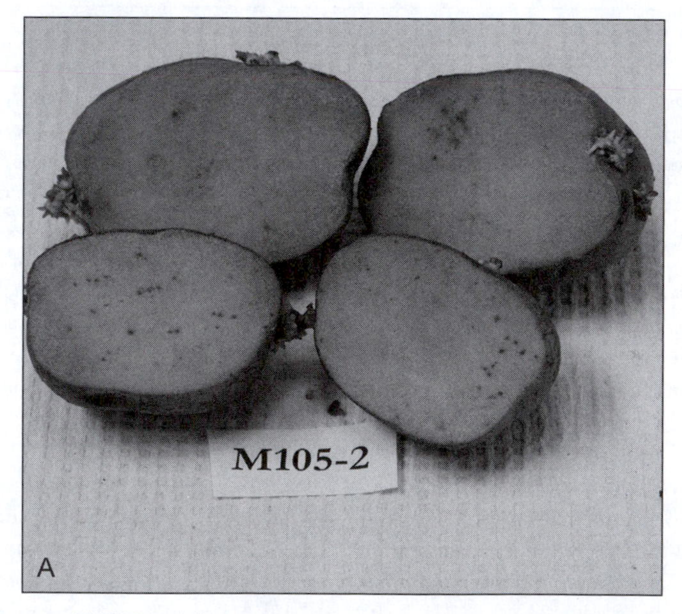

FIGURE 4.17 (**A**) Potatoes from chemically fumigated soil.

(*Continued*)

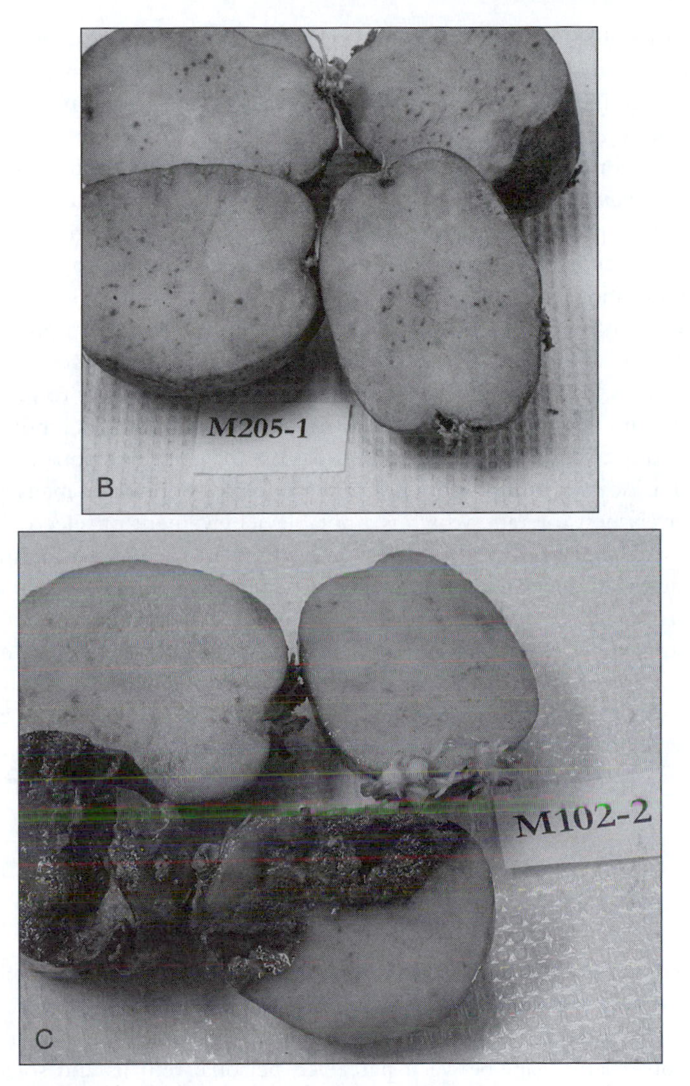

FIGURE 4.17 Cont'd (**B**) Potatoes from "biofumigated" soil where an oriental mustard cover crop was grown previously. (**C**) Potatoes from a fallow control soil.

farming systems. Seed predators such as beetles can help reduce weed pressure and are favored by growing borders or including some long-duration growth habit plants with annual crops. Pests are not, however, always suppressed in polyculture systems, and the complexity of managing multiple crops should not be underestimated. Growth stage, as well as species, influences pest population dynamics, and a diversity of age groups is often present in a polyculture. The presence of multiple crops can have a negative impact through the provision of secondary hosts for pests.

A "push–pull" system of diversification has been devised to suppress parasitic weeds such as striga (*Striga hermonthica* and *Striga asiatica*), a devastating pest on cereals in SSA (AATF, 2006). An intercrop of *Desmodium uncinatum* is planted between maize plants to exude root chemicals that induce dormant striga seeds to germinate and then die for lack of an appropriate host root (this is the "pull"). At the same time, *Desmodium* produces allelopathic compounds that are toxic to striga seedlings, providing the "push" in the system. Over time, the impact of agroecology-based interventions can have a cumulative effect that moves a system into a more resilient mode of operation.

There are many reports of low levels of pests in indigenous cropping systems, as reviewed by Morales (2002). This may be related to the preventative aspects of polycultures, where natural regulation of potential pests tends to prevent outbreaks. In some cases, it is also possible that smallholder farmers report low pest incidence due to high tolerance for pest damage, as partially damaged grain can be used for local consumption. The main staple crops grown in many places are also relatively pest tolerant, which is a notable achievement of selection by plant breeders and generations of farmers. In contrast, marketed horticultural crops often require production to a high-quality standard. There are many examples of farmers having to overcome severe pest problems when first growing a new crop. Considerable patience is required to address pest dynamics and interaction with an introduced species, as diversification and pest management strategies may take years, or indeed decades, to develop. Research over long timescales may be necessary to understand the processes involved in the biological management of pests.

PUTTING SUSTAINABLE AGRICULTURE IN PRACTICE

Soil protection and natural resource regeneration are at the foundation of a biologically smart, efficient, and resilient farming system, but these require tremendous investment by individuals and by communities. Often, soil and water management involves large gaps between perceived personal returns and societal-wide costs and benefits. Environmental problems suffer from this challenge overall, as change is often slow and difficult to understand. Cause and effect may not be easy to ascertain, as in global climate change. Soil conservation projects in particular often experience mixed results, as incentives do not always match the realities of food-insecure farmers. Subsidizes may be required if they are to address the long-term goals of protecting resources for future generations and protecting public goods today.

Lessons can be drawn from recent changes in farming practice; in some areas, reduced tillage systems have been adopted through a combination of lower farming costs (fewer equipment passes) and perceived soil-building benefits. Farmers had to overcome social norms that equated good farming practice with clean (e.g., no residue) fields and often spent years adapting their equipment and practices before adopting no till. In many areas, reduced tillage has not been adopted or is only used for specific crops. There are few examples of resource-poor farmers adopting reduced tillage, which is not surprising given the lack of economic incentive or access to herbicides. Recent interest has emerged in southwest Mali in using a preemerge herbicide to facilitate the direct planting of peanuts into weed residues, without tillage. This system may be driven by labor constraints and interest in a reduced requirement for weeding rather than by perceived environmental benefits. Peanut production has expanded in areas where this technology is being experimented with, at the initiative mostly of women farmers. The innovation is driven in part by interest in new market opportunities, along with family nutrition (Snapp and Weltzien, unpublished data).

Rural livelihoods and sustainable practice in a globalized economy involve increasing dynamic linkages among farming enterprises and evolving markets. This is illustrated by the aforementioned experience in southwest Mali. Rural family members move in and out of diverse livelihoods strategies, and market opportunities change rapidly—globalization forces are changing market demand at regional and international levels and are altering urban food preferences. Responding to this rapidly changing environment and developing sustainable responses require labor and often complex negotiations. There are conflicting time pressures on smallholders to conduct on-farm work and to pursue wage labor and other off-farm activities. Further, at the center of family and labor decisions is the issue of who benefits. Sustainable development requires attention to equity: new technologies should complement current efforts and not shift burdens unduly.

A case study of farmer adoption of grain legumes in northern Malawi is discussed in Chapter 9. Here the gendered aspects of intrahousehold discussions are highlighted to illustrate the shifts in labor requirements that often accompany biologically based technologies (Box 4.5).

The case study illustrates the close linkages between resource conservation and farmer decisions regarding when and where to invest energy. Intensification or extensification strategies may be appropriate, depending on the environmental and social contexts. Participatory action research acknowledges that farmers are at the center of this complex decision making and require support to enhance local adaptation, innovation, and knowledge. In participatory approaches to NRM, resource and energy flows are documented through farmers and researchers working together. Information generated is used to derive policy and management recommendations for sustainable production. Participatory NRM has been used in diverse contexts from Bolivia to Kenya (de Jager *et al.*, 2004). Interventions such as compost production and crop diversification have been shown to enhance the portfolio of sustainable options for the management of energy and nutrients (Sayer and Campbell, 2004).

BOX 4.5 Household Responsibilities and Residue Management in Northern Malawi

Legume residues are more effective at enhancing soil fertility if they are incorporated into the soil. However, residue incorporation is a laborious task, particularly if it is carried out at the agronomically optimum time when the soil is dry and very hard. Residue incorporation quickly became a contentious issue within households that adopted new legumes for soil improvement in a northern Malawi watershed (Bezner-Kerr *et al.*, 2007).

Traditionally, in this region, crop residues were incorporated as part of weeding operations during the growing season or as part of land preparation at the start of planting, which commenced with the rainy season. Burning of residues after harvest was also quite common as a means to reduce labor and to suppress weeds in the arduous process of hand-hoe agriculture. The new recommended management—to obtain maximum soil fertility benefit from residues—was postharvest incorporation of legume residues. This could be construed as a late weeding operation (the implication being that women should be responsible, as traditionally women lead on weed management) or as an early land preparation operation (the implication being that men should be responsible, as traditionally men lead on planting preparation). In order for a more sustainable cropping system to function in this northern Malawi watershed, complex negotiations were required on household task sharing of new duties in residue incorporation (Fig. 4.18).

FIGURE 4.18 There are many farming tasks, and careful consideration is required of new labor demands such as residue incorporation for soil improvement, shown here.

Centuries of farming have been maintained in China and other regions of the world showing the potential for long-term sustainability. At the same time, rising populations, poverty, and shifting land use are drivers of rapid change in natural resource management. The World Development Report (2008) documents the complexity of resource management, as farmers often ameliorate some areas within a farmscape while depleting other areas.

A rapidly changing world requires that sustainable agriculture focuses on supporting local knowledge generation to provide rural people with options rather than set recommendations and to support innovation (see Table 4.1). Biologically sound principles need to be taught and adapted to different circumstances. Biodiversity, resource efficiency, economic production, and resilience are all components fostering a sustainable trajectory. Farmers can improve the efficiency of nutrient, water, and energy cycles if knowledge is adequate and if the cultural–economic context is supportive.

REFERENCES AND RESOURCES

AATF (2006). "Empowering African Farmers to Eradicate *Striga* from Maize Croplands." The African Agricultural Technology Foundation, Nairobi, Kenya. http://www.aatf-africa.org/publications/Striga-eradicate.pdf.

Altieri, M. (1999). Applying agroecology to enhance the productivity of peasant farming systems in Latin America. *Environ. Dev. Sustain.* **1,** 197–217.

Alves, B. J. R., Boddey, R. M., and Urquiaga, S. (2003). The success of biological nitrogen fixation in soybean in Brazil. *Plant Soil* **252,** 1–9.

Arim, O. J., Waceke, J. W., Waudo, S. W., and Kimenju, J. W. (2006). Effects of *Canavalia ensiformis* and *Mucuna pruriens* intercrops on *Pratylenchus zeae* damage and yield of maize in subsistence agriculture. *Plant Soil* **284,** 243–251.

Bänziger, M., Edmeades, G. O., Beck, D., and Bellon, M. (2000). "Breeding for Drought and Nitrogen Stress Tolerance in Maize: From Theory to Practice." CIMMYT, Mexico.

Bennet, E. M., and Balvanera, P. (2007). The future of production systems in a globalized world. *Front. Ecol. Environ.* **5,** 191–198.

Bezner-Kerr, R., Snapp, S. S., Chirwa, M., Shumba, L., and Msachi, R. (2007). Participatory research on legume diversification with Malawian smallholder farmers for improved human nutrition and soil fertility. *Exp. Agric.* (in press).

Buckles, D., Triomphe, B., and Sain, G. (1998). "Cover Crops in Hillside Agriculture: Farmer Innovation with Mucuna International Development Research Centre, Ottawa, Canada." http://www.idrc.ca/en/ev-9307-201-1-DO_TOPIC.html.

de Jager, A., Onduru, D., and Walaga, C. (2004). Facilitated learning in soil fertility management: Assessing potentials of low-external-input technologies in east African farming systems. *Agric. Syst.* **79,** 205–223.

Drinkwater, L. E., and Snapp, S. S. (2007). Nutrients in agroecosystems: Rethinking the management paradigm. *Adv. Agron.* **92,** 163–186.

Landis, D., Soule, J., Gut, L., Gage, S., and Smeenk, J. (2007). Agricultural landscapes and ecologically based farming systems. *In* "Building a Sustainable Future: Ecologically Based Farming Systems"

(S. Deming, L. Johnson, D. R. Mutch, L. Probyn, K. Renner, J. Smeenk, S. Thalmann, and L. Worthington, eds.), pp. 13–31. Michigan State University, East Lansing, MI.

Ludwig, H. D. (1968). Permanent farming on Ukara: The impact of land shortage on husbandry practices. *In* "Smallholder Farming and Smallholder Development in Tanzania" (H. Ruthenberg, ed.), pp. 87–135. Weltforum, Afrika-Studien 24, München.

Morales, H. (2002). Pest management in traditional tropical agroecosystems: Lessons for pest prevention research and extension. *Integr. Pest Manag. Rev.* **7,** 145–163.

Netting, R. McC. (1993). "Smallholders, Householders, Farm Families and the Ecology of Intensive Sustainable Agriculture." Stanford University Press, Stanford.

Pengelly, B. C., and Maass, B. L. (2001). *Lablab purpureus* (L.) Sweet: Diversity, potential use and determination of a core collection of this multi-purpose tropical legume. *Genet. Res. Crop Evol.* **48,** 261–272.

Pieri, C. J. M. G. (1992). "Fertility of Soils: A Future for Farming in the West African Savannah." Springer-Verlag, Berlin.

Pound, B., Snapp, S. S., McDougall, C., and Braun, A. (eds.) (2003). "Uniting Science and Participation: Managing Natural Resources For Sustainable Livelihoods." Earthscan, London.

Robertson, G. P., Paul, E. A., and Harwood, R. R. (2000). Greenhouse gases in intensive agriculture: Contributions of individual gases to the radiative forcing of the atmosphere. *Science* **289,** 1922–1925.

Sayer, J., and Campbell, B. (2004). "The Science of Sustainable Development: Local Livelihoods and the Global Environment." Cambridge University Press, Cambridge.

Snapp, S. S., Date, K., Kirk, W., O'Neil, K., Kremen, A., and Bird, G. (2007). Root, shoot tissues of *Brassica juncea* and *Cereal secale* promote a healthy potato rhizosphere. *Plant Soil* **294,** 55–72.

Snapp, S. S., Rohrbach, D. D., Simtowe, F., and Freeman, H. A. (2002). Sustainable soil management options for Malawi: Can smallholder farmers grow more legumes? *Agric. Ecosyst. Environ.* **91,** 159–174.

Vernooy, R., and McDougall, C. (2003). Principles for good practice in participatory research: Reflecting on lessons from the field. *In* "Uniting Science and Participation: Managing Natural Resources for Sustainable Livelihoods" (B. Pound, S. S. Snapp, C. McDougall, and A. Braun, eds.). Earthscan, London.

World Development Report (2008). "Agriculture for Development," Chapter 8. http://siteresources.worldbank.org/INTWDR2008/Resources/2795087-1184361034870/Ch8.pdf.

Young, A. (1989). "Agroforestry for Soil Conservation." CAB International, Wallingford.

Zhu, Y., Chen, H., Fan, J., Wang, Y., Li, Y., Chen, J., Fan, J., Yang, S., Hu, L., Leung, H., Mew, T. M., Teng, P. S., Wang, Z., and Mundt, C. C. (2000). Genetic diversity and disease control in rice. *Nature* **406,** 718–722.

INTERNET RESOURCES

http://afsic.nal.usda.gov. Links to sustainable and alternative agriculture systems information in the United States and beyond.

Agroforestry systems examples at the Web site http://www.css.cornell.edu/ecf3/web/new/af/afSystems.html.

The Center for Cover Crops Information and Seed Exchange in Africa: http://www.ppath.cornell.edu/mba_project/CIEPCA/home.html.

International Cover Crops Clearinghouse based in Honduras Web site: http://cidicco.hn/newcidiccoenglish/bulletins.htm.

Centre for information on low external input and sustainable agriculture: http://www.leisa.info/index.php?url=index.tpl.

Organic agriculture searchable database. Organic Eprints is an international open access archive for papers related to research in organic agriculture. The Danish Research Centre for Organic Farming (DARCOF) and the Research Institute of Organic Agriculture (FiBL) in Switzerland are managing an open access Organic Eprints archive: http://orgprints.org/.

http://www.worldagroforestrycentre.org/. The World Agroforestry Centre has information on many perennial–annual farming systems by agroecology around the globe.

CHAPTER 5

Agricultural Change and Low-Input Technology

Robert Tripp

Summary

Low external input agriculture has become an important focus of development programs aimed at strengthening crop management capacities for resource-poor farmers. Strategies that eschew the use of synthetic fertilizers and pesticides often involve particular philosophical or political stances about the development process, but they also provide valuable insights into the nature of agricultural change. By examining experiences with promoting this kind of technology, we can draw some

Agricultural Systems: Agroecology and Rural Innovation for Development

general lessons about agricultural development. Our case studies illustrate adoption patterns for low-input technology that are, perhaps surprisingly, little different from those of adoption for conventional technology. Better-resourced farmers are favored, as are those with links to markets. The fact that an increasing amount of farm labor (even on very small holdings) is hired draws attention to the increasing differentiation of the countryside. Extension methods require examination, given that innovative group methods are often successful for encouraging participants to test new technologies, but are rarely the spark that ignites wider diffusion invites a review of extension methods. The experience examined here with an uptake of low external input technology indicates that we need a realistic examination of the incentives for investing in new agricultural technology (of any kind) and a robust and coordinated set of strategies for providing information to farmers.

INTRODUCTION

A systems approach to agricultural development emphasizes the importance of taking advantage of as many resources as possible to improve farm productivity. The role of plant breeding often captures significant attention, including both the contributions of modern crop varieties and farmers' development of landraces. The role of crop management is equally important but often less apparent. As with plant breeding, advances in crop management can occur through the use of both external resources and local ingenuity, and the source of innovation is sometimes used to distinguish agricultural development strategies. In particular, the use of external inputs (principally synthetic fertilizers and pesticides) may be discouraged or proscribed, leading to the promotion of various versions of "low external input agriculture."

This chapter reviews experiences with the promotion of technologies that support low external input agriculture. The focus of the discussion is not on a particular development philosophy, but rather on the lessons that can be learned from endeavors to incorporate new biological resources and innovative techniques in crop management for resource-poor farmers. Because most of the attempts to encourage this type of technology have been part of projects that emphasize farmer organization and capacity building, the lessons that emerge should be useful beyond the bounds of a particular type of technology and provide direction for more general strategies of small-farm development.

The chapter is organized as follows. The first section provides an introduction to low external input technology (LEIT) and reviews some of the evidence for who is likely to take advantage of this kind of innovation. Because LEIT assumes that a primary input for farm improvement is household labor, the next section reviews the role of labor in choices about agricultural technology. LEIT also emphasizes the importance of developing farmer skills and knowledge, and the next section reviews the nature of farmer knowledge and how LEIT projects attempt to strengthen farmer skills. The final section summarizes findings and discusses the design of more effective agricultural development strategies.

LOW EXTERNAL INPUT TECHNOLOGY

The technologies that concern us in this chapter are a collection of crop management inputs and techniques for soil conservation, soil fertility enhancement, crop establishment, and pest control. They are distinguished principally by what they are *not*: manufactured, "artificial" inputs introduced to the farming environment. The delineation of these technologies may serve either a restrictive or an integrative purpose. The restrictive interpretation promotes LEIT as a way of insulating farmers from external input markets in the service of socioeconomic and environmental ends. It may be the basis of development philosophies that promote the use of local resources in order to promote environmentally sustainable small-scale farming that emphasizes self-sufficiency, such as "low external input sustainable agriculture" (LEISA) (Reijntjes *et al.*, 1992), or it may promote active participation in the market by offering a distinctive, environmentally friendly brand, such as organic agriculture (Figure 5.1). However, an integrative interpretation sees LEIT as an underutilized resource that is an essential element in broad strategies for agricultural development based on agroecological principles. It incorporates LEIT along with appropriate "external" inputs in strategies such as integrated pest management and integrated plant nutrient management that promote sustainable agriculture. This interpretation sees LEIT as an important set of tools, but does not attempt to base a theory of economic development or social justice on a particular type of technology. Both interpretations have made productive use of LEIT and have successes to their credit. This chapter does not attempt to judge between the two interpretations, but the approach adopted here is highly compatible with a positive, integrative, and pragmatic view of LEIT.

FIGURE 5.1 New market opportunities provide an incentive for farmers to adopt alternative agriculture practices, as shown here for organic cabbage production in Thailand.

Examples of LEIT

Examples of the technologies included under this rubric are given in Box 5.1. The technologies include physical crop management (such as methods of terracing, tillage, and planting) and the use of biological resources (such as intercrops, mulches, and biocontrol agents). Although all of these are based on labor, implements, or biological inputs that might be locally available, they are not necessarily "indigenous." Many of the innovations in LEIT are based on the transfer of a plant species or other organisms from one environment to another, or the elaboration of novel crop management techniques. In addition, although the inputs may be theoretically available within the farm household or community, there are markets for some of these inputs (such as biocontrol products or manures), and the major input, labor, can be a market commodity as well, as shown later.

The technologies included within LEIT range from the mundane and traditional to the novel and exotic. Similarly, the amount of attention given to the promotion of LEIT varies considerably, from modest behind-the-scenes attempts to strengthen farmers' skills in crop management to well-publicized campaigns around particular innovations such as vetiver grass (National Research Council, 1993) or the SRI (Stoop *et al.*, 2002).

As much of LEIT represents iterative modifications to crop management practices, it is not always easy to assess progress, but there have been enough concerted efforts that it is possible to distinguish some successes and failures. Some of the successes have been the result of formal development projects, such as integrated pest management (IPM) in irrigated rice (Tripp *et al.*, 2005); others have benefited from both project support and farmer innovation, such as the spread of the velvet bean cover crop in Central America (Buckles, 1995); and others have relied primarily on farmer initiative, such as improved soil conservation in dryland Kenya (Tiffin *et al.*, 1994). The failures range from the many attempts at crop management improvement that have been quietly devised, tested, and abandoned to high-profile initiatives such as the attempt to introduce alley cropping in tropical Africa (Carter, 1995).

The Adoption of LEIT

There have been few attempts to document the global extent of adoption of this class of technology, and efforts in this direction suffer from inevitable problems of definition and rigor. For instance, a comprehensive study of the uptake of "sustainable agriculture" practices includes many examples of herbicide-based conservation tillage (in a list of otherwise chemical-free technologies) and relies almost entirely on self-reports by project staff (Pretty and Hine, 2001). There have been a number of attempts at assessing uptake of this kind of technology at the local level, but even here the literature is far less densely populated than for the seed-fertilizer technologies of the green revolution. Two of the problems are (1) the gradual, iterative nature of many of these management

BOX 5.1 Examples of LEIT

Soil and Water Management

• Terraces and other physical structures to prevent soil erosion. These may be the result of large-scale external investment or may be developed, often over many seasons, by hoe or plow or by the arrangement of stones to form barriers.

• Contour planting, in-row tillage, tied ridging. Ploughed or hoed ridges are laid out along contours on slopes; tillage only along the cropped row (in-row tillage) develops miniterraces (Bunch, 1999); tied ridges created by plow or hoe help conserve moisture and nutrients.

• Hedgerows and living barriers. Trash lines along the contour gradually form a bund; various shrub and grass species planted as intercrops on the contour form a living barrier against erosion.

• Reduced tillage systems, conservation tillage. A number of techniques promote a reduction in tillage; most require alternative planting systems and innovative ways of controlling weeds; some of the most prominent include the use of herbicide.

• Mulches, cover crops. Mulches may be derived from crop stover in the field or cut and transported from elsewhere; cover crops are usually grown in association with the field crop and may serve various purposes, including weed control and fertility enhancement (Anderson *et al.*, 2001).

Soil Fertility Enhancement

• Manures and composts. Manure from grazing animals or transported from stalls; various composting techniques, including vermicompost.

• Biomass transfer and green manures. A crop is grown in a separate field and cut and carried to provide organic matter (Cooper *et al.*, 1996); green manures are leguminous crops planted with the field crop or in rotation.

Crop Establishment

• Planting pits. Small pits or basins dug throughout the field, used as planting stations where fertility and moisture are concentrated; often used for rehabilitating degraded land.

• System of rice intensification (SRI). Rice seedlings are transplanted earlier than normal and planted at wide spacing; the soil is kept well drained and irrigation is managed to provide short periods of wetting and drying (Stoop *et al.*, 2002).

• Intercropping. A traditional practice in many areas that can enhance weed control and soil fertility as well as reduce risks.

(*Continues*)

BOX 5.1 *(Continued)*

Controlling Weeds and Pests

- Intercrops and rotations. Weeds and insect pests may be controlled by selection of appropriate rotations or intercrops.
- Integrated pest management. Insect control based on an understanding of ecological principles and employing a wide range of biological and, when necessary, chemical methods to keep pest damage below an economic threshold.

changes (making it difficult to define who is an "adopter") and (2) the lack of incentives for those who sponsor typically short-term development projects featuring LEIT to measure outcomes. A recently completed systematic study attempted to address this knowledge gap (Tripp, 2006). To overcome the difficulties outlined earlier, this study included a thorough review of the literature on the adoption of LEIT and conducted fieldwork that examined the long-term consequences of three prominent LEIT projects; it is the basis for many of the conclusions reported in this chapter.

Perhaps the most surprising conclusion from this review of LIET adoption patterns is that those who tend to take up and use such technology exhibit many of the characteristics associated with adopters of conventional technology. That is, they tend to be farmers with relatively more resources (land, education, access to finance) and those who rely more on agricultural markets. There are of course many variations and exceptions in these cases, but the overall pattern is quite clear. This challenges the view of many LEIT supporters who see these technologies as an alternative path for those farmers left behind by conventional, external input-based agricultural development.

Research on adoption provides explanations for the similarities observed in uptake patterns. Rogers' (1995) review of the diffusion of all types of innovations is one of the most thorough surveys available, and one of its principal conclusions is that the most common consequence of technology diffusion is to favor the better off, thus widening socioeconomic gaps. This is of course not an inevitable conclusion, and evidence shows that well-planned agricultural development efforts can at least limit such biases, but it should be clear that there is little evidence that LEIT follows a radically different pathway. However, there is no need to be unnecessarily despondent about the consequences of more conventional technology. Some of the early observers of the green revolution in Asia predicted a widening gap (Frankel, 1971), and there is evidence of undeniable instances of inequality (Freebairn, 1995; see Chapter 1). Yet long-term studies of areas that benefited from the seed-fertilizer technology associated with the green revolution provide a more hopeful picture of benefit sharing (Hazell and Ramasamy, 1991; Lanjouw and Stern, 1997).

Some examples from the literature illustrate instances in which LEIT adoption favors better-resourced farmers, even where poorer households have at least equivalent access to the innovations. A study of farming practices in an area of western Kenya showed that wealthier households are more likely to use both external inputs (hybrid maize, inorganic fertilizer) and low-input soil management techniques (fallows, compost, terraces) (Crowley and Carter, 2000). In Niger, the introduction of planting pits has helped rehabilitate degraded land and has contributed to an emerging land market, but the purchasers are concentrated among a rural elite (Hassane *et al.*, 2000). The initial experience with the SRI in Madagascar indicates that adopters are more likely to be surplus rather than deficit rice producers, with more land and often with better off-farm sources of income (Moser and Barrett, 2003). The examples are not confined to developing countries. A movement toward "restorative agriculture" in the United States in the early 19th century (featuring a shift from extensive cultivation toward more careful soil management and the use of manure) was distinguished by the fact that "the majority of improving farmers held a fortune somewhere above middling, including merchant-squires of great wealth.…Those who incorporated restorative methods almost always lived close enough to market towns to turn surplus into cash" (Stoll, 2002, p. 28).

A strong relationship also exists between commercial agriculture and the uptake of LEIT (see Box 5.2 for an example from Honduras). Cramb *et al.* (2000) compared the experiences of several soil conservation efforts in the Philippines and found that adoption is highest in an area where proximity to large urban markets gives farmers an incentive to conserve soil. The spontaneous adoption of planting pits in Zambia is higher among cotton farmers than among those who grow only food crops (Haggblade and Tembo, 2003). There is evidence that many instances of adoption of soil conservation measures (particularly terracing) for food crops in parts of eastern Kenya are related to a village's proximity to markets and, in some cases, to the ability to invest windfall profits from high coffee prices in labor for terrace construction (Zaal and Oostendorp, 2002). The opportunity for expanded commercial production may also be an incentive for the adoption of LEIT.

These examples are far from conclusive, but they certainly indicate that LEIT is not immune from distributional biases. It is of course misleading to simply classify farmers as "wealthy" or "poor" because resource and motivational distinctions in rural communities are multidimensional. A study of a soil conservation program in the Philippines showed that it tended to benefit a "clique of yeoman farmers" (Brown and Korte, 1997, p. 14), but not the village elite. Project management can also make a difference. Projects often begin in areas, and with farmers, that are more accessible. For instance, the early experience of farmer field schools (FFS) that introduced IPM in Indonesia concentrated on "the better informed and more affluent farmers" (Röling and van der Fliert, 1994, p. 103). A study of the impact of agroforestry innovations in Kenya found no relationship between wealth and adoption in the pilot villages that received extra attention for reaching disadvantaged groups,

BOX 5.2 The Adoption of Soil Regeneration Technology in Honduras

In the 1980s and 1990s, several nongovernmental organization (NGO) projects operated in Honduras promoting sustainable hillside farming, with particular attention to methods for soil fertility enhancement and soil and water conservation. A study revisited two areas in central Honduras where previous slash and burn agriculture has evolved to permanent cropping on hillside plots and farmers plant twice a year, with maize as the principal first season crop, followed by beans in the second season. The NGO projects promoted farmer experimentation and featured techniques such as in-row tillage and the use of cover crops. Although farmer leaders were identified from among project participants, the projects were sufficiently flexible that a wide range of farmer participation was possible.

An examination of who participated in the projects shows that those farmers with links to agricultural markets, particularly those with irrigation and commercial vegetables, were more likely to take an interest in the activities. Indeed, earlier project activity had helped many of these farmers enter into commercial vegetable production for the first time.

	Participants (n = 79)	Nonparticipants (n = 46)	Statistical significance
Education (years)	2.7	2.3	No
Age	44.7	51.7	<.05
Percentage farms with irrigation	60	35	<.05
Percentage farms with vegetables	43	22	<.05
Maize yield (kg/ha)	1105	806	<.05
Total chemical fertilizers (kg)	163	145	No

Not surprisingly, these initial differences in interest among the farmers are reflected in the final record of adoption, 5 or more years after the projects were completed in the study villages. The majority of the new technologies were applied to cash crops rather than subsistence crops.

Technology	Percentage of farmers using (n = 172)
In-row tillage with maize	16
In-row tillage with beans	17
In-row tillage with vegetables	45
Organic fertilizers on maize	12
Organic fertilizers on vegetables	28

Source: Richards and Suazo (2006).

but found a bias toward better-off farmers in villages included in a subsequent round of less intensive promotion (Place *et al.*, 2003).

The fact that LEIT seems to exhibit many of the diffusion characteristics of other agricultural technologies is disappointing for those who hope to find a direct way to reach the rural poor. At the same time, it indicates that any conclusions reached about the relevance, organization, and promotion of the technology are relevant to a wide array of strategies in agricultural development. Our interest is in who is able to take advantage of new technology and the implications for effectively supporting rural development. The next section looks more closely at the interactions between farm labor and technology adoption, and the following section examines how knowledge about technology is generated and made available.

LABOR

The fact that many examples of LEIT rely on some investment of labor for their implementation (and usually require some additional time for learning and adaptation) has been used by supporters of the technology as evidence of its relevance for poor households with few resources in addition to labor and has been used by detractors of the technology to characterize LEIT as impractical and labor intensive. Neither of these extremes represents a particularly useful assessment. This section examines the nature of the labor investment for LEIT, the sources of labor in small-farm agriculture, and the prevalence of off-farm labor.

Labor Requirements

Labor is a fundamental determinant of the acceptability of a technology to farmers. Moreover, the labor component cannot simply be assessed in terms of hours invested per hectare. The timing of labor during the season, the skill requirements (including the possibilities of learning to manage a technology more efficiently), and the difference between one-off investments (e.g., to establish a terrace) and recurrent requirements (e.g., to monitor pest damage) all must be taken into account.

It is inaccurate to characterize LEIT as necessarily labor intensive. Some examples of LEIT require no more labor than the farmer's present practice, whereas some types (such as certain variants of conservation tillage, or IPM that reduces the use of insecticides) are attractive precisely because they save labor. However, it remains true that the success of LEIT is often dependent on the efficient organization of labor supply. In some cases the crucial factor is an initial investment in labor for the establishment of LEIT (such as a soil conservation measure); once established, the labor requirements then fall, sometimes below those of conventional management. In a number of instances, however, LEIT implies a permanent increase in labor, whereas in other cases the crucial factor is not necessarily the investment in physical labor but rather the time for learning new skills that can be applied to the farm or for monitoring performance.

Many studies confirm the importance of routine labor demands as a determinant of adoption for LEIT. A number of observers have remarked on the problems that some IPM methods impose on farm labor patterns; for instance, farmers often find they do not have time to devote to frequent scouting for pests that some IPM techniques require. Carter's (1995) review of alley farming points to its high labor requirements at various times in the cropping calendar as one of the major factors explaining the very limited success of the technology.

Other aspects of labor are relevant to the use of LEIT. Even when labor availability per se is not an impediment, there are a number of instances where the new practices require additional skills. The skills may be needed for the one-time application of a particular technique (such as the establishment of contour ridges with an A-frame), but often they are necessary each season in order to make adjustments and adaptations and to promote further innovation. These skills may require significant time to acquire and, in this sense, may be counted as "labor" costs; Pretty (1995) sees them as part of "transition costs," that is, investment for transition from conventional to alternative agriculture. "Lack of information and management skills is, therefore, a major barrier to the adoption of sustainable agriculture (Figure 5.2). During the transition period, farmers must experiment more and so incur the costs of making mistakes, as well as acquiring new knowledge and information" (Pretty, 1995, p. 96).

The time to engage in learning and experimentation is less likely to be available to poorer members of the community, contributing to the patterns of adoption that have been observed. In establishing FFS for IPM in Zanzibar, it was "necessary to work with groups of farmers who were willing to learn, able to experiment, had enough flexibility to change and were prepared to commit themselves for one or more seasons. This automatically excluded the poorest farmers who had little physical and financial buffer for experimentation" (Bruin and Meerman, 2001, p. 67). It also is a significant barrier to participation by women who are responsible for a "doubleshift" of household and farm duties, particularly female-headed households. Issues of inequities in participation are discussed in more depth in Chapter 10.

The techniques of the SRI not only require more labor, but also new skills. Moser and Barrett (2003) pointed out that the labor requirements of SRI cannot be fulfilled easily with hired labor because the unfamiliar tasks associated with SRI require close supervision. One observer suggested that realizing the potential of SRI will depend on "agricultural professionalism" in mastering the various skills required (Stoop, 2003). Similarly, the adoption of conservation farming in Zambia requires advance planning and careful execution of tasks. "It requires a change of thinking about farm management under which the dry season becomes no longer a time primarily reserved for beer parties and socializing but rather an opportunity for serious land preparation work. Anecdotal evidence from our field interviews suggests that retired school teachers, draftsmen and accountants make good CF farmers" (Haggblade and Tambo, 2003, p. 39).

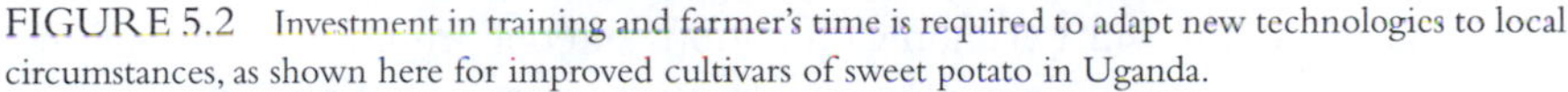

FIGURE 5.2 Investment in training and farmer's time is required to adapt new technologies to local circumstances, as shown here for improved cultivars of sweet potato in Uganda.

Hired Labor

The green revolution in Asia is a particularly important case of technological change leading to increased labor use. The development of short-stature rice and wheat varieties, in conjunction with increased use of fertilizer and irrigation, made possible significant increases in yield that reversed the trend toward increasing food grain imports. A review of the adoption of modern rice varieties in a number of Asian countries found, with one exception, a significant increase in labor use per hectare (David and Otsuka, 1994).

A very significant part of the increased labor use during the green revolution was hired mostly from landless or near landless households. Indeed, this added employment was one of the most significant contributions of the technology. Despite the fact that LEIT is often envisioned as a way to take advantage of supposedly surplus household labor, a significant proportion of the additional labor for managing these innovations is also hired. In the Honduras case (see Box 5.2), for instance, half the labor used to construct the miniterraces for in-row tillage was hired. The case for seeing LEIT as a distinct strategy is more difficult to maintain when labor is itself increasingly an external input. The growing importance of hired labor for many types of agricultural tasks in communities that are often viewed as composed of static, self-sufficient peasant households is a reminder of the importance of understanding farmers as dynamic actors. The relationship between labor demand and LEIT adoption is illustrated for a case from Kenya in Box 5.3.

The availability and price of hired labor depend to a large extent on the wider economy. Lipton with Longhurst (1989) estimated that every 10% increase in yield in the early years of the green revolution was responsible for a 4% increase in labor use, whereas the corresponding response 20 years later was associated with only a 1 to 1.5% increase. This is associated with shifts to labor-saving technology, which is in turn a response to rising rural wage rates, occasioned by growth in the non-farm economy. The role of wage rates in the uptake of LEIT is illustrated by an analysis of the use of green manures in irrigated rice production in Asia (Ali, 1999). Much of the behavior of rice farmers toward green manure options can be understood by looking at the relationship between labor costs and fertilizer price in individual countries. Green manures and fertilizer are alternative sources of nitrogen, and the cost of green manures is mostly dependent on the labor for their management. When the ratio of labor cost to fertilizer price is above a certain level, farmers abandon the use of green manure.

Off-Farm Labor

The influence of the wider economy on options and opportunities for agricultural technology development is also illustrated by the importance of off-farm labor for many so-called farming households. The labor requirements of any new technology may be interpreted differently by various types of farmers, depending not only on their access to household labor (or ability to hire labor) but also on their alternative employment opportunities. These alternatives can draw them away from the farm but also may provide extra cash resources to hire local labor. The availability of nonagricultural income opportunities may lower the amount of labor available on the farm. However, where agriculture is not profitable enough to offer a sufficient wage but there are relatively few outside alternatives, farm labor shortages and rural unemployment may, paradoxically, coexist.

Rural economic realities challenge any oversimplified images of the peasant farmer. Van der Ploeg (1990) provided a striking illustration of these realities in a description of a highland Peruvian farming community, where only 9% of the farmers work solely on their own farms. The poorest group of farmers combines subsistence production with wage labor in order to survive. Other farmers invest part of their time in wholesale trading and look to this as their principal economic strategy, seeking to reduce labor and other investments in the farm. Another group is involved in enterprises, such as petty trade, which offer less scope for expansion, and they channel much of their earnings to further intensifying agricultural production. Although it still may be possible to combine own-farm labor with off-farm migration after the cropping season, the concentration of rural resources is making this less of a possibility. Many farms may simply not be viable, and an absence of external

BOX 5.3 Labor Deployment and the Adoption of Soil Conservation in Western Kenya

Kenya's National Soil and Water Conservation Program operated from 1988 to 1998 and featured a catchment approach, where communities were encouraged to learn about and establish soil and water conservation techniques. Elected local catchment committees served as major actors in the project. A study examined the aftermath in a set of communities in high- and low-potential areas of Nyanza Province, western Kenya. The principal subsistence crop is maize, but beans, banana, groundnut, sweet potato, and sorghum are also grown. The project featured the promotion of vegetative and unplowed strips, simple terraces, and retention ditches.

Soil conservation technology often requires high labor inputs. The study found a total of 23 different types of soil and water conservation activity undertaken by farmers, but the technologies with lower labor requirements (such as grass strips and unplowed strips) were adopted much more commonly than those that required more labor (such as terraces). In addition, the strips take relatively little space and Napier grass can be grown on them to feed to cattle. Although many farmers adopted some kind of conservation structure, their spacing on the slope was less than one-third the recommended density for effective erosion control, another possible indication of labor constraints.

An examination of the nature of farmers who adopted the conservation technology also shows how labor opportunities influence interest in technology. In high potential areas (where the majority of the adoption took place), adopting farmers had more labor available (measured by household labor resources), relied more on crop sales, were less likely to be involved in nonfarm business, and hired more labor.

	Adopters ($n = 41$)	Nonadopters ($n = 13$)	Statistical significance
Landholding (acres)	6.8	3.2	No
Number of cattle owned	3.0	2.0	No
Percentage crop sale as most important source of income	59	15	<0.01
Percentage business of petty trade as most important source of income	7	39	<0.01
Percentage low labor availability	9	31	<0.1
Percentage hire labor for weeding	68	46	No
Percentage-derived income from casual labor	12	14	No

Source: Longley *et al.* (2006).

opportunities means that farmers end up seeking day labor with wealthier neighbors, as in the *ganyu* labor system in Malawi (Whiteside, 2000). Such systems can contribute to the downward spiral of small farms, as opportunities for laborers occur at precisely the times when they should be working on their own farms.

KNOWLEDGE

Most examples of LEIT involve an understanding of principles of crop management and the adaptation of innovations to local circumstances rather than the mere application of standard recommendations. This level of knowledge intensity is sometimes put forward as a distinguishing characteristic of LEIT, although it is debatable whether farmers' successful experience with "green revolution" technology was in fact merely a matter of following instructions, nor indeed whether new "external" technologies such as transgenic crops are any less knowledge intensive (Tripp, 2001). Nevertheless, it is certainly true that LEIT relies on building farmer knowledge systems. This section examines the resources available for that purpose.

Farmer Knowledge

Most efforts at LEIT are based on strengthening local knowledge and supporting farmers' experimental capacities. Both of these arenas have been the subject of considerable study and occasionally of controversy. A considerable amount has been written about indigenous technical knowledge (ITK), confirming that farmers often have a detailed understanding of their environment. For instance, the Hanunoo cultivators of the Philippines recognize many more distinctions in the plants growing in their region than do systematic botanists (Conklin, 1957). Traditional soil classification in Tlaxcala, Mexico, is a comprehensive yet flexible system that allows farmers to understand both fertility and soil management properties (Wilken, 1987). Such traditional knowledge is often overlooked in agricultural development efforts and farmers may not be encouraged to utilize this resource. For example, Murwira *et al.* (2000) reported how farmers in a rural development project in Zimbabwe were at first reluctant to discuss their use of traditional practices for pest control for fear of ridicule; it was only when the project confirmed that such knowledge could have significant value that farmers were willing to include the information in the project's activities.

Yet these capacities to observe and classify do not always lead to practical knowledge. For instance, Bentley (1989) argued that Honduran farmers' ITK is best developed for describing plants and plant growth and is less adequate for understanding insect behavior or the origins of plant disease. These deficiencies may at times contribute to harmful practices, such as overdependence on pesticides. Thus ITK cannot always be seen as a basis for further innovation or used as a certain defense against environmental mismanagement. In discussing the introduction of IPM in Vietnam, Heong and Escalada (1999) discussed the importance of changing farmers' beliefs about pest damage and their tendency to overestimate yield loss. Similarly, Orr (2003) stated that formal knowledge of insect pests and their habits is a priority if IPM is to be introduced to African farmers, and Neill and Lee (2001) suggested that the diffusion of resource-conserving technology in Honduras will be sporadic until farmers have a better understanding of system dynamics and the specific details of nutrient cycles and nitrogen fixation.

Taken together, these examples imply that the generation of LEIT cannot simply rely on providing support to farmer systems of knowledge, but also requires skilled and sustained technical input from extension or other service providers (Figure 5.3). Studies in Kenya and Zambia found that agroforestry-based soil fertility technology was used more extensively in villages that had served as pilot projects and thus had access to considerable technical backstopping (Place *et al.*, 2002). A number of NGOs that originally promoted conservation farming in Zambia have now retreated, in part because of a shortage of the management and agronomic skills among their staff required for the promotion of this complex technology (Haggblade and Tembo, 2003).

FIGURE 5.3 Farmer experimentation is often catalyzed by participating in on-farm research or extension activities, as illustrated here by an A-frame innovation developed by a Malawi farmer to speed up contour measurements and ridge alignment for soil conservation.

Another subject of debate is the extent to which farmers experiment. A study of farmer experimentation in Africa (Sumberg and Okali, 1997) identified two conditions for an "experiment": a farmer's initial observation of conditions or treatments and the observation or monitoring of subsequent results. The researchers found a wide difference among individuals and sites in the propensity to engage in experimentation. They also distinguished between "proactive" experimentation in which there was a conscious effort by the farmer to create or control certain conditions for the purposes of observation and "reactive" experimentation, which had no systematically chosen objectives. Their study found that men are more likely than women to report experimentation and that these men are likely to have more education and previous involvement with extension. They also found that experimentation was more frequent among people who saw themselves as full-time farmers, frequently (although not exclusively) engaged in commercial production.

A fairly wide variety of strategies have been used to build on farmer knowledge and innovation in the promotion of LEIT. Some of these involve expanding opportunities for farmer interchange, often based on formal group learning methods.

Group Methods

When LEIT involves the introduction of fairly complex technologies that require local adaptation, farmers often profit from "cross visits," opportunities to observe such technologies in practice in other communities and to discuss techniques and challenges with those farmers that have more experience. The *Campesino a Campesino* movement in Nicaragua, devised by the National Union of Agricultural Producers during the Sandanista regime, aims to help small farmers acquire environmentally sound production techniques. It makes considerable use of exchange visits, where members of one community visit those in another to learn about innovations. The movement has found that farmers are particularly effective at communicating their experiences to their counterparts (Anderson *et al.*, 2001; Holt-Giménez, 2001). Farm tours and visits are important to promoting understanding and farmer experimentation in widely different socioeconomic contexts, as witnessed by their importance in introducing IPM to Texas cotton farmers (Leslie and Cuperus, 1993).

Low external input technology activities are often organized around group methods, and there is a wide range of examples. Many LEIT projects include opportunities for social learning (the transmission of information in a social context). In addition, there have been advances in developing farmer participation in problem diagnosis and technology development. Several formal methodologies exist. An additional characteristic of many LEIT efforts is an aspiration that once useful principles or techniques are identified and introduced on a small scale, local farmers can take increasing responsibility for their diffusion, and LEIT projects often make provision for this strategy.

Group methods are of course not the exclusive preserve of LEIT activities; as Pretty (1995) pointed out, many Asian countries promoted uniform green revolution packages through local groups (often formed or mandated for that purpose). However, the group approach is particularly relevant to low external input strategies for several reasons (Pretty, 1995). First, resource management often requires more than the efforts of individual farmers; for instance, communities need to agree on common strategies for challenges such as soil erosion control, and some pest management techniques require coordination among neighboring farmers. The effective management of common property resources can play an important role in environmentally sound development strategies. In addition, low external input strategies often go beyond the goal of introducing particular technologies and seek to build, or rebuild, local institutions. Finally, to the extent that LEIT requires the application of principles and the local adaptation of technology, it usually makes sense to organize group activities, not only because they are more efficient but also because they promote the interchange of ideas and experiences.

An example of group activity for improving common property management is the Landcare movement, which originated in Australia and represents a productive collaboration between farmers and conservationists (Roberts and Coutts, 1997). Landcare groups comprise local farmers and other members of the community interested in land management. They meet to discuss common problems and engage in a range of activities, including teaching, training, and trials. Successful interventions have been carried out by Landcare groups in areas such as soil conservation and wetlands management. Many of these activities are eligible for state funding, and Landcare groups also participate in the development of integrated catchment management plans, under the aegis of the state government. The strategy has been transferred elsewhere and is used to promote resource conservation on individual farms. One example is a region of the southern Philippines, where a Landcare association and local groups have been formed (Cramb, 2005). The local groups include farmers, community leaders, and extension agents and are able to support their activities through a decentralized Philippine government funding initiative related to environmental protection.

One of the most well-known innovations for introducing LEIT is the farmer field school. The development of an effective IPM strategy for rice in Asia is based on work in Indonesia that led to the emergence of the FFS concept. In order to appreciate the rationale behind lowering pesticide use and allowing ecological processes to reestablish their regulation of pest activity, farmers needed to learn more about pest–plant and pest–predator interactions; the FFS met this challenge (Kenmore, 1996). The FFS fosters discovery learning by facilitating farmer opportunities to observe and discuss important ecological relationships in the field. A FFS typically includes about 20 farmers and a trained facilitator who meet once a week during the cropping season. Farmers spend part of each session observing pest and predator behavior, drawing diagrams of the relationships they uncover, and debating their implications. Farmers are encouraged to make "insect zoos" (collecting certain pests to observe their behavior or life cycles) and to conduct simple experiments (such as mechanically cutting a certain proportion of leaves early in the crop cycle to mimic early insect damage and to observe the recuperative capacity of the plant). Each FFS also includes exercises in group dynamics to promote a group spirit and to foster collaboration.

In the 1990s, more than 75,000 FFS were conducted in Asia (Pontius *et al.*, 2002). The FFS strategy is now being applied in a wide range of settings for purposes well beyond the extension of IPM. The Food and Agriculture Organization (FAO) is promoting the concept of integrated production and pest management through FFS in a number of African countries where farmers are exposed to a range of principles in crop production (Okoth *et al.*, 2003). There are worldwide examples of the FFS approach being applied to programs in livestock health and production, soil fertility, community forestry, gender equity, HIV/AIDs prevention, and other topics (LEISA, 2003; CIP-UPWARD, 2003).

The Sustainability of Groups

Group formation may be useful in the context of a project, but the sustainability of such groups is often in doubt. The rapid spread of conservation tillage in Brazil was due in part to the formation of "Friends of the Land" clubs that facilitated farmer-to-farmer exchange of experience, provided support to farmers experimenting with the technology, and allowed access to outside expertise. (Many of these clubs were assisted by support from the chemical companies that were selling the herbicides used for conservation tillage.) However, once farmers became familiar with the techniques, interest in the clubs tended to wane unless other activities were included (Landers, 2001). Winarto (2002) described how an NGO in Indonesia that successfully introduced IPM to rice farmers helped form an alliance of farmers' associations that campaigned to remove pesticides from the government credit package and lobbied for lower prices and more timely delivery of fertilizer. However, there is little evidence that the majority of FFS formed in Indonesia for introducing IPM survive beyond the initial season of training.

Although group approaches can promote widespread community involvement in technology generation, they may also be susceptible to capture by local elites. Projects promoting the construction of stone bunds in Burkina Faso relied on the formation of village groups.

> Influential village members, such as the better-off, local chiefs, and the so-called enlightened (those allegedly well-versed in modern ways, often returned migrants), tend to dominate these groups. While there is usually an atmosphere of free discussion, it is true that the dominant ones are able to rely on the group to help them develop their own farmland. The less well-off do participate in the activities of the group, although they rarely benefit directly. Quite often, their own farms are neglected. (Atampugre, 1993, p. 106)

Other Techniques for Technology Generation

In addition to taking advantage of group dynamics, LEIT projects use a range of techniques for promoting on-farm technology generation. Farmer participatory research has been a part of many formal agricultural technology development strategies at least since the emergence of farming systems research (FSR) in the 1970s. Biggs (1989) prepared a classification of types of farmer participation in public FSR programs, and others have helped draw attention to wider possibilities for farmer involvement in technology generation (Chambers *et al.*, 1989). A particularly useful guide for helping organize practical farmer experimentation in the context of community-based projects was developed from experience with introducing resource-conserving technology in Central America (Bunch, 1982). Participatory action research approaches such as the "mother and baby" trial design have been shown to promote farmer experimentation (Figure 5.4). Braun *et al.* (2000) compared the development of local agricultural research committees (a technique used in several Latin American countries for developing community-level experimentation capacity) with the experiences of FFS and found that each method has particular strengths and advantages.

An exceptionally comprehensive set of guidelines for developing options for integrated soil fertility management is found in the participatory learning and action research (PLAR) methodology. The objectives are to develop useful soil fertility management innovations based on local resources and knowledge and, where appropriate, external inputs. PLAR was developed through a series of field

FIGURE 5.4 An informal experiment labeled by the farmer who designed it is shown from Central Malawi, where she is testing complex legume mixtures and planting arrangements.

experiences in Africa (Defoer and Budelman, 2000). The process is based on a series of activities with farmers over the course of several growing seasons. It is an example of joint learning and experimentation, involving an interdisciplinary team of facilitators and farmers, and demands a considerable amount of time from all participants.

The Diffusion of Information

Intensive methodologies such as FFS or PLAR, which direct considerable facilitation resources toward a small number of communities, are often defended on the basis that once extension services have gained experience with the concepts in an initial period, they can adjust the methodologies to local resources. A complementary expectation is that farmers who have been through this process can serve as catalysts to help neighboring villages initiate their own work. Extension services can make sure that experienced farmers have the resources to offer advice to others. A number of national FFS programs in Asia have devoted resources to training and supporting farmer facilitators who can take the place of extension agents in leading FFS and thus extend the methodology at a lower cost; however, the success of these efforts has been the subject of recent critical analysis (Quizon *et al.*, 2001). Beyond these large-scale efforts, many other initiatives in promoting low external input agriculture include farmer-led extension or farmer promoters in their strategies (Bunch 1982; Holt-Giménez, 2001). In some cases, projects pay a small stipend to local farmers who have been identified and trained to play a formal extension role. In other cases, the projects simply try to increase contact between participant farmers and their neighbors in the hopes that this will facilitate the diffusion of the innovations. Farmer-to-farmer exchanges and field visits have been successful in enhancing learning and diffusion of information (Figure 5.5).

In many cases there appears to be an assumption that the experience of working in a group and generating LEIT will serve as a spark that motivates farmers to share their technological and organizational experience with others. The participant observation that Winarto (2004) carried out with a FFS group in Java provides insights into the challenges of communicating knowledge gained in a FFS (Figure 5.6). Although there is certainly evidence of social learning, with participants debating, sharing observations, and learning from each other, their desire to communicate this new-found knowledge with other farmers was often thwarted. A farmer's status in the community determined the degree to which others would listen to him (e.g., a young participant had no success in getting his father to modify his practices). New terms (such as *natural enemy*) did not easily find a place in conversation, and the experiences of the FFS were difficult to communicate to others. Despite the persistent proselytizing of some of the IPM farmers, only modest progress was achieved in changing nonparticipants' views on insecticides. Box 5.4 summarizes the successes and failures of FFS in Sri Lanka.

FIGURE 5.5 A farmer-to-farmer exchange visit in Northern Malawi is shown where farmers gain first-hand experience with institutional innovations such as nutrition education through farmer-led recipe days.

IMPLICATIONS FOR RURAL DEVELOPMENT STRATEGIES

The technologies described under the rubric of LEIT make indispensable contributions to improving small-farm productivity, and the innovative methods used to promote the generation and diffusion of this type of technology help strengthen farmers' capacities. However, the relatively limited spread of LEIT, and uncertainties about the cost effectiveness of many of the methods used to support it, challenges any aspirations to use this experience as a model for pro-poor technological change. Interactions with farm labor and local knowledge, and the only modest achievements of LEIT projects on the ground, illustrate why it is unwise to hope that rural development strategies based on particular types of agricultural technology can bring about fundamental social or political change. More fundamental issues stand in the way of technology-led rural development. This section discusses two of those issues: the incentives that motivate farmers to seek new technology and the organization of development projects. The discussion is based on the experiences of promoting LEIT but should have relevance to a wider range of agricultural development strategies.

Incentives for Technology Generation and Adoption

In order for LEIT to make a significant impact on the countryside, farmers must have appropriate incentives to experiment with, adapt, and gain control of new technology. This review has shown that the supposedly "simple" technologies of

FIGURE 5.6 Social learning at a farmer field school in Sri Lanka.

LEIT do not often reach the poorest farming households. One of the principal explanations is the nature of income generation for the rural poor. The custom of referring to all landholding rural households as "farmers" masks the complexity of rural livelihood strategies and often overstates the importance of agriculture to the income streams of these households (see Chapter 2 for an in-depth discussion of livelihood strategies). For example, less than one-third of maize growers in Kenya or Zambia are net grain sellers (Jayne *et al.*, 1999). The deployment of farm household labor is receiving increased attention due to growing interest in rural livelihood diversity (Ellis, 1998), although the phenomenon is not of recent origin. A situation where most farmers "were involved in several 'occupations'… [and] might take laboring jobs for other people or they might have expertise in

BOX 5.4 Farmer Field Schools for IPM in Sri Lanka

From 1995 until 2002 a program of FFS for introducing IPM and other crop management techniques to rice farmers in Sri Lanka was managed by the Department of Agriculture, with assistance from the FAO. A study was conducted in communities in the Southern Province among farmers with access to irrigated paddy land. The project concentrated on helping farmers lower the use of insecticides, particularly early in the season. It also supported the incorporation (rather than burning) of rice straw and promoted more rational fertilizer management through single nutrient fertilizers.

Each FFS could accommodate about 20 farmers on a first-come, first-served basis. An examination of the characteristics of participants reveals few differences with nonparticipants in terms of income source, age, or education; however, farmers who also worked as casual laborers were much less likely to participate.

The effects in terms of technology change are remarkable and apparently sustainable 5 or more years after participation in the FFS. A comparison of FFS farmers, neighbors in the same irrigation tract, and control farmers from a nearby irrigation tract in seven locations showed that only FFS farmers experienced significant change.

Practice	FFS farmers ($n = 70$)	Neighbors ($n = 70$)	Statistical significance of difference between FFS and neighbors	(Control farmers) ($n = 70$)
Insecticide applications (2002/2003 season)	0.6	1.7	<0.001	(1.7)
Report lower insecticide use (%)	80	49	<0.001	(49)
Use triple superphosphate (%)	83	51	<0.01	(54)
Use muriate of potash (%)	83	53	<0.01	(56)
Incorporate rice straw (%)	86	73	<0.1	(60)

A remarkable feature of these adoption patterns is the fact that there is virtually no difference in technology use between neighbors of FFS farmers and those in control villages several kilometers away. (The only possible exception is the incorporation of rice straw.) This is one indication of the fact that although the FFS was responsible for significant changes in farming practices, the message did not spread to other farmers.

(*Continues*)

BOX 5.4 (*Continued*)

Another indication of the low degree of information diffusion is the response of neighboring farmers about lessons learned from the FFS farmers. The only practice for which there is much evidence of communication is the incorporation of rice straw, a practice whose visibility may make it more amenable for farmer-to-farmer transmission.

Information	Number (and %) of neighbors who received information from FFS farmers (n = 70)	Number (and %) of neighbors who acted on the information (n = 70)
Use of insecticides only as last resort	8 (11%)	6 (9%)
Importance of beneficial insects	4 (6%)	4 (6%)
Use of straw as soil amendment	21 (30%)	19 (27%)
Use of single nutrient fertilizers	10 (14%)	5 (7%)

Source: Tripp *et al.* (2005).

a particular craft or skill which they combined with farming" (Overton, 1996, p. 36) is an increasingly common understanding of contemporary rural economies in developing countries, although this description is drawn from 16th century England.

The fact that farming is not a full-time occupation does not necessarily make it any less important or automatically reduce incentives for its improvement, but it must be acknowledged that in many situations households' interests in access to more efficient production technology may take second place to their concerns as consumers seeking affordable food prices.

A relatively high proportion of time devoted to nonagricultural activities and a declining role of farming as a source of cash income are factors that can predispose households to avoid participation in LEIT projects and can thwart efforts to use this technology for improving the livelihoods of resource-poor farmers. The interactions between off-farm income and agricultural activity can be complex; additional income sources can capture time and attention that might otherwise be devoted to farming; or income earned off the farm can be invested in improving farming capacity (Reardon *et al.*, 2000). The strategies elected depend to a great extent on the resources available to the household and the opportunities for remunerative agriculture. Probably the major lesson is that characterizing resource-poor farmers as a homogeneous class, susceptible to a uniform set of development initiatives, is bound to fail.

In addition, some difficult choices must be faced, and strategies articulated, regarding the future role of agriculture in rural development. The choices involve both an assessment of the agricultural potential of diverse rural areas and the capacities of different types of households. Opinions differ regarding the potential of so-called marginal areas for agricultural growth (Renkow, 2000). Policy makers must present clear strategies and, in some cases, be willing to defend decisions to divert agricultural investments from such areas (and to provide cogent alternative rural development strategies).

Within both more- and less-favored areas, a better differentiation of households is also required. De Janvry and Sadoulet (2000) have proposed that we should recognize four types of paths leading away from rural poverty: households may "exit" agriculture through migration or the development of rural employment opportunities; some may follow an "agricultural path" that connects them with agricultural markets; others can follow a "pluriactive path" that combines off-farm income with subsistence farming; and finally some households must be provided an "assistance path" through income or food transfers that allows immediate survival and eventual opportunities to follow other paths.

A more recent treatment by Dorward *et al.* (2006) denominates the first three of these options as "stepping out," "stepping up," and "hanging in," respectively. None of these schemes is a perfectly distinct category, but they are helpful in differentiating strategies with respect to the promotion of LEIT. Households that are "stepping out" of farming will not be major targets for agricultural technology generation and should instead be eligible for assistance that improves their skills and capacities as urban migrants or participants in a diversifying rural economy. We have seen that those who are "stepping up" their commitment to agriculture as a major source of income will have increasing participation in agricultural markets and are, in many respects, the most logical targets for technologies such as LEIT. Those who are "hanging in" are perhaps the most problematic. Their relative lack of attention to farming may make them more likely to misuse inputs such as pesticides and less likely to invest in resource conservation, hence making them particularly important candidates for LEIT. However, their diverse income strategies lower their incentives for participation in technology generation activities and often restrict their capacities to invest in new technology; they may require different approaches if LEIT is to make a contribution to their farming.

Strategies for Demand-Driven Research

The different motivations and pathways of rural households add complexity to the already difficult challenge of promoting LEIT. The way forward is not clear, but there is growing evidence that the strategy of many independent, short-lived projects is not productive. The problem extends beyond LEIT and calls into question the common strategy of donors in funding small pilot projects focused on specific technological innovations with the attendant demonstrations, group formation, participatory exercises, and assumptions that somehow there will be a spontaneous diffusion of results.

Overall, it is clear that many LEIT projects are limited in breadth and focus. They typically concentrate on a few specific technologies and cover a relatively limited number of communities. There is nothing wrong with working with a restricted technology set, as long as it responds to farmers' priorities rather than project mandates and as long as there is a capacity for evolution. There is also nothing intrinsically wrong with starting on a small scale, as long as there is some conception of the next steps that are necessary and a willingness to collaborate with similar efforts. The latter is a particular problem with LEIT and it is not uncommon to find several separate projects working on similar issues (e.g., soil fertility) in the same region with little or no communication, coordination, or joint learning. In addition, technology themes proliferate and farmers may come into contact (simultaneously or serially) with efforts in participatory plant breeding, group formation for fodder crop nurseries, and a farmer field school for IPM.

When there is evidence of at least modest success at the pilot stage, the next step is often to call for scaling up. This is a particularly imprecise term that usually does not define exactly what is expected (policy change, project replication, investment in an extension effort, developing new organizations) and is instead a symptom of poor planning in agricultural development assistance. It is crucial that governments and development agencies decrease their reliance on these piecemeal strategies. There is a need to think about institutions that allow more effective farmer demand and mechanisms that are more efficient in facilitating access to information and advice.

Although LEIT projects pride themselves on promoting "demand-driven" technology development, most are short-term expressions of particular donor priorities. More truly demand-driven activity will only come in response to effective political pressure from well-organized farmers. We have seen that most farmer groups promoted by LEIT projects are too narrowly focused to have any chance of a sustainable existence or widespread support. Developing strong, broad-based farmer organizations that can exert pressure for more effective public research and extension will have higher payoffs than small-group activity in response to a brief donor-driven project. There is a growing consensus that farmer organizations are a vital element for rural development, although heightened expectations and donors' desire to provide external assistance can jeopardize their progress (Chirwa *et al.*, 2005; see Chapter 11 on innovations in extension).

Farmer organizations will only be sustainable if they address major issues of concern to their members. It is important to recognize that although access to technology may be one of these, it is unlikely that technology generation, on its own, will be the basis of a significant growth in viable organizations. Most successful farmer organizations address the economic or political priorities of their members. For households with less participation in agricultural markets, other types of rural organization may be called for, but institutions that promote farmer voice and interaction can provide incentives that direct participants' attention to agricultural innovation. Organizations need to offer as many advantages to farmers as possible in order to elicit commitment and offer opportunities for varying levels of participation.

It is easy to overestimate the importance of technology generation as a basis for the development of farmer organizations. Such organizations are only likely to invest in technological innovation when there are good returns to farming. The rapid growth of local farmers' associations in mid-19th century England and of the Farm Bureaus in early 20th century United States took place when the respective agricultural economies were thriving (Campbell, 1962; Fox, 1979). In addition, it must be recognized that political organization in support of agriculture does not necessarily come from grassroots; many of India's most prominent farmers' organizations represent an elite of larger farmers interested in access to subsidized technology (Brass, 1995). Grassroots farmers' organizations are also likely to focus on economic issues rather than technology; the Granger movement in late 19th century United States provided many opportunities for the exchange of technical information, but its principal impetus was to defend the farming community from the growing influence of the railroads and grain merchants (Buck, 1913). Thus, even in circumstances of agricultural growth and opportunity, keeping a farmers' organization interested in technology generation is not easy.

Despite these limitations, strong farmers' organizations and other robust examples of rural civil society are necessary to generate sufficient demand for technology development. On the other side of the equation, public agricultural research and extension must be guided by clear policies on rural development and have the skills and resources to respond to farmers' requirements. Rural organizations will provide many opportunities for social learning and the development and transmission of information about new technology, but this is not sufficient. Other sources of information and debate, such as newspapers and FM radio, offer opportunities for discussion of the breadth of issues affecting rural residents. Media at their best can enlist the interest and attention of the diversity of households that should be able to take advantage of agricultural innovations. Finally, efficient and transparent input and output markets must be able to provide the information that farmers require.

CONCLUSIONS

This review of the experience of promoting low external input technology, which is sometimes seen as a radically alternative path for agricultural development, has found patterns of success and failure that are remarkably similar to those of other technology-based development efforts. This analysis is not designed to challenge the philosophical or political roots of some of the strategies that favor LEIT, but rather to point out that regardless of orientation, many of our efforts at improving the welfare of resource-poor farmers face similar challenges and are presumably amenable to similar reorientations of focus and purpose to make them more effective.

The adoption of most agricultural technology depends to a considerable extent on labor resources (although we have seen that LEIT itself is not necessarily labor intensive). Farmers deploy additional labor when they see that the returns are sufficient, although increasingly, even on very small holdings, much of that labor is hired, indicating that labor is now often an "external" resource purchased on the market. The use of hired labor is an indication of the diverse livelihoods of many rural households; some of these can balance on- and off-farm labor, whereas others are in a downward spiral that may end in an exit from farming. This diversity of income sources also helps explain the diversity of interest in new technology in supposedly homogeneous farming communities. The adoption of new technology also requires knowledge, and farmers' incentives for developing that knowledge are varied. Methods such as social learning are often quite effective for introducing new technology, but farmers require time, and hence adequate incentives, to participate. Project-led group formation may help develop new knowledge, but there is rarely motivation for maintaining narrowly focused groups. This must be overcome if groups are to engage in diffusion of new knowledge to a larger audience of farmers.

The reorientation of development strategies suggested by this review includes (1) a careful examination of the heterogeneity of the countryside and the attendant diversity of incentives for acquiring new technology and (2) reconsideration of development assistance strategies based on pilot projects and vaguely defined scaling-up strategies.

There is not one type of "resource-poor farmer" but many. This requires a differentiated strategy. Some farmers are already participating in agricultural markets or have the potential to do so and often have strong motivation to acquire new farming techniques. Others are balancing several, often insecure, sources of income with subsistence farming, and although their environmental footprint and household food insecurity argue for the provision of low-input technology, these households' attention is more difficult to capture. Many other rural households have so few agricultural resources that assistance should be directed toward the development of alternative sources of livelihood. A substitution of realism for romanticism in agricultural development must be accompanied by more responsible and coordinated programs, integrated with policy. We need urgent reexamination of the idea that an endless number of small donor-funded or government projects, even if based on imaginative techniques that involve farmers in technology generation, is going to promote meaningful technological change. More investment is required in modalities such as broad-based rural organizations and other means of rural communication. These will allow farmers to exert more political pressure on technology providers from both public and private sectors.

REFERENCES

Ali, M. (1999). Evaluation of green manure technology in tropical lowland rice systems. *Field Crops Res.* **61,** 61–78.

Anderson, S., Gündel, S., and Pound, B. with Triomphe, B. (2001). "Cover Crops in Smallholder Agriculture: Lessons from Latin America." ITDG Publishing, London, p. 112.

Bentley, J. W. (1989). What farmers don't know can't help them: The strengths and weaknesses of indigenous technical knowledge in Honduras. *Agric. Hum. Values* **6,** 25–31.

Biggs, S. (1989). "Resource-Poor Farmer Participation in Research: A Synthesis of Experiences from Nine National Agricultural Research Systems." OFCOR Comparative Study Paper No. 3, ISNAR, The Hague, p. 37.

Brass, T. (ed.) (1995). "New Farmers' Movements in India." Frank Cass, Illford, UK.

Braun, A. R., Thiele, G., and Fernández, M. (2000). "Farmer Field Schools and Local Agricultural Research Committees: Complementary Platforms for Integrated Decision-Making in Sustainable Agriculture." AgREN Paper No. 105, ODI, London.

Brown, D., and Korte, C. (1997). "Institutional Development of Local Organisations in the Context of Farmer-Led Extension: The Agroforestry Programme of the Mag'uugmad Foundation." AgREN Paper No.68, ODI, London.

Bruin, G., and Meerman, F. (2001). "New Ways of Developing Agricultural Technologies. The Zanzibar Experience with Participatory Integrated Pest Management." Wageningen University and Research Centre, Wageningen.

Buck, S. J. (1913). "The Granger Movement." Harvard University Press, Cambridge, MA.

Buckles, D. (1995). Velvetbean: A "new" plant with a history. *Econ Bot* **49**(1), 13–25.

Bunch, R. (1982). "Two Ears of Corn. A Guide to People-Centered Agricultural Improvement." World Neighbors, Oklahoma City, OK.

Bunch, R. (1999). Learning How to "Make the Soil Grow": Three Case Studies on Soil Recuperation Adoption and Adaptation from Honduras and Guatemala. *In* "Issues and Options in the Design of Soil and Water Conservation Projects" (M. McDonald, and K. Brown, eds.), Proceedings of a workshop, Llandudno, Conwy, UK.

Campbell, C. (1962). "The Farm Bureau and the New Deal." University of Illinois Press, Urbana, IL.

Carter, J. (1995). Alley Farming: Have Resource-Poor Farmers Benefited? Natural Resource Perspectives No. 3. ODI, London.

Chambers, R. Pacey, A., and Thrupp, L.A. (eds.) (1989). "Farmer First: Farmer Innovation and Agricultural Research." Intermediate Technology Publications, London.

Chirwa, E., Dorward, A., Kachule, R., Kumwenda, I., Kydd, J., Poole, N., Poulton, C., and Stockbridge, M. (2005). Walking tightropes. Supporting farmer organizations for farmer access. Natural Resource Perspectives No. 99, ODI, London.

CIP-UPWARD (2003). "Farmer Field Schools: From IPM to Platforms for Learning and Empowerment." Users' Perspectives with Agricultural Research and Development, Los Baños, Philippines.

Conklin, H.C. (1957). "Hanunoo Agriculture: A Report on an Integral System of Shifting Cultivation in the Philippines." FAO, Rome.

Cooper, P., Leakey, R., Rao, M., and Reynolds, L. (1996). Agroforestry and the mitigation of land degradation in the humid and sub-humid tropics of Africa. *Exp. Agric.* **32,** 235–290.

Cramb, R. (2005). Social capital and soil conservation: Evidence from the Philippines. *Aust. J. Agric. Res. Econ.* **49,** 211–226.

Cramb, R. A., Garcia, J. N. M., Gerrits, R. V., and Saguiguit, G. C. (2000). Conservation farming projects in the Philippine uplands: Rhetoric and reality. *World Dev.* **28,** 911–927.

Crowley, E., and Carter, S. (2000). Agrarian change and the changing relationships between toil and soil in Maragoli, western Kenya (1900–1994). *Hum. Ecol.* **28,** 383–414.

David, C., and Otsuka, K. (1994). "Modern Rice Technology and Income Distribution in Asia." Lynne Rienner, Boulder, CO.

Defoer, T., and Budelman, A. (eds.) (2000). "Managing Soil Fertility in the Tropics: A Resource Guide for Participatory Learning and Action Research." Royal Tropical Institute, Amsterdam.

De Janvry, A., and Sadoulet, E. (2000). Rural poverty in Latin America: Determinants and exit paths. *Food Policy* **25,** 389–409.

Dorward, A., Wheeler, R., MacAuslan, I., Buckley, C., Kydd, J., and Chirwa, E. (2006). Promoting agriculture for social protection or social protection for agriculture: Strategic policy and research issues. Discussion paper, Future Agricultures. http://www.future-agricultures.org/pdf%20files/SP_Growth_Final.pdf

Ellis, F. (1998). Household strategies and rural livelihood diversification. *J. Dev. Stud.* **35,** 1–38.

Fox, H. (1979). Local farmers' associations and the circulation of agricultural information in nineteenth-century England. *In* "Change in the Countryside: Essays on Rural England 1500–1900" (H. Fox, and R. Butlin, eds.). Institute of British Geographers Special Publication No. 10, London.

Frankel, F. (1971). "India's Green Revolution." Princeton University Press, Princeton, NJ.

Freebairn, D. (1995). Did the Green Revolution concentrate incomes? A quantitative study of research reports. *World Dev.* **23,** 265–279.

Haggblade, S., and Tembo, G. (2003). "Conservation Farming in Zambia." EPTD Discussion Paper, No. 108, IFPRI, Washington, DC.

Hassane, A., Martin, P., and Reij, C. (2000). "Water Harvesting, Land Rehabilitation and Household Food Security in Niger." Rome. IFAD.

Hazell, P., and Ramasamy, C. (1991). "The Green Revolution Reconsidered: The Impact of High Yielding Rice Varieties in South India." Johns Hopkins University Press, Baltimore, MD.

Heong, K., and Escalada, M. (1999). Quantifying rice farmers' pest management decisions: Beliefs and subjective norms in stem borer control. *Crop Prot.* **18**, 315–322.

Holt-Giménez, E. (2001). Scaling up sustainable agriculture: Lessons from the Campesino a Campesino movement. *LEISA Mag.* **17**(3), 27–29.

Jayne, T., Mukumbu, M., Chisvo, M., Tschirley, D., Weber, M., Zulu, B., Johhannson, R., Santos, P., and Soroko, D. (1999). "Successes and Challenges of Food Market Reform: Experiences from Kenya, Mozambique, Zambia and Zimbabwe." MSU International Development Working Paper No. 72, Michigan State University, East Landsing, MI.

Kenmore, P. (1996). Integrated pest management in rice. *In* "Biotechnology and Integrated Pest Management" (G. Persley, ed.). CAB International, Wallingford, UK.

Landers, J. N. (2001). "Zero Tillage Development in Tropical Brazil." FAO Agricultural Services Bulletin, 147, FAO, Rome.

Lanjouw, P., and Stern, N. (1993). Agricultural change and inequality in Palanpur 1957–1984. *In* "The Economics of Rural Organization: Theory, Practice and Policy" (K. Hoff, A. Braverman, and J. Stiglitz, eds.). Oxford University Press, New York.

LEISA (2003). Learning with farmer field schools. *LEISA Mag.* **19**(1), 1-36.

Leslie, A., and Cuperus, G. (eds.)(1993). "Successful Implementation of Integrated Pest Management for Agricultural Crops." Lewis Publishers, Boca Raton, FL.

Lipton, M., and with Longhurst, R. (1989). "New Seeds and Poor People." Unwin Hyman, London.

Longley, C., Mango, N., Nindo, W., and Mango, C. (2006). Conservation by committee: The catchment approach to soil and water conservation in Nyanza Province, western Kenya. *In* "Self-Sufficient Agriculture: Labour and Knowledge in Small-Scale Farming" (R. Tripp, ed.). Earthscan, London.

Moser, C., and Barrett, C. (2003). The disappointing adoption dynamics of a yield-increasing, low-external-input technology: The case of SRI in Madagascar. *Agric. Systems* **76**, 1085–1100.

Murwira, K., Wedgewood, H., Watson, C., Win, E. and Tawney, C. (2000). "Beating Hunger. The Chivi Experience." Intermediate Technology Publications, London.

National Research Council (1993). "Vetiver Grass: A Thin Green Line against Erosion." National Academy Press, Washington, DC.

Neill, S., and Lee, D. (2001). Explaining the adoption and disadoption of sustainable agriculture: The case of cover crops in Northern Honduras. *Econ. Dev. Cult. Change* **49,** 793–820.

Okoth, J., Khisa, G., and Julianus, T. (2003). Towards self-financed farmer field schools. *LEISA Mag.* **19**(1), 28–29.

Overton, M. (1996). "Agricultural Revolution in England." Cambridge University Press, Cambridge.

Place, F., Adato, M., Hebinck, P., and Omosa, M. (2003). "The Impact of Agroforestry-Based Soil Fertility Replenishment Practices on the Poor in Western Kenya." FCND Discussion Paper, No.160, IFPRI, Washington, DC.

Place, F., Franzel, S., DeWolf, J., Rommelse, R., Kwesiga, F., Niang, A., and Jama, B. (2002). Agroforesty for soil fertility replenishment: Evidence on adoption processes in Kenya and Zambia. *In* "Natural Resources Management in African Agriculture" (C. B. Barrett, F. Place, and A. A. Aboud, eds.). CABI, Wallingford, UK.

Pontius, J., Dilts, R., and Bartlett, A. (eds.) (2002). "Ten Years of IPM Training in Asia: From Farmer Field School to Community IPM." FAO, Bangkok.

Pretty, J. (1995). "Regenerating Agriculture." Earthscan, London.

Pretty, J., and Hine, R. (2001). "Feeding the World with Sustainable Agriculture: A Summary of New Evidence." SAFE-World Final Report, University of Essex, Colchester, UK.

Quizon, J., Feder, G., and Murgai, R. (2001). Fiscal sustainability of agricultural extension: The case of the farmer field school approach. *J. Int. Agric. Extens. Educ.* Spring 2001, 13–23.

Reardon, T., Taylor, J., Stamoulis, K., Lanjouw, P., and Balisacan, A. (2000). Effects of non-farm employment on rural income inequality in developing countries: An investment perspective. *J. Agric. Econ.* **51,** 266–288.

Reijntjes, C., Haverkort, B., and Waters-Bayer, A. (1992). "Farming for the Future." ILEIA, Leusden.

Renkow, M. (2000). Poverty, productivity and production environment: A review of the evidence. *Food Policy* **25,** 463–478.

Richards, M., and Suazo, L. (2006). Learning from success: Revisiting experiences of LEIT adoption by hillside farmers in central Honduras. *In* "Self-Sufficient Agriculture: Labour and Knowledge in Small-Scale Farming" (R. Tripp, ed.). Earthscan, London.

Roberts, K., and Coutts, J. (1997). "A Broader Approach to Common Resource Management: Landcare and Integrated Catchment Management in Queensland, Australia." AgREN Paper, No. 70, ODI, London.

Rogers, E. M. (1995). "Diffusion of Innovations," 4th Ed. Free Press, New York.

Röling, N., and van de Fliert, E. (1994). Transforming extension for sustainable agriculture: The case of integrated pest management in rice in Indonesia. *Agric. Hum. Values* **11**(2/3), 96–108.

Stoll, S. (2002). "Larding the Lean Earth." Hill and Wang, New York.

Stoop, W. (2003). The system of rice intensification (SRI) from Madagascar. Myth or missed opportunity? Unpublished report.

Stoop, W., Uphoff, N., and Kassam, A. (2002). A review of agricultural research issues raised by the system of rice intensification (SRI) from Madagascar: Opportunities for improving farming systems for resource-poor farmers. *Agric. Systems* **71,** 249–274.

Sumberg, J., and Okali, C. (1997). "Farmers' Experiments: Creating Local Knowledge." Lynne Rienner, Boulder, CO.

Tiffin, M., Mortimore, M., and Gichuki, F. (1994). "More People, Less Erosion." Wiley, Chichester.

Tripp, R. (2001). Can biotechnology reach the poor? The adequacy of information and seed delivery. *Food Policy* **26,** 249–264.

Tripp, R. (2006). "Self-Sufficient Agriculture: Labour and Knowledge in Small-Scale Farming." Earthscan, London.

Tripp, R., Wijeratne, M., and Piyadasa, V. H. (2005). What should we expect from farmer field schools? A Sri Lanka case study. *World Dev.* **33,** 1705–1720.

Van der Ploeg, J. (1990). "Labor, Markets, and Agricultural Production." Westview Press, Boulder, CO.

Whiteside, M. (2000). "*Ganyu* Labour in Malawi and Its Implications for Livelihood Security Interventions: An Analysis of Recent Literature and Implications for Poverty Alleviation." AgREN Paper 99, Overseas Development Institute, London.

Wilken, G. C. (1987). "Good Farmers: Traditional Agricultural Resource Management in Mexico and Central America." University of California Press, Berkeley, CA.

Winarto, Y. T. (2002). From farmers to farmers, the seeds of empowerment. *In* "Beyond Jakarta: Regional Autonomy and Local Societies in Indonesia" (M. Sakai, ed.). Crawford House, Adelaide.

Winarto, Y. T. (2004). "Seeds of Knowledge: The Beginning of Integrated Pest Management in Java." Yale Southeast Asia Studies, New Haven, CT.

Zaal, F., and Oostendorp, R. H. (2002). Explaining a miracle: Intensification and the transition towards sustainable small-scale agriculture in dryland Machakos and Kitui Districts, Kenya. *World Dev.* **30,** 1271–1287.

CHAPTER 6

Ecologically Based Nutrient Management

L. E. Drinkwater, M. Schipanski, S. S. Snapp, and L. E. Jackson

Agricultural Systems: Agroecology and Rural Innovation for Development

Summary

The management of nutrients is fundamental to agricultural productivity and viable rural livelihoods. Farmers have the dual goals of supporting crop and animal growth, while minimizing losses to the environment. Nutrient availability is a function of management practices and inherent characteristics of the environment (i.e., climate and soil type) and the organisms in that environment. In this chapter we explore how the application of ecological principles will provide managers with tools for developing coherent nutrient management strategies that optimize the complex processes governing nutrient cycling in agroecosystems. Sustainable management requires close attention to internal nutrient cycling, through building labile/available pools *and* nutrient pools that are more easily retained in the soil. Strategic choices are required in orchestrating the full suite of management decisions that contribute to effective nutrient management, including decisions such as which fertility sources to use and how much to apply in a given field, how to use rotation and intercropping in concert with these sources, which crop genotypes to grow, and how to best use interventions such as tillage or fire. Participatory techniques for assessing current nutrient management practices and developing innovative systems are invaluable in developing ecologically based nutrient management schemes.

NUTRIENT MANAGEMENT AS APPLIED ECOLOGY

Ecologically based nutrient management is an integrated approach that applies systems thinking to optimizing soil fertility for crop production and sustainability. Nutrient management falls within the purview of ecosystem ecology that aims to understand ecological processes, such as productivity and nutrient cycling, which are governed by intact ecosystems. An *ecosystem* is a dynamic complex of organisms and the physical environment with which they interact. Application of an ecosystem framework provides agriculturalists with a flexible systems approach that can be used to organize the complex, dynamic interactions among the organisms and their environments that govern agroecosystem functions. *Ecosystem functions* are those processes, such as nutrient cycling, water and energy flows, soil retention and production, or crop yield, that result from the interactions among ecosystem components; i.e., plants, decomposers, climate, soil environment, etc.

In applying principles and concepts from ecology, the scope of nutrient management is expanded to include a wide range of soil nutrient reservoirs and biogeochemical processes. For example, rather than focusing solely on soluble, inorganic plant-available pools, an ecosystem-based approach seeks to optimize organic and mineral soil reservoirs that are more efficiently retained in the soil, such as organic matter and sparingly soluble forms of phosphorus. This framework also expands the processes targeted by management beyond the specific process of crop nutrient uptake. Efforts are directed toward managing nutrient cycling processes occurring at a variety of spatial and temporal scales. Integrated management of the full array of ecosystem processes that regulate the cycling of nutrients and carbon in soil will improve productivity while also increasing nutrient use efficiency over the long term.

In this chapter we will focus on the integrated management of nitrogen, phosphorus, and carbon. Nitrogen and phosphorus are the two nutrients that most commonly limit crop production in agroecosystems. Biological processes play major roles in regulating the cycling of N and P, and, for this reason, the fate of these nutrients is strongly linked to the flow of carbon. We will discuss strategies for managing N, P, and C cycling processes, giving particular attention to the role of C in influencing the fate of N and P. Although we emphasize these major nutrients, many of the concepts we discuss are widely applicable to the other macro- and micronutrients important for plant growth. Because a basic understanding of these cycles is fundamental to ecological nutrient management, we will first briefly review the key features of these elemental cycles.

THE BASICS OF NUTRIENT CYCLING

To understand nutrient cycling, we must consider distribution, fluxes, and the regulatory mechanisms that make up an elemental cycle. Nutrients move from one *compartment* or *pool* to another. *Reservoir* is another term commonly used to refer to stores of nutrients in the soil. A compartment is usually defined by physical boundaries, while distinct pools can exist within a single compartment. For example, the soil compartment has several distinct pools of N. Plant uptake of NO_3^- results in the movement of N from the soil compartment (or more specifically, from the inorganic soil pool) to plant biomass. The distribution of nutrients among compartments and pools in agroecosystems varies in terms of the absolute amounts, depending on soil, climate, and biotic and management factors.

We refer to the rate of transfer from one pool to another as a *flux*. The flux of nutrients is often framed in terms of *source/sink* transfers when we want to emphasize the role of a particular process in regulating nutrient flows. A *source* simply refers to the pool where the nutrient came from whereas a *sink* is the pool actively taking up the nutrient. All fluxes are regulated by a process, which can be either *biotic* (controlled by living organisms; e.g., mineralization) or *abiotic* (controlled by chemical and/or physical mechanisms; e.g., precipitation). The flux from one pool to another often entails a chemical modification of the nutrient. The most common chemical modifications are organic/inorganic and oxidation/reduction reactions.

We use models to depict relationships between location, form, and transfer of nutrient cycles. These models can be adapted to represent nutrient cycling at any scale with varying degrees of detail. An ecosystem can be divided up into very few compartments; for example, the simplest nutrient cycle might only distinguish between plant and soil compartments. As more compartments and pools are added, the cycling model becomes more complex. To address nutrient flows at the landscape level, individual fields or farms and adjacent waterways would be the designated compartments. A very simple depiction of N flows is shown in Fig. 6.1A.

Only three compartments are shown with two biologically mediated processes that control the flux of N from soil organic matter (SOM) into the inorganic pool (Flux A, mainly controlled by microorganisms) and then from the inorganic N pool to plant biomass (Flux B, regulated by the plant). If mineralization and plant assimilation are equal (N moving in and N moving out of the inorganic N pool are the same), and if these two processes are the dominant fluxes regulating this pool, then the size of the inorganic N pool will remain a constant even thought NO_3^- is actually moving in and out. This situation is called a *steady state*. You can see that if we collected monthly soil samples and extracted inorganic N under steady state conditions, the NO_3^- pool will appear static, since the concentration remains constant through time. We would miss the dynamics that are actually taking place; i.e., N is moving in and out of this pool. This is one of the difficulties when using static measurements of pool size as indicators of nutrient availability. The limitation of static measurements is particularly prominent when standing pools of inorganic N are very small. Small standing pools of NO_3^- are usually interpreted as indicators of low N fertility.

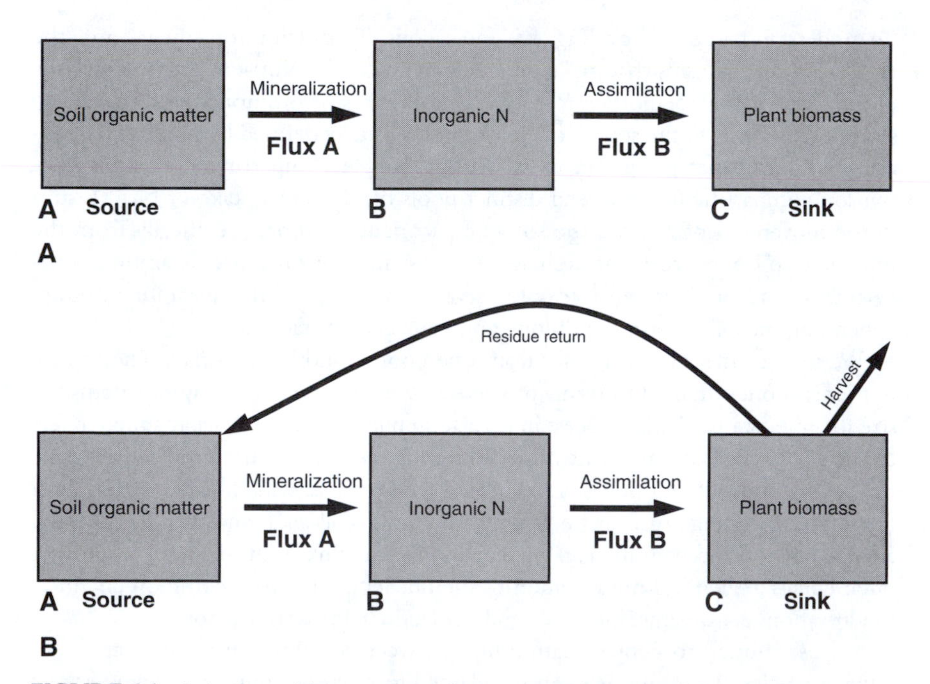

FIGURE 6.1 A simple model demonstrating the use of compartments and fluxes to depict nutrient flows. (**A**) Three compartments (A, B, and C) are shown with two biologically mediated fluxes. (**B**) The cycle is closed when plant residues are returned to the soil. Net export of N occurs through harvest.

However, if plant assimilation is keeping up with mineralization, you can have a very large flux in a very small inorganic N pool. This is often the case in fields where organic residues have been used as nutrient sources for many years.

We can close the cycle in this simple model by adding two more fluxes (Fig. 6.1B). These two new processes are also biotic in that they are the result of human management. In this model, harvest removes the N from our agroecosystem. We don't consider the fate of the harvested N that could be going to animals and/or humans, and the N remaining in crop residues is left in the field to become part of the SOM. In this case harvest is considered to be an *export*, since the harvested N leaves our system, while the other three fluxes are part of the *internal* N cycle.

When developing nutrient management strategies it is important to remember that the rates of different processes can vary by orders of magnitude. This impacts the distribution of nutrients among pools and results in compartments with widely varying turnover times. Turnover time is defined as the time it will take to empty a reservoir if the source is cut off and if sinks remain constant. In other words, fluxes out of the compartment continue, but the influx is shut off. Another useful way to compare the dynamics of different compartments is mean residence time (MRT) or the average amount of time a nutrient spends in the compartment before being transferred out. Mean residence time is calculated as the pool size/flux when assuming the pool is close to steady state conditions; i.e., in ≈ out. For example, to estimate the mean residence time of nitrous oxide in the atmosphere on a global basis, we calculate the total size of the pool and then divide by the estimated global rate of production:

$$[N_2O] = 300 \text{ ppb, Total } N_2O = 2.3 \times 10^{15} \text{ g}$$
$$\text{Rate of production: } 20 \times 10^{12} \text{ g yr}^{-1}$$
$$\text{MRT} = 2.3 \times 10^{15} \text{ g}/20 \times 10^{12} \text{ g yr}^{-1} = 110 \text{ years}$$

Both turnover time and MRT require detailed knowledge of fluxes and pool size that can be difficult to accurately measure in soils without the use of expensive tracer experiments. However, the MRTs for important soil nutrient pools vary widely. For example, the MRT for NO_3^- is about a day, while humic components of soil organic matter have MRTs of hundreds to thousands of years (Tan, 2003). Given the huge difference in the temporal dynamics of these pools, it is not necessary to take measurements in your agroecosystems to apply these useful concepts. Estimates from the literature can be very helpful as a starting point.

Nitrogen Cycling

In unmanaged terrestrial ecosystems, the soil N cycle is driven by SOM, which contains approximately 50% C and 5% N, of which typically <5% is in labile forms. In agroecosystems, nitrogen added as inorganic fertilizer or in organic residues from biological N fixation and various soil amendments such as composts or animal manures also plays a major role in driving the N cycle. The breakdown, or depolymerization of the large, complex molecules that make up organic residues, is facilitated by extracellular enzymes secreted mainly by soil fungi and prokaryotes. These exoenzymes catalyze the release monomers, such as amino acids and sugars, which are small enough to be transported into cells. These labile compounds are recycled and reused through microbial metabolism, faunal grazing of microbes, as well as microbial death and damage that are caused by stress, such as wet-dry or freeze-thaw cycles (Schimel and Bennett, 2004; Fig. 6.2). Plants also contribute to internal cycling via root exudation of a diverse array of organic compounds (Bais *et al.*, 2006).

Soil microorganisms play a dominant role in regulating soil N cycling. Mineralization occurs when heterotrophic microbes break down nutrient-rich organic monomers and release energy and NH_4^+. Ammonia can be assimilated by heterotrophs or used as an energy source by ammonia-oxidizing microbes to produce nitrite (NO_2^-) that is quickly converted to NO_3^- in the process of nitrification. During nitrification, some nitric oxide (NO) and nitrous oxide (N_2O) are also produced and lost from the soil (Godde and Conrad, 2000; Fig. 6.2). Alternatively, NH_4^+ can be lost from the soil through the emission of ammonia (NH_3) gas if soil pH is greater than 8. Nitrate can be lost from the system through several processes. Denitrification takes place when heterotrophic bacteria under oxygen limitation use NO_3^- as an alternative electron acceptor to produce N_2O and N_2. The leaching of NO_3^-,

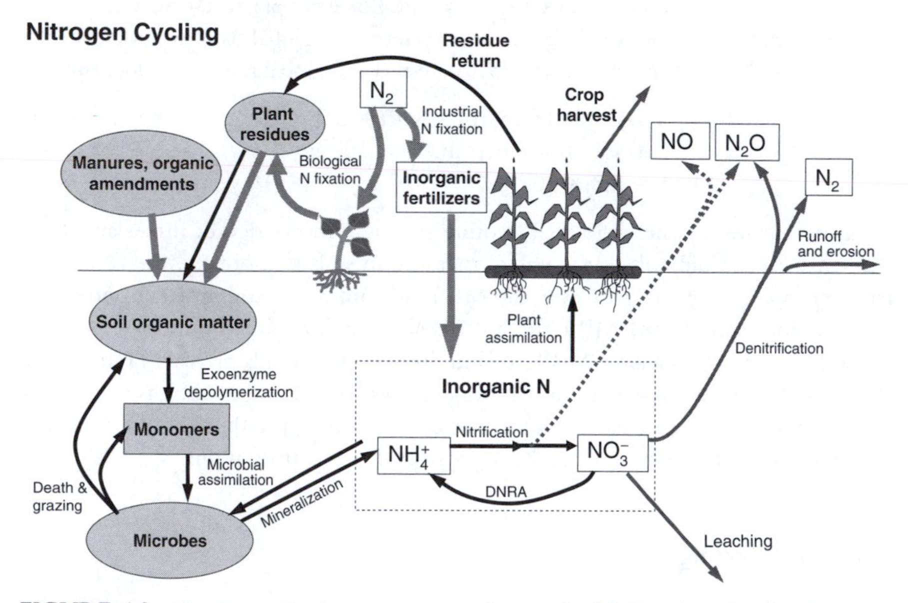

FIGURE 6.2 Nitrogen cycling in agroecosystems. See text for full discussion of cycling processes. New N is added through biological N fixation, synthetic fertilizers, or organic amendments such as manure or compost (gray arrows). The main pathways of removal are through harvested exports, leaching, and denitrification. Some gaseous losses also result from nitrification; however, this is thought to be a less significant flux (dotted arrows). Internal cycling processes occur through human management of residues, plant assimilation, and microbially mediated transformations.

which contaminates groundwater, occurs when rainfall exceeds evapotranspiration, especially in coarse-textured soils. Runoff also carries N in various forms to surface waters. A second anaerobic pathway that helps to retain NO_3^- in the soil and involves the conversion of NO_3^- to NH_4^+ includes the dissimilatory nitrate reduction to ammonium (DNRA). This pathway can compete with denitrification and can be the dominant dissimilatory reduction pathway in some tropical soils (Silver *et al.*, 2001).

Phosphorus Cycling

While N transformations are primarily controlled by microbially mediated processes, the soil P cycle is regulated by both biological and geochemical processes that compete with one another for the small amounts of soluble, inorganic P that are typically present in the soil solution (Cross and Schlesinger, 1995). A second major distinction shaping the P cycle is that it cannot be converted into gaseous forms that can be lost from the system. For convenience, the P cycle is portrayed as consisting of two subcycles reflecting the abiotic and biotic mechanisms (Fig. 6.3).

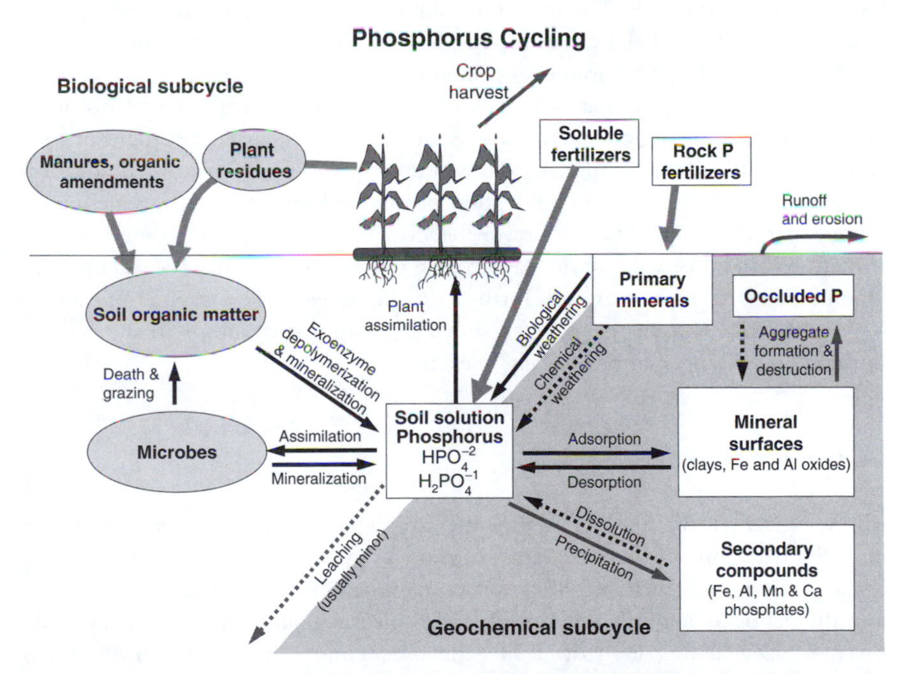

FIGURE 6.3 The phosphorus cycle consists of biological and geochemical subcycles. See text for full discussion of cycling processes. New P is added through soluble synthetic fertilizers, sparingly soluble amendments such as rock P, or organic amendments such as manure or compost (gray arrows). The main pathways of removal are through harvested exports, erosion, occlusion, and precipitation with small losses occurring through leaching in some systems. Internal cycling processes occur through human management of residues, plant assimilation, microbially mediated transformations, and geochemical processes.

The geochemical and biological subcycles are composed of processes that are distinct from one another, with the exception of the weathering of primary minerals, which is mediated by both biological and geochemical mechanisms (Schlesinger, 2005). Biological weathering occurs at rates many times faster compared to abiotic weathering processes. The biological transformations involving P are fewer compared to N and begin with reactions mediated by exoenzymes that release P. Phosphorus mineralization is the microbial conversion of organic P to orthophosphates ($H_2PO_4^-$ or $H_2PO_4^{-2}$, depending on soil pH) which can in turn be assimilated by either plants or microorganisms. The microbial P will become available over time as the microbes die or are grazed. As with N, soil organic matter and newly added organic residues can serve as an important source of P.

The geochemically mediated sinks for orthophosphates compete with biological assimilation and include two types of inorganic reactions; these are precipitation/dissolution and sorption/desorption processes. Precipitation/dissolution reactions involve the formation and dissolving of precipitates. Precipitation reactions occur with dissolved iron, aluminum, manganese (acid soils), or calcium (alkaline soils) to form phosphate minerals. The rate of dissolution is negligible for these precipitates with the exception of the calcium phosphates. Calcium phosphates such as apatite account for 95% of P found in primary minerals of the Earth's crust and are commonly referred to as *sparingly soluble P*, since these minerals can be dissolved by chemical and biological weathering. Apatite is the primary constituent of rock P, which can be added to infertile soils as a slow source of P. Sorption/desorption reactions involve sorption and desorption of ions and molecules at the surfaces of mineral particles. Adsorption is a reversible chemical binding of P to soil particles. In some soils, adsorbed phosphate may become trapped on the surface of soil minerals when a Fe or Al oxide coating is formed on the mineral. The trapped phosphate is then described as being occluded. For all practical purposes, P that becomes occluded is no longer agronomically relevant.

Carbon Cycling

All biologically mediated cycling processes are dependent on C, either for energy or as the backbone of biomolecules that must be synthesized for life to exist. Soil organic matter is defined as all carbon-containing soil constituents and is therefore the major biologically relevant soil reservoir for N and P in most arable soils. Because SOM is the result of all life, the biochemistry of SOM constituents is complex, reflecting the diverse array of compounds produced by plants, microbes, and larger soil organisms. The chemical composition and the accessibility of the OM to decomposing organisms (i.e., the actual size of the OM and whether or not it is protected by soil minerals through occlusion or surface interactions) regulate the rate of decomposition, with the former being more important in the early stages of decomposition and the latter exerting more influence later in the process

(Lutzow *et al.*, 2006). For practical purposes, SOM is conceptualized as a series of pools with varying flux rates that reflect differences in chemical composition and the degree of physical accessibility to microorganisms (Fig. 6.4). Decomposition of OM is mediated primarily by bacteria and fungi that release the majority of the C as CO_2 via respiration while incorporating a small portion of the C into cellular structures through biosynthesis (growth and reproduction). Growth, reproduction, and death, combined with interactions among soil organisms as part of the soil

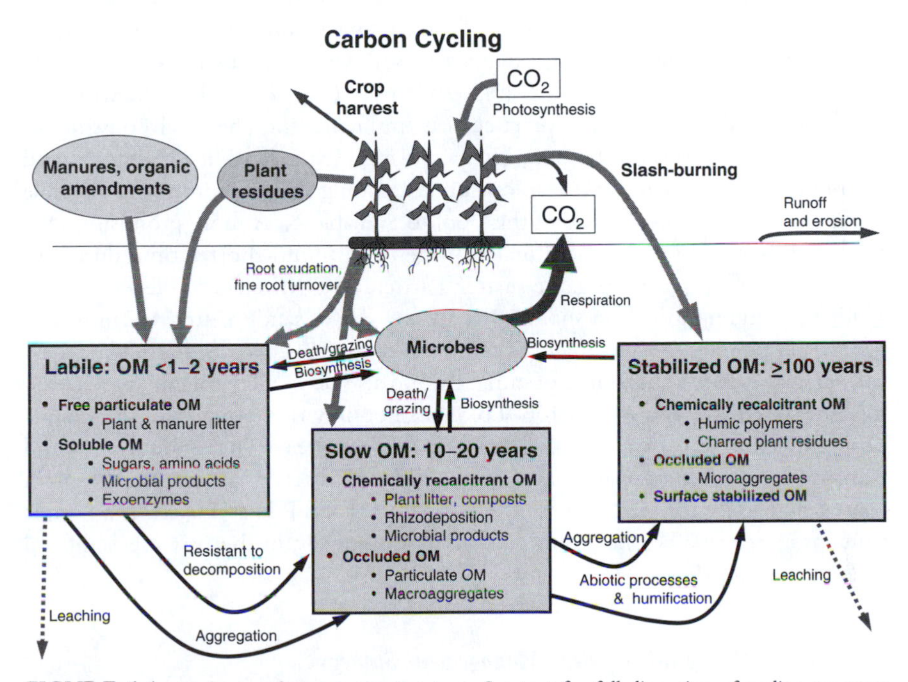

FIGURE 6.4 Carbon cycling in agroecosystems. See text for full discussion of cycling processes. The level of SOM is determined by the balance between photosynthesis or new OM additions and decomposition. Decomposition encompasses two distinct processes that reflect the dual function of C: 1) respiration (energy) and 2) biosynthesis (growth and reproduction). Biosynthesis results in C from the various substrates actually being incorporated into microbial biomass, while respiration results in the release of CO_2 into the atmosphere. In this diagram we separate out OM pools based on their approximate rates of turnover. The stable OM pool is by far the largest, usually accounting for >80% of SOM. The only route to stabilized OM that is directly under management control is through charcoal production. The vast majority of OM in the stabilized pool has undergone some form of microbial processing. In addition to the biological processes of respiration/biosynthesis, there are numerous abiotic mechanisms that are not well understood but are thought to contribute to stabilization of OM including adsorption, adventitious chemical reactions, and physiochemical interactions between clay particles and organic compounds (Lutzow *et al.*, 2006). Thus, OM can become stabilized through three mechanisms: 1) chemical resistance to decomposition; 2) inaccessibility to decomposers or exoenzymes; and 3) interactions with metal ions and surfaces. Aggregate formation that results in occluded OM is mediated by both soil organisms and abiotic processes.

food web, such as grazing, predation, and parasitism, regulate the flow of C and accompanying nutrients such as N and P.

In thinking about how to impact soil OM through management, it is essential to consider exactly which forms of organic matter you are aiming to influence. The bulk of SOM is present in the soil as stabilized OM, including polymers such as humins, humic acids, and fulvic acids, which are mainly present in intimate associations with mineral soil particles (Schulten and Reinhold, 1992). The MRT values for the humic fraction vary considerably, from 250–1900 years, clearly beyond the time frame of planning for agricultural management. Nevertheless, this pool represents a sizable N reservoir, and the elemental composition of stable OM, usually called *humus*, is fairly consistent across soil types and climatic zones with C and N contents of 50–60% and 2–4%, respectively (Tan, 2003). Recent work has demonstrated that agricultural practices can influence the chemical constituents of humified OM in ways that may alter its N-supplying capability (Schulten and Reinhold, 1992), suggesting that as our understanding increases, it may be possible to manipulate the contribution of this pool to available N. A more promising OM fraction that responds to agricultural management within shorter time durations (1–10 years) is particulate organic matter. Particulate OM refers to pieces of plant residues, including roots and shoots, that are the size of sand (53 μm to 2 mm) and has MRTs ranging from a single growing season to 10–20 years. Particulate OM can serve as a significant source of nutrients and also plays a key role in aggregation in some soils. Lastly, the soil microbial biomass not only is important for its decomposing function but also serves as a labile pool of nutrients. The amount of N and P in soil prokaryotes is nearly equal to the amount in terrestrial plants (Whitman *et al.*, 1998). For cultivated systems, the estimated N and P in soil bacteria amounts to an average of 630 kg ha^{-1} and 60 kg ha^{-1}, respectively, in the first meter of soil (Whitman *et al.*, 1998).

Applying an Ecological Nutrient Management Strategy

The plethora of processes controlling the cycling of nutrients in agroecosystems presents ample opportunities for enhancing the flows of N, P, and C. The overarching strategy guiding ecosystem-based nutrient management is distinct from the conventional approach that has focused primarily on fertilizer management for the past 50 years. Table 6.1 compares the two management schemes. The underlying theory guiding conventional nutrient management emphasizes developing optimum delivery systems for soluble inorganic fertilizers and managing the crop to create a strong sink for fertilizer by removing all other growth-limiting factors. The primary difficulty with this strategy is that soluble inorganic forms of N and P are fast cycling and are subject to multiple pathways of loss. As you might predict based on the nutrient cycling diagrams, when the pool of soluble inorganic N or P is greatly increased, undesirable fluxes also increase, and a proportion of these added nutrients is lost. So, while this approach has resulted in greater yields, it has

TABLE 6.1 Characteristics of the Conventional Agronomic Approach (Balasubramanian *et al.*, 2004; Havlin, 2004; Doberman and Cassman, 2004) Compared to an Ecosystem-Based Approach to Nutrient Management (Modified from Drinkwater, 2004)

	Agronomic framework	Ecological framework
Goals	Maximize crop uptake of applied N, P to achieve yield goal and reduce environmental losses	Achieve optimal yields and maintain soil reservoirs while balancing nutrient additions and exports as much as possible
Nutrient management strategy	Manage crop to create a strong sink for fertilizer by removing all growth-limiting factors and by providing an optimum delivery system (Balasubramanian *et al.*, 2004)	Manage agroecosystem to increase internal cycling capacity to 1) maintain nutrient pools that can be accessed through plant- and microbially-mediated processes and 2) conserve N and P by creating multiple sinks in time and space
Nutrient pools actively managed	Inorganic N and P	All N and P pools, organic and inorganic
Processes targeted by nutrient management	Crop uptake of N and P	Plant and microbial assimilation of N and P; C cycling; N, P, and C storage; other desirable processes that conserve nutrients
Strategy toward microbially mediated N transformations	Eliminate or inhibit as much as possible	Promote processes that conserve N; reduce processes that lead to losses (e.g., denitrification) by maintaining small inorganic N and P pools
Strategies for reducing NO_3 leaching, P occlusion/precipitation	Increase crop uptake of added N; use chemicals that inhibit nitrification	Minimize inorganic pool sizes through management of multiple processes: cover cropping, additions of N and P w/organic residues
Assessment of NUE	Based on fertilizer uptake of the crop in one growing season	Based on budgeting framework, reflect agroecosystem level retention, multiyear
Typical experimental approaches	Short-term, small-plot, empirical, factorial experiments dominate	Participatory, systems approaches, on-farm research is important; spatial and temporal scales of experiments are determined by the processes of interest

also resulted in poor nutrient use efficiency, and major losses of fertilizers to the environment are widespread (Drinkwater and Snapp, 2007a). Soil degradation is also a secondary consequence of these intensive, fertilizer-driven cropping systems, mainly due to the use of intensive tillage combined with reduced inputs of organic residues and bare fallows.

Ecological nutrient management seeks to overcome these drawbacks leading to nutrient loss and soil degradation. The goal of ecological nutrient management is to achieve optimal yields while maintaining soil nutrient reservoirs, balancing nutrient exports with additions, and minimizing losses of nutrients and soil to the environment. In agroecosystems with poor or degraded soils an additional goal is to restore soil fertility and agroecosystem functions.

To implement this comprehensive goal, ecological management must target a variety of nutrient reservoirs and cycling processes. The basic strategy is to enhance or conserve nutrient pools that can be accessed through plant- and microbially mediated processes by creating sinks for inorganic N and P that will promote nutrient retention and internal nutrient cycling of nutrients. The nutrient reservoirs that are targeted include labile and humified soil organic pools, microbial biomass, and sparingly soluble P. Management aims to promote processes that conserve these pools while minimizing those that lead to nutrient losses. For example, practices that enhance plant and microbial assimilation of N and P and other processes leading to N and P storage are favored. While flux through the inorganic N and P pools may be very large in these systems, a central objective of management is to reduce the size of these pools that are the most susceptible to loss. Examples of management practices that are compatible with this approach include diversifying nutrient sources, cover cropping and intercropping, and legume intensification for biological N fixation and P-solubilizing properties. The particular suite of cropping practices used are site specific and reflect the environmental characteristics of the agroecosystem (climate and soils), the crops that are being grown, the resources available to the farmer, and the livelihood goals of the household.

THE ROLES OF PLANTS AND MICROORGANISMS IN CYCLING NUTRIENTS

Cycling Processes Influenced by Plants

Effective use of plant diversity in agroecosystems requires some understanding of the roles played by different crop species in nutrient cycling. Plants and their associated microbes regulate innumerable ecosystem processes that ultimately control ecosystem fluxes of C, N, and P (Hooper and Vitousek, 1997). Intentional use of plant diversity based on the capacity of a species to enhance particular ecosystem processes is an important strategy in ecological nutrient management. Agroecosystem plant species diversity can be increased either by introducing additional cash crops or noncash crops (e.g., cover crops and intercrops) selected to serve specific ecosystem functions.

The most easily defined plant functional roles are those relating to phenology, productive potential, and above- and belowground architecture. *Phenology* refers to plant life cycle characteristics, such as germination, growth, flowering,

Which legume growth type?

Plant phenology varies from short-duration, determinant to long-duration, indeterminant (flowers repeatedly)

FIGURE 6.5 Examples of legumes with differing phenology. Legumes can be integrated into cropping systems using a number of different strategies, depending on their life cycles.

and reproduction, which are controlled by climatic conditions and seasons. For example, the functional role of legumes varies with phenology (Fig. 6.5). Many legume species used for grain production are short-duration annuals with determinant flowering and a high harvest index (the proportion of aboveground biomass that is harvestable product). Green manure legume species are at the other end of the spectrum. They provide large amounts of nutrient-rich residues and are generally short- or long-lived perennials with indeterminant flowering and low-to-no harvest index. Differences in the seasonal niche of plants can be used to expand the amount of time a field is covered with actively growing plants, increasing nutrient uptake in space and time and reducing nutrient losses (McCracken *et al.*, 1994; Snapp and Silim, 2002). Increased plant growth has cascading effects on internal cycling processes in agroecosystems. For example, if bare fallow periods are replaced with a cover crop, rhizodeposition provides C to the soil microbial community for a greater part of the year, increasing the potential for assimilation of nutrients into the microbial biomass (Drinkwater *et al.*, 1998). The tremendous variation among plant genotypes in root/shoot partitioning and root architecture can be exploited to complement cash crop characteristics and optimize plant-mediated processes belowground. For example, root biomass makes greater contributions to SOM than shoots, which tend to decompose more rapidly (Puget *et al.*, 2000).

Plant species characteristics, such as biochemical composition of litter and root exudates, fine root turnover, and the characteristics of the rhizospheric environment, influence ecosystem function through their impact on processes related to decomposition such as net mineralization of nutrients, aggregate formation, and humification.

Striking plant species effects have been documented for decomposition dynamics and net mineralization of N and P (Wedin and Tilman, 1990; Fierer *et al.*, 2001), aggregate formation (Tisdall and Oades, 1979; Angers and Mehuys, 1989; Haynes and Beare, 1997), availability of nutrients such as Ca, Mg, and P from mineral sources (Marschner and Dell, 1994; Johnson *et al.*, 1997; Neumann and Romheld, 1999; Kamh *et al.*, 1999), and microbial community composition (Kennedy, 1999). Many of these observed plant species' impacts on nutrient cycling processes are actually mediated by microorganisms associated with the roots.

In addition to these impacts on nutrient cycling, plant biodiversity can also enhance disease suppression (Abawi and Widmer, 2000), reduce weed competition and herbicide requirements (Gallandt *et al.*, 1999), and foster beneficial arthropod communities (Lewis *et al.*, 1997). Inclusion of all of these functions is integral to agroecological management of crop production. One useful approach is to compile information on the functional traits of potential cover crops (Table 6.2). Decisions about rotation and intercropping that impact plant species composition can contribute to a reduction of the need for agrochemical inputs (Drinkwater and Snapp, 2007a & b).

Plant–Microbial Interactions

While plants themselves can directly impact biogeochemical processes through nutrient assimilation and the quantity and quality of litter and root exudates, many influences on nutrient cycling are the result of plant–microbial interactions. The *rhizosphere* is the region of soil that is immediately adjacent to the plant root and is the site of plant–microbial interactions (Fig. 6.6). The importance and extent of plant–microbial interactions that take place in this microenvironment have not been fully appreciated in the past (Drinkwater and Snapp, 2007b).

Mycorrhizal Fungi

Arbuscular mycorrhizae (AM) fungi, which are endosymbionts, are the most important fungal symbiont in agroecosystems. Plant-mycorrhizal associations are the major mechanism for phosphorus uptake in over 80% of plant species. Colonization of roots by mycorrhizal fungi provides the plant with a well-distributed and extensive absorbing system in soil and a greater chance of encountering fertile microsites not available to roots alone. The ability of mycorrhizal fungi to access small soil pores (Drew *et al.*, 2004), and their ability to quickly respond to localized nutrient patches (Tibbett, 2000; Cavagnaro *et al.*, 2005), increases the plant's access to these nutrients. This is of particular significance in soils of low nutrient status, and for immobile nutrients, such as NH_4^+ and PO_4 (Ames *et al.*, 1983; Menge, 1983; Hetrick, 1991). Also, under drought stress, the role of mycorrhizal uptake of NO_3^- becomes more important, since the NO_3^- supply to the roots via mass flow is reduced (Nichols *et al.*, 1985). The N uptake mechanisms are largely unknown, but NH_4^+ is preferentially used. For example, corn plants colonized by *Glomus aggregatum* took up ten times more N from a $^{15}NH_4^+$

TABLE 6.2 Example of Cover Crop Functions That Can Be Evaluated

	Forbs			Legumes				Grasses		
	Brassicas	Bell beans	Medics	Rose clover	Strawberry clover	Vetches	Barley	Oats	Orchard grass	Tall fescue
Function										
Adds N to soil		X	X	X	X	X				
N retention	X						X	X	X	X
Erosion control	X			X			X	X	X	X
Weed suppression			X						X	X
Improves soil structure and water infiltration							X	X	X	X
Inhibits nematodes	X			X						
Attracts beneficial insects	X					X	X	X	X	X
Opens up heavy soils	X	X								

Note that the legumes supply new soil N while Brassicas and grasses excel at N retention. This is a simple yes (X) or no (blank) assessment; however, a more detailed evaluation could provide a ranking or some other more quantitative information. Modified from Eviner, 2001.

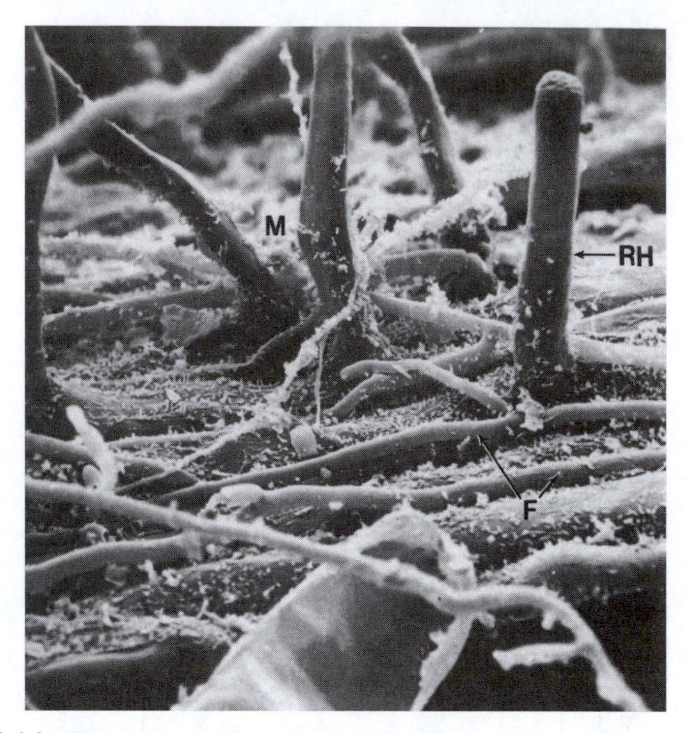

FIGURE 6.6 Electron micrograph of the root surface. Dense bacterial colonization can be seen on the root surface as well as fungal hyphae (F), root hairs (RH), and mucigel (M) from root exudates. (From Foster *et al.*, 1983. Ultrastructure of the Root-Soil Interface.)

patch than from a $^{15}NO_3^-$ patch (Tanaka and Yano, 2005). While AM fungi increase the recovery of ^{15}N from decomposing plant residues in soil, it is unclear how much they rely on organic N or if they accelerate organic matter decomposition (Hodge, 2004).

Background soil fertility and species diversity can influence the role of mycorrhizal contribution to nutrient cycling in agroecosystems. The species type and extent of mycorrhizal diversity can greatly influence nutrient uptake efficiency, ecosystem function, and net primary productivity (van der Heijden *et al.*, 1998). Under the nutrient-rich conditions that occur in industrialized agricultural systems, formation of mycorrhizal associations may become a cost to the plant, as the plant is able to satisfy its own nutrient requirements (Johnson *et al.*, 1997). Agricultural production practices appear to have inadvertently reduced diversity, function, and efficiency in plant–mycorrhizal symbiosis (Daniell *et al.*, 1998). In a meta-analysis of AM and ectomycorrhizal studies, colonization generally declined in response to N and P fertilization, although N effects on AM abundance were less strong than those for P (Treseder, 2004). One explanation is that mycorrhizae may be less important in facilitating plant uptake of NO_3^-, due its availability via mass flow, except in very N-limited ecosystems. In an organic farming system, a mycorrhiza-defective tomato mutant had 12% lower N content than the mycorrhizal wild-type, and there was more soil NO_3^- as well (Cavagnaro *et al.*, 2006), indicating that AM are important in farming systems where fungicides and P fertilizers are not used. Manipulation of mycorrhizal populations to develop more efficient plant–symbiont combinations is in its infancy, but strategies that can be pursued include use of sparingly soluble rock P, reduced tillage, and integration of auxiliary plants that are highly mycorrhizal.

Biological N Fixation

Plants lack enzymes that can convert N_2 gas into a usable form, and, as a result, most plants in natural ecosystems rely on the N released via microbial decomposition of soil organic matter. Some bacteria, known as *diazotrophs*, do produce the enzyme nitrogenase that catalyzes the reduction of N_2 into NH_3, a plant-available form of N. This microbially mediated process is referred to as biological nitrogen fixation (BNF). Because N_2 is a stable molecule with a strong triple bond, BNF is an energy-intensive process. Carbon is the primary energy source for the range of diazotrophs that carry out BNF. The high energy demand of BNF may explain why it is not more universally found in plant–microbe associations and why N cycling in natural ecosystems is driven primarily by the recycling of previously fixed N through mineralization and immobilization of N from organic matter pools. Globally, biological nitrogen fixation in unmanaged ecosystems is estimated to convert about 150 Tg of N_2 gas into plant-available N every year (Vitousek, 1997). In managed ecosystems, only an estimated 33 Tg N/year is fixed by cultivated legumes, while more than 100 Tg N/year is fixed nonbiologically through the fossil-fuel based production of synthetic nitrogen fertilizers (Galloway, 2003).

Diazotrophs and plants have evolved different degrees of association that facilitate the transfer of photosynthetically derived carbon from plant to bacteria to support BNF. Most diazotrophs are heterotrophic and rely on plant-derived carbon to support BNF. Symbiotic diazotrophs that fix nitrogen within nodules of leguminous plants are the most familiar example of BNF and are typically referred to collectively as *rhizobia* (Fig. 6.7). In most cases, the legume–rhizobia symbiosis is highly specific. Complex chemical signaling has evolved between legume species and specific rhizobial strains to initiate nodule formation. Symbiotic rhizobia in the nodule receive a direct supply of C in exchange for N fixation for plant growth. The nodule provides physical protection for the rhizobia while increasing the capture of the fixed N by the plant.

More recent work has identified numerous other diazotrophs that are associated with plant roots, either externally or internally within intracellular root spaces. These associative diazotrophs utilize labile carbon root exudates as an energy source to support BNF. The plant has less control over the fate of the fixed N in this situation, compared with the N fixed within a root nodule. The N fixed by associative diazotrophs is incorporated into the bacterial biomass. This N becomes available

FIGURE 6.7 Leguminous plants are important in cropping systems worldwide. (**A**) In the Potosi region of Bolivia, Tarwi (*Lupinus mutabilis* Sweet) is a multipurpose legume that serves as a fertility source and grain crop. (**B**) The root nodules on Tarwi are large and numerous.

for plant uptake when grazers feed on these bacteria and trophic interactions in the rhizospheric food web result in a net N release (Clarholm, 1985). Due to the rapid turnover of microbial biomass in comparison with the much longer life cycle of the plant, significant quantities of associatively fixed N end up in the associated plant (Ladha and Reddy 2003). Some free-living diazotrophs, such as cyanobacteria, are autotrophs, and they are capable of both fixing carbon via photosynthesis and fixing N via BNF. Many cyanobacteria, in spite of their relative self-sufficiency, form symbiotic partnerships with plants, and the plants again are eventually the beneficiaries of the fixed N (Yoneyama *et al.*, 1987).

The ecology of biological nitrogen fixation is complex, and many things must be considered to optimize this valuable process. For example, at the field scale, BNF is regulated by interactions between plant species, climate, and soil type. Similarly, at microscales, BNF is regulated by plant–microbe–microsite environmental interactions. The complexity of the ecology of BNF is reflected in the high variability found in BNF rates in natural and agroecosystems (Ojiem *et al.*, 2007; Walley *et al.*, 2001). The soil environment exerts influence on BNF through direct and indirect effects on the plants and microorganisms involved in N fixation. High soil temperatures (>27–40° C), water stress, soil acidity (pH < 5), low soil P availability, and Al toxicity—all common conditions in certain tropical systems with highly weathered soils—can limit rhizobial growth and nitrogenase activity (Hungria and

Vargas, 2000; Graham and Vance, 2003). The availability of molybdenum (Mo), a key component of the nitrogenase enzyme, can also be an important limiting factor in BNF in some soils (Ohara, 2001). Because BNF only supplies new N to agroecosystems while recycling other nutrients, integration with other soil amendments is critical to both the ability of the system to support the nutritional demands of BNF and to maintain longer-term nutrient balances.

Partnerships with the Rhizospheric Community

Because the rhizosphere is the site of increased C availability, there are numerous other kinds of interactions that involve free-living or rhizospheric microorganisms that are not obligate symbionts. Since crop plants mainly take up NH_4^+ and NO_3^- rather than amino acids or other monomers, mineralization is important for the N supply to plants in the absence of inorganic N fertilizer additions. While C limitation rarely occurs in the rhizosphere (Cheng *et al.*, 1996), decomposers in bulk soil are usually C limited (Koch *et al.*, 2001) and less numerous (Cardon and Whitbeck, 2007). Plants can stimulate mineralization of organic substrates by supplying labile C to decomposers in the rhizosphere (Clarholm, 1985; Cheng *et al.*, 2003; Hamilton and Frank, 2001). A priming effect of increased mineralization mediated by plant–microbial interactions has been demonstrated for N (Clarholm, 1985; Hamilton and Frank, 2001), P (Gavito and Olsson, 2003; Helal and Dressler, 1989), and S (Vong *et al.*, 2003; Kertesz and Mirleau, 2004).

The role of these microbial–plant interactions in stimulating N mineralization is the most often studied. The rate of decomposition and N mineralization varies with plant species (Cheng *et al.*, 2003), rhizospheric community composition (Clarholm, 1985; Ferris, 1998; Chen and Ferris, 1999), and nutrient availability (Tate *et al.*, 1991; Liljeroth *et al.*, 1994). The release of nutrients for plant uptake appears to be enhanced by the involvement of secondary consumers feeding on the primary decomposers due to differences in the stoichiometry[1] between the two trophic levels (Clarholm, 1985; Ferris, 1998; Chen and Ferris, 1999; Fig. 6.8). This trophic cascade provides a mechanism for the primary producers to influence nutrient mineralization, similar to the so-called "microbial loop" in aquatic ecosystems, where primary producers have been shown to increase excretion of soluble C under nutrient-limiting conditions (Elser and Urabe, 1999). There is some evidence suggesting that terrestrial plants can influence the rate of net N mineralization through this mechanism, based on their need for nutrients, by modifying the amount of soluble C excreted into the rhizosphere (Hamilton and Frank 2001; Fig. 6.9).

Greater reliance on SOM as a nutrient source increases the importance of microbially-mediated processes such as decomposition and mineralization. The tight coupling that occurs in the rhizosphere between net mineralization of N and P and

[1]Stoichiometry refers to the ratio of elements to one another in different organisms; e.g., N:P or C:N.

FIGURE 6.8 A swarm of protozoa grazing on red fluorescent bacteria in the rhizosphere. The rhizosphere is the home of numerous organisms that influence nutrient cycling and plant access to nutrients through food web interactions.

plant assimilation reduces the potential for nutrient losses. Inorganic nutrient pools can be extremely small in ecosystems while high rates of plant growth are maintained if N mineralization and plant assimilation are spatially and temporally connected in this manner (Jackson *et al.*, 1988). The identity of the SOM pools that are accessed through this mechanism remains unknown; however, phytoremediation studies show that decomposition of chemically recalcitrant substrates is accelerated

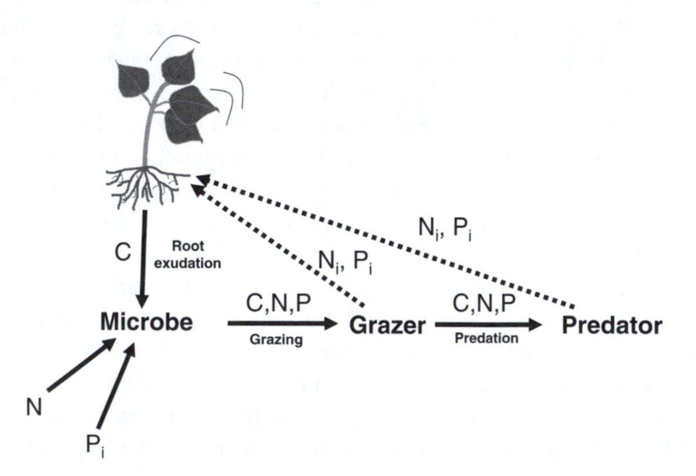

FIGURE 6.9 Feeding interactions across trophic levels increase net mineralization of N and P in the rhizosphere. Plants supply C to microbes that take up N and P from sources that are not available to plants; grazing of these microbes increases the rate of mineralization.

in the rhizosphere as compared to bulk soil (Reilley *et al.*, 1996; Siciliano *et al.*, 2003). This evokes the intriguing possibility that plants may be able to promote access to humified organic matter pools through partnerships with rhizospheric microorganisms.

Microbial Mediation of Nutrient Cycling

Microorganisms represent a substantial portion of the standing biomass in agricultural ecosystems and contribute to the regulation of C sequestration, N availability and losses, and P dynamics. The size and physiological state of the standing microbial biomass are influenced by management practices, including rotational diversity (Anderson and Domsch, 1990), tillage (Holland and Coleman, 1987), and the quality and quantity of C inputs to the soil (Kirchner *et al.*, 1993; Wander and Traina, 1996; Lundquist *et al.*, 1999; Fliessbach and Mäder, 2000). The mechanisms that control community structure and functional characteristics of belowground ecosystem processes remain largely unknown; however, this is an area of active research and much progress has been made in the last decade.

While some plants are able to produce and secrete enzymes required for P mineralization (Vance *et al.*, 2003), release of nutrients from organic compounds is largely carried out by heterotrophic microorganisms (Paul and Clark, 1996). Microbial production of extracellular enzymes that can attack polymers and release small, soluble molecules is an important mechanism contributing to the internal cycling of N, P, and S (McGill and Cole, 1981; Paul and Clark, 1996). Microbial community composition and metabolic status determine the balance between C released through respiration and C assimilated into biomass during decomposition as well as the biochemical composition of that biomass and the net release of plant-available nutrients. Decomposers in soils with greater plant species diversity or greater abundance of C relative to N have reduced energy requirements for maintenance and therefore convert a greater proportion of metabolized C to biomass (Anderson and Domsch, 1990; Fliessbach *et al.*, 2000; Aoyama *et al.*, 2000). Changes in microbial community structure can lead to increased C retention if the management practices result in fungal-dominated decomposer communities (Holland and Coleman, 1987).

Microbial Control of N Cycling

The relative abundance of C and N strongly influences the rates of competing microbial processes and offers opportunities for farmers to optimize N cycling through manipulating microbial metabolism. For example, plant litter with a high C:N ratio initially increases microbial N immobilization and decreases NH_4^+ and NO_3^- availability to plants. As microbial decomposition of these residues continues and cascading effects on grazers and other trophic levels increase, net N release increases (Booth *et al.*, 2005).

When large amounts of inorganic N are added, as in industrialized cropping systems, inorganic N pools expand beyond the capacity of crop and microbial uptake so that pathways of loss, such as denitrification and leaching, are increased. Soluble fertilizer additions appear to stimulate preferential decomposition of some SOM pools, including particulate OM (Neff *et al.*, 2002).

In agroecosystems with low N fertility, plants and soil microbes compete for NH_4^+ and NO_3^-. In short-term studies (i.e., one to several days) microbes take up more inorganic ^{15}N than plants, presumably because they have higher substrate affinities, larger surface area to volume ratios, and faster growth rates than plants (Hodge, 2004; Schimel and Bennett, 2004). But after a month or so, plants contain an increasing proportion of the added ^{15}N, because the gradual release of microbial ^{15}N into the soil becomes available for root uptake, and plants hold on to N longer than microbes (Harrison *et al.*, 2007).

Increased soil stocks of labile C substrates contribute to N conservation through both aerobic and anaerobic pathways (Silver *et al.*, 2001; Burger and Jackson, 2003). Studies in agricultural soils simulating conditions in bulk soil indicate that the major fate of NH_4^+ is nitrification (Shi and Norton, 2000; Burger and Jackson 2003). Competition between heterotrophs and nitrifiers for NH_4^+ is strong, resulting in very small NH_4^+ pools, and nitrification rates can be two- to threefold greater than NH_4^+ immobilization (Burger and Jackson, 2003). Nevertheless, soils receiving greater C additions support a larger, more active microbial biomass, resulting in a greater proportion of NO_3^- assimilation and reduction of standing NO_3^- pools (Burger and Jackson, 2003).

Carbon abundance also influences dissimilatory NO_3^- reduction pathways in ways that support N conservation and reduce environmental impacts. In one study, denitrification in soils receiving organic N amendments reduced the proportion of N lost as N_2O (Kramer *et al.*, 2006). Carbon abundance can also favor a second anaerobic pathway, dissimilatory nitrate reduction to ammonium (DNRA) (Silver *et al.*, 2001). This process was thought to be limited to extremely anaerobic, C-rich environments such as sewage sludge and estuarine or lake sediments (Maier *et al.*, 1999) but has recently been detected in a broad range of unmanaged terrestrial ecosystems (Silver, 2003, pers. comm.) and in agricultural soils (Yin *et al.*, 2002). Silver *et al.* (2001) reported average rates of DNRA were threefold greater than denitrification in humid tropical forest soils. The resulting reduction in NO_3^- availability to denitrifiers and leaching may contribute to N conservation in these ecosystems (Silver *et al.*, 2001). In rice paddies, soils with greater levels of SOM due to additions of straw mulch had threefold greater DNRA when compared to soils where straw was removed and endogenous SOM was reduced (Yin *et al.*, 2002).

Microbially Mediated P Transformations

As with N, SOM levels and C abundance influence microbially mediated processes that control the uptake of P by microbes as well as mineralization and biological weathering. For example, microorganisms solubilize sparingly soluble inorganic P through several mechanisms if they have adequate C substrates for growth and reproduction but are lacking P (Illmer *et al.*, 1995; Oberson *et al.*, 2001). Direct excretion of phosphatase enzymes is one mechanism of phosphate solubilization. Another is local acidification through organic acid excretion, such as occurs in the soil fungus *Penicillium radicum* when isolated from a low P rhizosphere of unfertilized wheat (Whitelaw *et al.*, 1999). In this system, phosphate solubilization from insoluble or sparingly soluble complexes with calcium, colloidal aluminum, and iron was related to titratable acidity and gluconic acid concentration. Organic acid excretion not only alters pH but also may chelate Al_3^+ or other cations, directly further enhancing the solubilization of phosphate (Erich *et al.*, 2002).

The assimilation of inorganic phosphorus by microbes may protect phosphorus from geochemical adsorption reactions with soil particles through microbial turnover and organic matter mineralization processes that are synchronized with plant and further microbial uptake. Indirect evidence for this is the enhanced levels of microbial P and cycling of P from inorganic to organic in plant forms associated with managed systems that had enhanced soil biological activity and legumes present (Oberson *et al.*, 2001). Labeled glucose and residue studies have recently shown that biomass P turnover is rapid, approximately twice as fast as C (Kouno *et al.*, 2002). This indicates that the potential for microbial P pools to support plant P requirements may have been markedly underestimated.

CONCEPTS AND STRATEGIES FOR MANAGING COMPLEXITY

Using Spatial and Temporal Scales to Organize Management Decisions

The use of both temporal and spatial scales to organize nutrient flows into a logical structure is fundamental to developing a coherent set of management strategies that act together to achieve the goals of ecological nutrient management. The spatial scales we must consider range from microns to the plant, field, and farming community or regional scales. We can think of these spatial scales as nested within one another, so that at any level we are able to identify the location of the processes we are aiming to manage. For example, the use of rock phosphate as a source of P involves processes occurring at the micron, plant, and field scales (Fig. 6.10).

In using a sparingly soluble P source, the farmer is aiming to modify the solubilization of P, a microscale process that is mediated by microorganisms and the rhizosphere of some plant species. The background soil environment and climate affect processes occurring at every level, including the farmer's decisions. Assuming that P is a limiting factor in this field, plant productivity will be impacted by field-scale management decisions and the resulting rate of P solubilization. The field-scale

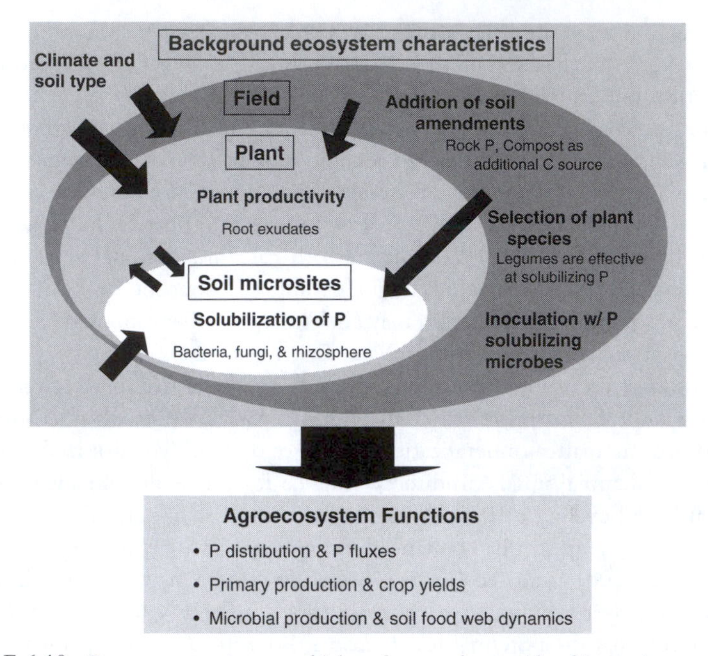

FIGURE 6.10 Processes occurring at multiple scales must be considered in nutrient management decision making. Here the processes that impact the decision to use rock P and the ultimate outcomes are illustrated. See text for full discussion.

decisions that will directly influence this process are 1) choice of amendments at the field scale, 2) selection of plant species, and 3) inoculation of P-solubilizing microbes. (This may be an option in the future!) Interactions across these scales impact one another; i.e., field management impacts the plants and soil microbes; P solubilization influences plant productivity and creates a feedback because increased productive capacity increases the ability of the plant to stimulate P solubilization through direct and indirect means. Adding rock P in conjunction with planting a legume can be particularly effective because legumes are able to access sparingly soluble P. If rock P was added without consideration of plant species or use of an additional C source (such as compost or manure) to support microbial P solubilization, then it is possible that there may not be a detectable improvement in crop yields in the first growing season because microbial activity is limited by access to C (energy), not P. The inclusion of a supplemental C source is particularly important if a nonmycorrhizal crop is to be planted. This example illustrates how systematic analysis of processes occurring at different scales can help in planning management interventions.

Just as interactions across spatial scales were important in conceptualizing the key processes in the rock P example, interactions among processes occurring at different rates is also a useful organizing principle. This is particularly important when a major change in an agroecosystem's management regime is implemented, such as increased inputs of organic residues or a change in tillage regime. There are major differences in the process rates and the flux through various pools. As a result, the MRTs of relevant nutrient pools range from minutes for transient inorganic N forms such as nitrite to centuries for humified pools of SOM. Generally speaking, the spatial and temporal scales of ecological processes are commonly linked. Small-scale or local processes are often ephemeral and rapid. Examples are nutrient transformations controlled by microorganisms such as nitrification or mineralization and nutrient uptake by a single fine root. These rapid, small-scale processes and interactions can be highly variable in space and time, but in aggregate, they determine agroecosystem functions at the field scale. For example, two competing biological processes that occur very rapidly yet play a significant role in regulating the amount of N lost from a field on an annual basis are the flux of NH_4^+ into fine roots versus the conversion of NH_4^+ to NO_3^- by nitrifying microorganisms. If the former predominates, then the NH_4^+ available to nitrifiers is reduced, and NO_3^- pools remain small. On the other hand, when nitrifiers have access to ample NH_4^+, then NO_3^- production is elevated, resulting in larger NO_3^- pools and increased losses of N.

Managing agroecosystems to modify reservoirs with longer MRTs usually requires planning for management that occurs over longer time frames compared to small-scale processes and pools that are cycling faster. For example, soil degradation and restoration result from slow changes that accrue over decades rather than years and that represent the sum of many shorter-term processes and events. Yet it is these longer-term processes that are critical to the long-term sustainability of agroecosystem production.

One approach that has been used in the case of SOM is to focus efforts on OM pools with MRTs that can be influenced by management in shorter time frames. Because of the different MRTs of the soil organic pools that are impacted by management changes, the shift to new steady-state conditions will occur over multiyear, decadal, and even longer time scales, depending on the MRT of the particular SOM pool. Agroecosystems that are undergoing changes in ecological processes are considered to be "in transition" (Liebhardt *et al.*, 1989). During this transition period, there are clear signs of directional change. For instance, when soluble fertilizers are replaced with organic nutrient sources, subsequent shifts in C and N cycling impact crop yields and the distribution of SOM pools (Wander *et al.*, 1994; Liebhardt *et al.*, 1989). In the short term, organic inputs will have a greater impact on faster cycling processes. To impact slower cycling, SOM reservoirs require that nutrient management strategies consider time frames of 5–10 years. Fig. 6.11 illustrates how a green manure incorporation impacts SOM pools with differing MRTs and their contribution to crop N supply.

While many ecological processes that govern nutrient availability fall somewhere along the continuum from small-scale and fast to large-scale and slow, there are exceptions. In agricultural systems, it is not uncommon to have large-scale processes that occur very rapidly. Management interventions such as crop harvest, burning,

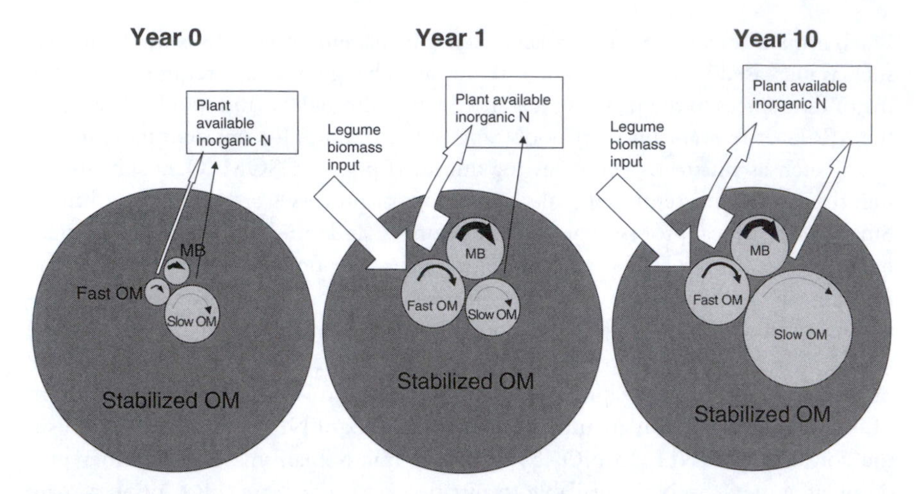

FIGURE 6.11 The effect of legume additions on SOM pools and N availability over time. Four SOM pools are represented in each diagram: microbial biomass (MB); fast-cycling organic matter pool primarily composed of recent litter additions (fast OM); slow-cycling organic matter pool primarily composed of partially decomposed litter (slow OM); and the much larger soil humus pool, which is unavailable for microbial decomposition and plant N uptake. Year 0 represents a relatively degraded soil with low SOM levels and low N supplying capacity from decomposition of fast OM with small contributions from slow OM. In Year 1, legume biomass is incorporated into the cropping system. Biomass enters the fast OM pool and drives a rapid increase in the size and cycling rate of MB. Labile C and N compounds are decomposed rapidly by microbes, resulting in a quick burst of N availability. The slow OM pool is not immediately impacted by the legume addition. Less available C and N compounds eventually become part of the slow OM pool on a decadal time scale. By Year 10, the size of MB and fast OM pools are maintained, and a larger slow OM pool also contributes to the N-supplying capacity of the soil. The net effect is N availability for crop uptake and small gains in total SOM to support longer term nutrient-supplying capacity.

and tillage are examples. These events cause rapid, large-scale changes and result in dramatic shifts in virtually all the smaller scale ecosystem processes from one moment to the next. A well-known example is the pulse in soil respiration that occurs after tillage.

Evaluating Management Impacts on Agroecosystem-Scale Nutrient Flows

To efficiently gauge the impact of farm management on longer term soil fertility and sustainability, it is crucial to consider the flow of nutrients across boundaries of management units as well as the larger landscape in which they are embedded. To analyze the movement of nutrients across field and farm boundaries, net flows of nutrients can be estimated by constructing nutrient budgets. This approach is used in ecosystem ecology to compare fluxes into and out of a defined compartment.

In order to find out whether the balance of these fluxes is positive or negative, a compartment can be as small as a patch of organic residue in the soil (Hodge, 2004) or as large as the entire atmosphere (Schlesinger, 2005). Over the last ten years, the value of nutrient budgeting as a tool for analyzing nutrient flows in agroecosystems and agricultural landscapes has become apparent, and the approach has been widely applied at a variety of scales. Depending on the questions that are being addressed, the scale of land unit used can be individual fields, farms, watersheds, or even whole regions and countries.

To conceptualize how management is affecting the nutrient status of a field or other management unit, we treat the field as a compartment and focus on inputs and exports across the field boundary. The simplest nutrient budgets emphasize the flow of nutrients that are controlled by the manager, such as fertility inputs and harvested exports (Fig. 6.12). These fluxes are usually the dominant flows that regulate the transfer of nutrients into and out of a field or farm. These simple mass balances provide a starting point for managing smaller scale processes that are regulating the fate of nutrients in agroecosystems. While this balance does not quantify internal

FIGURE 6.12 Major nutrient flows in a smallholder cropping system in the Potosi region of Bolivia where animals are an integral part of fertility management. Farmers harvest nutrients from rangeland through grazing their animals on marginal lands (dotted arrows). Internal transfers of P also occur when harvested crop residues serve as forages for animals (lower gray arrow). Manure is used primarily on fields that are closest to the homestead, although some is transported to farther fields (black arrows). The manure contains nutrients that have been captured from communal rangelands as well as recycled from cropping fields. Some nutrients leave the agroecosystem as harvested crops that are exported to markets (gray arrows to the right).

nutrient cycling processes or environmental losses resulting from these internal processes, it provides useful information for assessing whether surplus or inadequate nutrients are being added and thus is useful in developing nutrient management strategies. All things being equal, environmental losses are directly related to the level of N and P availability. Soils with excess applications will lose more through microbially mediated processes compared to soils that do not have surplus nutrients (Aber *et al.*, 1989).

Construction of a field-scale mass balance entails calculating the difference between inputs and harvested exports over the course of a rotation cycle (Drinkwater *et al.*, 1998). All fertilizers, soil amendments, and N-fixing crops must be accounted for as inputs. One of the most challenging aspects of using this budgeting approach is determination of N inputs from leguminous cover crops. A common practice for legumes is to measure N in standing biomass for green manures as an estimate of N from BNF and to consider no net gain or loss of N for leguminous grain crops (Drinkwater *et al.*, 1998). The exports are all harvested crops or animals, including grains, forages or crop residues removed, or manure or animal biomass removed. This simple budgeting method can be very useful as an indicator of directional change and the relative efficiency of divergent nutrient management strategies (Box 6.1). Negative balances indicate that deficits are accruing and that nutrients are being extracted from the soil (Box 6.2). In this case, if nutrient management

BOX 6.1 How Do Different Nutrient Management Strategies Affect N Mass Balance in Grain Systems?

Fifteen-year N balances for three distinct grain production systems: 1) MNR–Integrated system with grains, forages and legumes with animal manure returned to the field, 2) Cash grain systems with leguminous green manures as the only N source, 3) Cash grain system based on soluble N fertilizer inputs (modified from Drinkwater *et al.*, 1998). All systems include maize and soybean while only the MNR and LEG also grow wheat and leguminous green manures.

These simple input-output balances show that the LEG system is running close to steady state; that is, inputs are roughly equal to harvested exports while the MRN and CNV have accrued comparable surpluses of N over this 15-year period. These differences in N balance are driven mainly by the inputs rather than yields since the harvested N in these three cropping systems is similar. If we include data on soil N using samples conducted at the beginning of the experiment and then 15-year later, we can detect a significant increase in soil N for the MNR while the CNV system shows a significant decline in soil N for the same time period. The small increase shown for the LEG system is not statistically significant.

BOX 6.2 Intercropping of Pigeon Pea Reduces Erosion and Increases Grain Yields and P Recycled Through Active Soil OM Pool

The table below shows, net exports of P in yields and through erosion at two sites with differing erosion rates in Songani, Southern Malawi. Erosion P losses were estimated after Stoorvogel and Smaling, 1990 where erosion rates were estimated as follows: 1) site 1 (2% slope) erosion was estimated at 5 ton/ha/yr and 2) site 2 (30% slope) at 20 ton/ha/yr (Snapp *et al.*, 1998). Based on percentage ground cover, we estimated that erosion and the resulting P loss were reduced by 25% when maize was intercropped with pigeon pea intercrop, compared to monoculture maize. Long duration pigeon pea extends the period of soil cover over a 4 month period of intermittent rains.

	Yield Maize	Yield P'pea	P harvested in grain	Erosion P loss	Total P lost	P recycled in crop residues (internal P cycle)
	$t\ ha^{-1}$	$t\ ha^{-1}$	$kg\ ha^{-1}$	$kg\ ha^{-1}\ yr^{-1}$	$kg\ ha^{-1}\ yr^{-1}$	$kg\ ha^{-1}\ yr^{-1}$
Maize, Low Erosion Site 1	1	0	2	2.3	− 4.3	7.6
Maize, High Erosion Site 2	0.5	0	1	9	− 10.0	3.8
Maize + Pigeon Pea, Site 1	1.1	0.4	3.4	1.7	− 5.1	16.4
Maize + Pigeon Pea, Site 2	0.5	0.3	1.9	6.8	− 11.7	9.8

Because there are no inputs of P for this maize crop, all P balances at the end of the season are negative, indicating that a net export of P has occurred. Phosphorus lost through erosion is threefold greater in the steeper field. The addition of pigeon pea into this system increases the export of harvested P while also reducing P lost through erosion. However, because of the increase in harvested yields, overall P removal is accelerated with pigeon pea + maize. As a result, although erosion losses are reduced at each site, the need to add P through soil amendments is increased by intercropping. The last column reports the P content of crop residues from maize or maize + pigeon pea and shows how the inclusion of the legume more than doubles the amount of P that will be recycled back into labile OM pools.

practices are not modified, soil fertility will continue to be depleted and production will decline. Chronic surpluses may indicate that overapplication is a problem; however, to fully determine whether or not these surpluses are being retained in the field, additional analysis of soil stocks (i.e., Box 6.1) and potential loss pathways such as erosion (Box 6.2) will need to be evaluated.

Integrating Background Soil Fertility into Nutrient Management Planning

We have discussed how nutrient cycling in agroecosystems reflects interactions between the environment, management, and the organisms present in the system. While management practices can exert a strong influence on shorter term outcomes such as crop nutrient uptake and yields, the particular effect of identical management strategies will vary across farms, depending on climate, soil type, and the legacy of past management decisions. These inherent characteristics of the agroecosystem need to be considered in developing overall nutrient management strategies. Fig. 6.13 illustrates how management practices can have different results depending on the initial fertility status of a site. In this diagram, we have laid out three different management scenarios for two fields that differ in terms of the initial fertility status. The cause of this difference in soil fertility is inconsequential; it could be due to either soil type difference or past management history.

For a low fertility field (1a), small soil N, P, and organic matter reservoirs support low productivity. With crop exports, small soil reservoirs continue to shrink, leading to a downward spiral of soil degradation. If legumes alone are incorporated into a cropping system in this field (1b), BNF will only provide a small benefit, as legume growth will be limited by a small and increasingly shrinking soil P reservoir. The presence of legumes in the system can increase the availability of soil P (Bah, 2006), but this will only increase the rate of P depletion over the long term. In this scenario, legumes are not likely to increase productivity, but they may help maintain soil N reservoirs and SOM status. If incorporation of legumes is paired with modest P fertilizer additions (1c), BNF can make much larger contributions to overall productivity. If P fertilizer is in organic form, such as manure, SOM reservoirs can also be increased. It is at higher levels of productivity that retaining crop residues becomes more economically feasible for farmers, reinforcing the maintenance of the SOM reservoir. As the organic matter reservoir increases, the capacity of the soil to retain N and P in relatively available forms increases.

For a medium fertility field (2a), modest crop production can be sustained over the short term by the mining of existing soil N and P reservoirs. With the incorporation of legumes into this field (2b), BNF can provide substantial benefits because biomass production is not limited by P and other nutrient availability. Legumes can improve the P status of the soil by moving P from less to more available soil pools (Bah, 2006). Long-term dependence on just legume BNF input will eventually

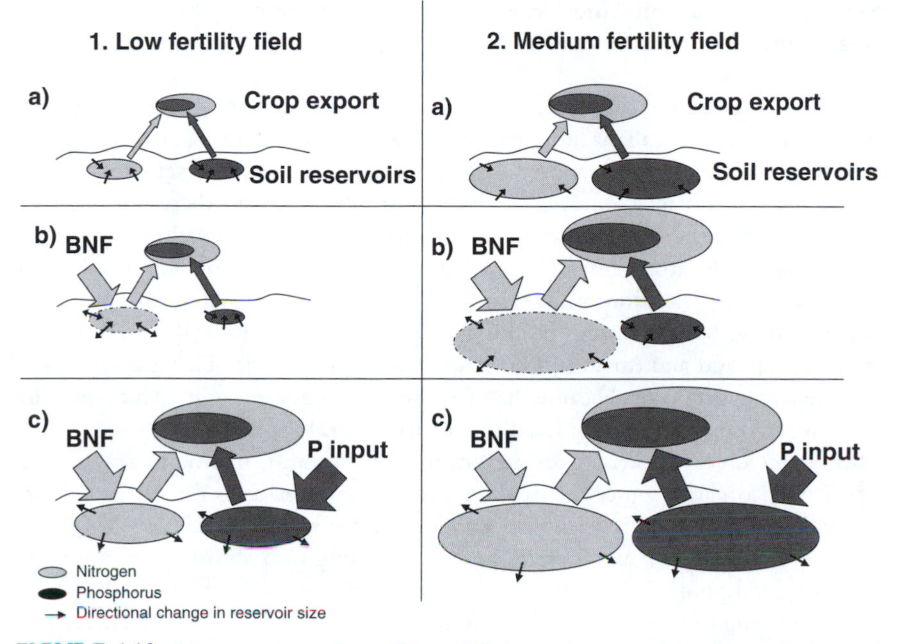

FIGURE 6.13 Management practices will have differing consequences depending on background soil fertility. A (1) low fertility background and (2) medium fertility background are compared for three management options: a) absence of any fertility additions; b) inclusion of N-fixing crop in the system where at least part of the N fixed is retained in the system; and c) N-fixing crop combined with modest addition of rock P. The effect of these three management regimes is illustrated in terms of relative flows of N and P into soil pools and crop harvest, changes in soil pools, and crop N and P content. See text for full discussion.

lead to P depletion, and this medium-fertility soil could shift to a low-fertility status, as in Fig. 1b. Again, if incorporation of legumes is paired with modest additions of P fertilizer (2c), BNF can make much larger contributions to productivity over the long term. As in 1c, the form of P fertilizer and the quantity of crop residues retained affect the longer term SOM reservoir.

Synthetic N fertilizers, as an alternative to BNF, could help boost productivity in any of the non-P limiting scenarios. However, inorganic N fertilizers can exacerbate soil P depletion (Lupwayi and Haque, 1999) and do not contribute to longer term nutrient cycling capacity of the system if they are not coupled with C inputs.

The size of nutrient exports relative to crop residues retained in the field determines the directional change in soil reservoir size. Legume grain crops, for example, tend to export almost as much N as they fix (Alves *et al.*, 2003). As yields increase with increasing soil fertility, grain crops can export more N than they fix (Ojiem *et al.*, 2007). Incorporation of green manures as intercrops or relay crops can help balance N exported in a cash crop with N fixed by the legume (Lupwayi and Haque, 1999).

Strategic Use of Soil Amendments and Cover Cropping to Enhance Linkages between Cycles

We have discussed the role of plants and microbes in connecting N and P cycling with C flows and emphasized how these linkages support internal cycling capacity and promote nutrient retention. Here we provide specific examples of how use of various nutrient sources can either promote or impair these linkages. For example, while the initial crop uptake of inorganic, soluble fertilizers is greater than crop uptake from other forms of amendments such as organic residues or rock phosphate, retention of these soluble forms in the ecosystem through conversion to SOM is reduced, resulting in greater environmental losses (Bundy *et al.*, 2001; Ladd and Amato, 1986; Drinkwater *et al.*, 1998). This greater loss of soluble fertilizers occurs because the processes that sequester soluble nutrients are saturated (Azam *et al.*, 1985; Ladd & Amato, 1986, Hodge *et al.*, 2000), leaving NO_3^- vulnerable to leaching/denitrification. In contrast, microbial assimilation of N from organic sources is two- to fourfold greater than for N from inorganic fertilizer, leading to increased storage of legume-derived N in SOM pools. Likewise, soluble surplus P sources pushes P cycling into absorbed precipitation and occluded pools.

In cropping systems where soluble fertilizer is part of the overall nutrient management strategy, fertilizer additions should be managed to enhance assimilation in biologically regulated sinks. In rotations, inorganic sources can be preferentially applied to those crops with higher nutrient use efficiency. Improving the spatial/temporal connections between fertilizers and senescent crop residues appears to increase the retention of older SOM fractions (Clapp *et al.*, 2000), suggesting it may be advantageous to add small portions of fertilizer when high C crop residues are being incorporated. A review of three long-term trials from temperate countries suggests that the fate of soluble P from fertilizers depends on whether P was added primarily as an organic source or inorganic source, as well as soil characteristics (Blake *et al.*, 2000). In theses studies, P use by plants was much more efficient if applied in balance with C.

Targeted use of animal manures facilitates plant and microbial uptake of P through a range of mechanisms. These include direct competition for adsorption sites by manure compounds, enhanced release of P from sparingly soluble pools through altered pH and soluble C addition, and enhanced microbial activity (Erich *et al.*, 2002; Laboski and Lamb, 2003). Where manure is utilized at sustainable, moderate levels and livestock are distributed extensively across the landscape, organic P sources appear to be inherently less vulnerable than inorganic fertilizer sources to loss from occlusion, erosion, or leaching (Powell *et al.*, 1999). While manure additions also contribute to N fertility and SOM pools, in the long term soil nitrogen status will depend in large part of the proportion of land devoted to symbiotically fixing plant species. Use of animal manures serves as a mechanism to recycle N and P back to cropping fields where forages were produced.

Likewise, use of sparingly soluble inorganic P inputs should be combined with strategies to link P solubilization with C flows. Application of sparingly soluble sources of P to crops (e.g., most legumes) that can assimilate P into biological pools is an efficient strategy to bypass desorption, precipitation, and occlusion of P (Oberson *et al.*, 1999). In degraded soils, additions of rock P may be needed to be combined with the use of shrubby short-lived mycorrhizal plants that have been shown to reduce erosion, build organic matter, and assimilate N and P into plant-accessible N and P pools. Two legumes species, pigeon pea and lupin, are notable for providing these multiple ecoservices and have also been shown to access sparingly soluble phosphorus pools. Interestingly, these crops are commonly integrated with nutrient-demanding crops in indigenous cropping systems. For example, pigeon pea is grown as an intercrop with maize in India, and lupin is grown as a rotational crop just before potato in the Andes. Use of legumes to transfer P from mineral forms to labile OM pools increases the amount of P that is actively cycling via biological processes and can contribute to increased P uptake by subsequent crops that may not have the ability to access sparingly soluble P. An example is provided in Box 6.2, where the amount of P recycled in crop residues increased nearly threefold when pigeon pea was intercropped with maize.

The consistent theme uniting all of these strategies is to evaluate the possible fates of various nutrient sources and to also consider how to link use of these sources with enhanced C cycling. Furthermore, the greatest potential for effective nutrient cycling is realized when soil amendments are combined with use of biological N fixation.

Biological N Fixation: A Key Source of Nitrogen

Effective management of biological N fixation is central to ecologically based nutrient management. The most familiar example of symbiotic nitrogen fixation is the close association between legumes and rhizobial bacteria (*Rhizobium, Mesorhizobium, Sinorhizobium,* and *Bradyrhizobium*), although associative and free-living diazotrophs are potentially important in several monocot crops.

Symbiotic N Fixation

Legumes can be incorporated into crop rotations either intercropped with nonlegumes or in sequential (relay) rotations. A disadvantage of relay cropping is that mineralization of N may not coincide with the subsequent crop N demand. Beneficial effects of relay cropping systems include the addition of organic matter and mineralization of N from residual legume biomass that can support the growth of subsequent, nonlegume crops. Grain legumes, such as soybeans, are typically grown as monocultures in rotation with nonlegume grain crops such as maize. Grain legumes are the most common type of legume in cropping systems because they provide essential

TABLE 6.3 Average and Upper Range of Biological N Fixation Contributions to Tropical Cropping Systems

Associated crop	Average N fixed (kg N/ha year)	Upper range of N fixed (kg N/ha year)
Rice-*Cyanobacteria*[a]	30	up to 80
Azolla-*Anabaena* in rice[c]	32	—
Sugarcane-*Acetobacter*[b]	—	up to 150
Grain legumes[c]	77	up to 200
Green manure legumes[c]	85	up to 300
Pasture legumes[c]	78	up to 250
Leguminous trees and shrubs[c]	150	up to 275

[a]From Roger and Ladha, 1992 (as cited in Reis 2000).
[b]From Boddey *et al.*, 1995.
[c]From Giller 2001. Legume nitrogen fixation values do not include belowground biomass and are, therefore, underestimates.

human and livestock protein sources in a form that is easily stored and transported. Grain legumes can fix up to 200 kg N/ha-year (Table 6.3). However, most of this N is exported off the farm in the protein-rich seeds, resulting in low or negative net soil N balance. Most estimates of N fixation, however, do not include root biomass, which can be 16–77% of total plant N (Table 6.4). Root biomass is difficult to measure; however, from the limited data available, it is clear that legume species can vary greatly in root to shoot ratios. Perennial species tend to have a higher root:shoot ratio than annual species (Antos and Halpern, 1997). This is generally supported by recent belowground N results (Table 6.4), where perennial legumes tend to have a

TABLE 6.4 Measured Legume Belowground N Biomass as a Percentage of Total Plant N

Legume	Primary use	BGN as % of total plant N	Source
Chick pea (*Cicer arietinum*)	grain	29	Turpin *et al.* (2002)
Fava bean (*Vicia faba*)	grain	25	Khan *et al.* (2003)
Fava bean (*Vicia faba*)	grain	17	Mayer *et al.* (2003)
Field pea (*Pisum sativum*)	grain	16	Mayer *et al.* (2003)
Jack bean (*Canavalia ensiformis*)	green manure/forage	39	Ramos *et al.* (2001)
Mucuna (*Mucuna aterrima*)	green manure	49	Ramos *et al.* (2001)

higher belowground N as percent of total plant N (average of 43%) than the annual grain legumes (average of 32%). Environmental conditions can also influence root biomass and root architecture. Generally, plant allocation to roots increases under drought conditions. If estimates of root biomass are included, grain legumes can provide modest positive N balances even with high grain N exports.

Intercropping systems incorporate legumes into agroecosystems by creating spatial rather than temporal separation between legumes and nonlegumes. Examples of an annual intercropping system include maize–pigeon pea mixtures (Snapp *et al.*, 2003). Legume intercrops can supply a slow but steady supply of N for the nonlegume crop. Intercrops can also compete for other nutrients and water. One of the major constraints to the adoption of legumes in cropping systems is the opportunity cost of taking land out of production in either space, as part of an intercrop, or in time as part of a legume relay cropping rotation. For this reason, successful adoptions are more likely when legumes serve multiple purposes of producing a net positive N balance while still producing consumable products or livestock forage. Pigeon pea is one such example of a green manure crop that produces a high-protein vegetable product while maintaining a positive N balance (Ghosh *et al.*, 2007).

In contrast with grain legumes, green manures are grown for the primary purpose of improving soil N status and are typically incorporated into the soil at a maximal stage of biomass production. Tropical green manures, such as *Canavalia, Crotalaria,* and *Mucuna*, commonly fix over 100 kg N/ha year, all of which is retained in the system, resulting in more positive N balances than grain legumes. Green manures as relay crops are more commonly used in temperate systems because of lower land pressures and because they can be grown during the colder winter months when crop production is not possible. In tropical systems, relay green manures are less common due to high land pressures, limited labor supply, the ability to produce crops year-round in some regions, or the lack of water to support green manure growth during the dry seasons between cropping seasons. Intercropping of green manure crops to supply N to a simultaneously growing cash crop have been adopted in some systems. The aquatic fern *Azolla* and its symbiotic association with the cyanobacteria *Anabaena* provide an example of a green manure that is used exclusively as an N source when intercropped in lowland rice systems. With 80–95% of *Azolla* N derived from fixation, rice–*Azolla* intercrops can fix approximately 30 kg N/ha (Eskew, 1987; Kikuchi *et al.*, 1984; Yoneyama *et al.*, 1987). Some constraints to more widespread adoption of *Azolla* are pest pressures, P limitation, and limited irrigation availability in some regions (Giller, 2001).

Farmers that have limited land, labor, and other resources are interested in "dual purpose" legumes, which have an intermediate phenology. That is, they provide a product, such as leaf, vegetable, or grain, while at the same time providing long-term benefits through residues that suppress weeds and build soil fertility (Fig. 6.14). There is a trade-off, as carbohydrates and nutrients invested in residues provides less resources for yield potential; thus, residue biomass is inversely related to harvest index across legume species (see Chapter 3, Fig. 3.12). Examples of dual purpose,

FIGURE 6.14 A Bolivian farmer shows off his fava bean crop. The previous potato crop failed due to unfavorable climatic conditions, leaving behind P from the manure application that is normally applied to potatoes but not to bean crops. As a result, the fava beans produced a very large biomass.

low harvest index legumes include long-duration pigeon pea, forage soybean, and mucuna. Such species provide returns to farmers in the short term—and thus the economic feasibility of adoption—while simultaneously contributing to ecosystem services.

Over the long term, dual-purpose plant types contribute to resilient cropping systems. This is both through the soil-building properties of high-quality residues, and the inherent ability of indeterminant growth types to recover from pest epidemics. Plant breeding efforts have historically focused on producing high-yield potential phenotypes. Examples include the development of new varieties of pigeon pea and cowpea that are extra early and extra short duration. These crops often incorporate high harvest index traits, having the unintended consequence of reducing biomass available for fodder, weed suppression, and soil fertility enrichment. Producing a wider range of dual-purpose genotypes with intermediate phenology and experimenting with intercrops of short- and long-duration crops are approaches that require careful consideration in the future.

Alley cropping involves the use of woody or shrub perennial legumes between "alleys" of nonlegume crops. Prunings from the legumes are used as livestock forage and/or added to the soil as an N source for the nonlegume. Inclusion of perennials in cropping systems provides important ecological benefits due to their extensive rooting systems that persist across multiple cropping seasons. Perennials can reduce soil erosion, access deeper soil pools of nutrients and water, provide a critical microbial habitat between annual cropping seasons, and increase SOM. *Leucaena* and *Gliricidia* are two common leguminous alley crop species. *Leucaena*, intercropped with sorghum, increased sorghum yields by 73% as compared to sorghum grown without N fertilizer, and yields were 43% greater than with a low-rate N fertilizer application (Ghosh *et al.*, 2007). Alley cropped legumes can fix between 200 and 300 kg N/ha-year (Giller, 2001). Some of the challenges in the adoption of alley cropping systems include the competition of the legume with the cash crop for moisture in dry years, the pruning labor required, and the use of land for a non-cash crop. Selection of species that have complimentary rooting systems with cash crops (i.e., a deep-rooted perennial legume cropped with a shallow-rooted annual) and species that grow at a manageable pace to supply N while not requiring excessive pruning inputs are important considerations in the selection of legume species for alley cropping.

Non-Symbiotic N Fixation

Associative and free-living diazotrophs are commonly found in the rhizosphere of gramineous species, such as rice, maize, sugar cane, and tropical pasture grasses. Field measurements of the contributions from associative and free-living diazotrophs reveal extreme variability with contributions of 0–80 kg N ha^{-1} to crop growth (Bremer *et al.*, 1995; Peoples *et al.*, 2001; Table 6.3). With the rapid development of improved molecular methods, we are only beginning to scratch the surface of identifying the variety of diazotrophs responsible for fixing N in soils, and we still do not have methods that provide accurate measures of how much N free-living and associative diazotrophs are fixing in different agroecosystems (Buckley *et al.*, 2007).

One of the most promising areas of current biological N fixation research is in understanding the ecology and importance of associative N fixation with key grain species such as rice, maize, and sorghum. Watanabe (1979) found that 80% of bacteria in rice roots are capable of fixing N. *Acetobacter*, a non-obligate diazotroph commonly found in sugar cane roots, can fix up to 150 kg N/ha/year (Boddey *et al.*, 1995). *Azospirillum* and *Herbaspirillum* are examples of diazotrophs commonly found associated with rice, sugar cane, maize, and sorghum roots. The non-specificity of diazotroph–plant interactions makes it difficult to intentionally manage in comparison to the more specific legume–rhizobial symbiosis.

Management practices that affect the availability of soil C should significantly impact the potential for BNF. For example, the retention of the C in straw from a wheat crop with a yield of 2 t ha^{-1} could theoretically fuel the production of 50–150 kg N ha^{-1} if utilized by diazotrophs to drive N fixation (Kennedy and Islam, 2001). In addition, crop selection and breeding can affect BNF potential because plant species differ greatly in the quantity and quality of root exudates produced.

Agroecosystem Scale Nutrient Use Efficiency (NUE)

A central theme of any fertility management regime is the idea of evaluating the efficiency of nutrient inputs. In our experience, understanding and promoting nutrient efficiency are the key concerns of resource-constrained farmers. They are much more important than determining the rate of nutrient application that will maximize agronomic return. This is because smallholder farmers with very limited assets need to optimize returns to their modest investments, rather than optimizing profitability per se. An efficiency approach is a different way to think about nutrient management compared to the majority of soil fertility management research and fertilizer recommendations developed around the globe, which focus on optimizing plant yields. When nutrient use efficiency is considered within industrial agriculture management regimes, it is usually measured as yield per nutrient input from fertilizer; i.e., kg maize/kg fertilizer N. In other words, the efficiency of a nutrient source is evaluated based on the estimated contribution to yield for a single growing season.

There are several drawbacks to this approach. First and foremost, the focus on the single process of plant assimilation of the nutrient input leaves out many processes that retain nutrients for crop use in subsequent years and are beneficial for long-term improvement of soil fertility. Furthermore, this metric is limited to a single growing season, so the fate of these fertilizer inputs over longer time frames is not factored into the assessments of NUE. You can see that reliance on this metric as an indicator of NUE leads to management decisions that are driven solely by consideration of immediate yield outcomes, while more complex outcomes such as longer term benefits to soil fertility or retention of nutrients in SOM do not factor into nutrient management strategies. An additional consequence is that organic amendments such as green manures or composts that contribute to building SOM are judged to be inefficient nutrient sources and therefore inferior to inorganic, soluble fertilizers. One consequence of the wide application of this single NUE metric to drive nutrient management decisions is that farmers find themselves on a "fertilizer treadmill" where their farming systems have become dependent on high inputs of soluble fertilizers simply to maintain acceptable yields (Drinkwater and Snapp, 2007).

A more comprehensive, ecologically based model for NUE assessment takes into account diverse nutrient fates over a longer time scale than a single growing season. From this holistic perspective nutrient use efficiency is defined in terms of the retention of nutrients within the agroecosystem, usually at either the field or farm scale, in conjunction with plant-production related outcomes. Therefore, we distinguish between crop-scale NUE and agroecosystem-scale NUE. Crop-scale NUE, or yield/fertilizer input, is certainly one useful measure to consider in the context of nutrient management decisions; however, use of this metric cannot support integrated management. Agroecosystem-scale NUE can be estimated using the simple input–output mass balance approach we discussed earlier. This requires information on rotation, fertility inputs, and crop yields for at least one rotation cycle. Clearly, there are many sources of error in these simple budgets; however, we have found them to be a useful starting point for developing strategies to improve nutrient management in a variety of agroecosystems. In the future, it may be possible to use natural isotopic ratios of $^{15}N/^{14}N$ as indicators of agroecosystem-scale NUE. While NUE is a useful concept, it should only be used as one of the many factors that contribute to the development of field-specific nutrient management planning.

Integrating Nutrient Management with Other Farming System Decisions

In addition to the processes that are directly linked to nutrient cycling, nutrient management practices have cascading effects on other agroecosystem processes, making it advantageous to integrate nutrient management planning with tillage, pest management, and marketing and livelihood goals. A farmer's perspective on the decision of how best to manage a fertilizer source use is illustrated by the "what to do with a goat's worth of proceeds" dilemma described in text Box 6.3. The question facing many smallholder farmers is how to optimize returns from the modest proceeds raised by selling one goat. Should this be invested in fertilizer, in improved seed, in hiring labor to carry out extra weeding, or in some combination of these strategies? Trade-offs need to be considered. Is it worthwhile to invest in fertilizer for parts of the farm where an extra weeding operation cannot be undertaken, due to labor or financial constraints? Integrated nutrient management occurs within the context of investment decisions such as these, which are made on a whole-farm basis. This further complicates farmer decision making, as an investment in fertilizer or compost at high rates in one field may preclude nutrient investment in other fields. An ongoing question is the extent to which returns

BOX 6.3 The Goat Dilemma: How Should Revenues from the Sale of a Goat Be Used?

A farmer sells her goat at the start of the planting season. Should she A) use the proceeds to buy fertilizer to apply at the recommended rate of 45 or more kg N per ha, which has been shown to be profitably applied to a maize crop? or B) use the proceeds to apply a moderate dose such as 17 kg N fertilizer per ha over a larger area? She also needs to consider if she can afford to apply fertilizer and hire extra labor to weed the crop intensively. Her decisions need to take into consideration the value of concentrating the fertilizer in fields where she usually obtains high yields, versus a strategy that includes application of the fertilizer to low yield potential fields that might help enhance the yield output from the entire farm.

can be enhanced through targeting fertilizer to the highest performing fields, or through spreading fertilizer throughout a farm to obtain the highest efficiency possible with low rates of fertilizer.

The interaction among these allocation decisions was studied using simulation modeling and on-farm research in southern Africa to evaluate combinations of weeding intensity and N fertilizer rates (Dimes *et al.*, 2001). In these systems, N was the limiting nutrient and, therefore, N fertilizer additions should have increased maize yields. However, yield increases were not achieved unless an extra weeding was carried out in fields receiving N fertilizer. For these site-specific management decisions, the most promising strategy is expected to vary, depending on the heterogeneity of resources across a farm and the background rate of fertility; e.g., what production is obtainable without fertilizer, based on a minimal investment in planting, weeding, and harvest. To illustrate how allocation of resources to fertilizer applications and labor for weeding interact with the inherent productivity at the farm scale, we have compared the impact of three different management scenarios on maize yields (Fig. 6.15). Scenarios of targeted and homogeneous applications are explored for a farm with two maize production fields that vary in yield potential, one being low (0.5 t/ha without fertilizer) and the other being high

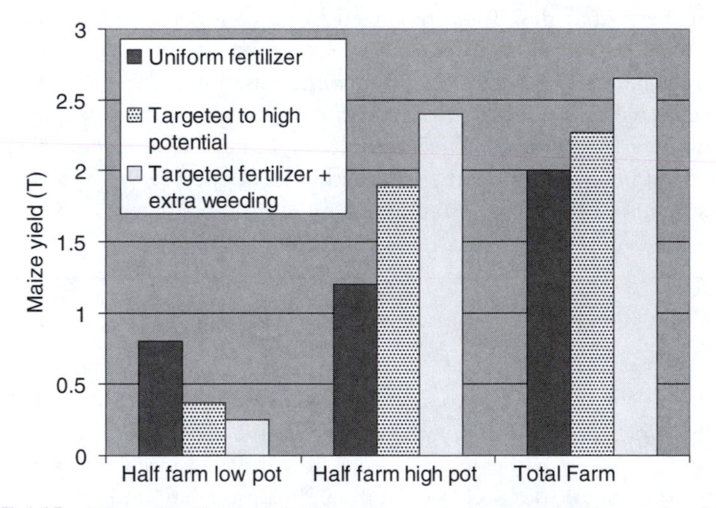

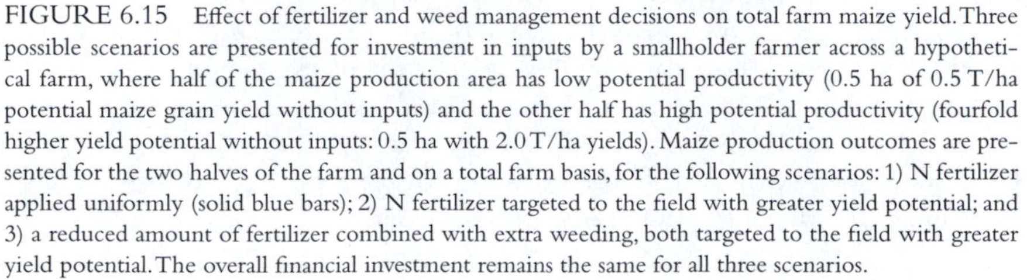

FIGURE 6.15 Effect of fertilizer and weed management decisions on total farm maize yield. Three possible scenarios are presented for investment in inputs by a smallholder farmer across a hypothetical farm, where half of the maize production area has low potential productivity (0.5 ha of 0.5 T/ha potential maize grain yield without inputs) and the other half has high potential productivity (fourfold higher yield potential without inputs: 0.5 ha with 2.0 T/ha yields). Maize production outcomes are presented for the two halves of the farm and on a total farm basis, for the following scenarios: 1) N fertilizer applied uniformly (solid blue bars); 2) N fertilizer targeted to the field with greater yield potential; and 3) a reduced amount of fertilizer combined with extra weeding, both targeted to the field with greater yield potential. The overall financial investment remains the same for all three scenarios.

(2.0 t/ha without fertilizer). In Scenario One, uniform application of a 25 kg of fertilizer per ha rate across the farm lead to the lowest yield potential overall, although the poor yield potential site had higher yield than in other scenarios. Scenario Two targeted a higher dose of fertilizer to the high yield potential site, combined with a lower rate at the low yield potential site, and had a significant positive effect on the overall production of maize grain from the farm. In Scenario Three, trading off some fertilizer for an extra weeding, which is again targeted to the higher potential site, produced the largest maize yield overall for the same level of investment across the farm. Notice that in this third scenario where fertilizer resources and weeding efforts are directed towards the more productive half of the farm, yields in the other half with poorer soils are exceedingly low.

The take-home message from this example is that trade-offs occur across a farm and the outcomes of management decisions will vary, depending on the particular situation on that farm. Yield from the low-potential fields on a smallholder farm may be at such a low level that the grain produced and response to input are minimal and are not able to significantly influence overall productivity of the farm. Abandoning part of the farm as a minimal investment site and intensifying production on higher potential sites may be a useful strategy in some circumstances. If input resources are limited, for example, farmers may not be in a position to apply all of the inputs that economic returns would justify. It is important to take into consideration the background level of fertility, the interactions of fertility and other inputs at different sites across the farm, and, overall, the response of staple grains to complementary investments over the short and long term, including weeding and SOM building practices.

DEVELOPING SITE-SPECIFIC ECOLOGICAL NUTRIENT MANAGEMENT SYSTEMS

Clarify Goals of Nutrient Management

A first step in managing the nutrient cycling to support agricultural goals is to identify nutrient management goals for your agroecosystem. What are the yield and fertility objectives? What is the relative balance between fertility and food or nutritional goals? Is there a perceived problem that needs to be addressed? Initially, the goals do not need to be prioritized or evaluated for whether or not they can be reasonably achieved. Refinement of goals will be easier after a concept map is developed.

Concept Map of Nutrient Flows

Drawing a conceptual diagram of nutrient flows, compartments, and processes regulating those flows similar to some of the diagrams we have used in this chapter can be a useful exercise. The use of conceptual models as communication and planning tools has proven to be a useful tool for facilitating communication and planning in groups with diverse perspectives (Heemskerk *et al.*, 2003). A conceptual model is a visual representation of the system to be studied. Conceptual models are particularly useful in planning interdisciplinary agricultural systems research because they require the team to graphically represent the problem to be addressed within a larger systems context. Ideally, to be useful as a planning tool, a conceptual model should:

- Describe a system that encompasses the research questions/management issues but has clear boundaries;
- Explicitly define the components of the system and how they interact with one another; i.e., it should identify the factors that directly or indirectly contribute to production, environmental outcomes, or nutrient flows;
- Provide a logical framework for the problems or questions to be addressed;
- Be simple enough to be understood by scientists from a variety of disciplines and stakeholders; and
- Be developed and agreed upon by all stakeholders and researchers.

Diagrams of agroecosystem nutrient flows can serve as a vehicle for achieving several outcomes that are prerequisites for successful implementation of ecologically based nutrient management. This process is important for:

1) *Facilitating information exchange*—Ensures that farmers and researchers have an agreed-upon understanding of nutrient management practices while also helping scientists to share information about important soil processes that control nutrient availability with farmers.

2) *Organizing a complex system*—By laying out the relationships among the interacting processes that are occurring at different spatial and temporal scales, trade-offs and linkages between management strategies become apparent.

3) *Moving the local nutrient cycling knowledge system forward*—The process of agreeing upon a diagram that represents diverse perspectives helps to identify knowledge gaps while also promoting the incorporation of innovations and new knowledge into the shared knowledge structure.

Resource Inventory

As part of the information gathering stage, it is important to define the agroecosystem characteristics that provide the backdrop for nutrient management decisions. These include:

1) *Background environment*—Soil types, soil fertility status, climate
2) *Cropping system characteristics*—Crops that are grown; rotation and proportion of land that is usually in each crop; relationship between crops, forage, and animal production; field sizes, locations; management intensity

3) *Nutrient input sources*—Identification of the sources of nutrients that are locally available and constraints that impact their use by farmers
4) *Relationship to other management practices*—How do other management issues such as weed control and tillage systems impact nutrient management?
5) *Fate of crops*—Importance of distinguishing between crops grown for family consumption and those aimed at markets; identity of markets; relative value of cash crops.

There are numerous resources available outlining methods that can be used in characterizing agroecosystems and in problem diagnosis (Gonsalves *et al.*, 2005).

Revisit and Refine Goals

With the above information in hand it will be possible to prioritize, evaluate trade-offs, and identify which goals are easily achievable. At this point a useful step might be to distinguish between long-term and short-term goals. If farmer-identified problems are the catalysts for this evaluation, then the range of possible solutions should be evaluated using the conceptual diagram and information that has been gathered.

Quick Assessment of Consequences of Current Nutrient Management Practices

Before moving forward to develop nutrient management strategies to achieve the goals (or solve the problems) that have been identified, prioritized, and analyzed, construction of simple input–output balances is a further step that can be used to analyze the current management. This approach has proven useful in pinpointing the most important weaknesses in nutrient management systems that are currently being used by farmers. In the United States, application of this tool has indicated that organic vegetable growers were overapplying compost, leading to environmentally unsound levels of soil P. In Andean systems this approach demonstrated that P management practices in fields closer to the community provided sufficient P and were compatible with increased use of legumes for N fixation, while fields that were farther from communities did not receive adequate P to benefit from legume intensification (Vanek, S., pers. comm., 2007). Further study of these systems revealed that potassium was being extracted at rates that far exceeded inputs, indicating that over the long term, potassium limitations may reduce yields. An additional example is the resource allocation map, specifically designed to track nutrient flows at the farm or community scale where transfers across fields, rangeland, and corrals are important. Readers should visit the Web site for this textbook for updates on tools that are being developed to facilitate the use of nutrient budgeting in developing management strategies.

BOX 6.4 Mapping Farm and Community Scale Nutrient Flows

Participatory research approaches have illustrated that farmer resource management can be improved through maps of agroecosystem nutrient resource flow, also called resource allocation maps (RAMS; Defoer 2002). Farmers and researchers together develop the maps and use them to record, monitor, and analyze data and decision making, which enhances understanding of soil fertility status, nutrient transfers, and degree of recycling associated with management options. Information gathered in this way is of value at different levels. This includes local participants who may be able to better assess where losses are potentially high on their farm, and thus where opportunities to recycle should be concentrated to improve overall nutrient efficiency. The RAMS approach illustrates the exciting potential of approaches that act as an interface between a "hard system" of knowledge (resource flow budgeting which can be used for modeling and comparisons with other systems) and a "soft system," integrating knowledge gained from collaborating with farmers and improved understanding of farmer perception of losses, gains, and transformations within and across a farm. Participatory research that integrates qualitative and quantitative approaches may provide new insights into designing sustainable agricultural systems that are not only efficient from a bio-engineering perspective but also relevant to real world farmers.

At a community or small watershed scale resource mapping is also being pursued as a means to enhance understanding and recycling of resources on a larger scale than the farm. In Nicaragua for example, participatory micro-watershed studies were initiated through community meetings of stakeholders, where resource mapping, transect analysis, and indicator-based assessment were used to evaluate current status and opportunities for improvement.

Livestock-crop integrated systems are ideal ways to concentrate and transfer nutrients, as animal manure is collected by corralling animals at night and during the day pasturing them over a wide area. A cow pastured on 4 hectares can provide sufficient nutrients to support half a hectare of nutrient-demanding crops such as maize. Thus livestock transforms widely spread, relatively unavailable nutrient sources from wild or semi-improved pastures or even urban streets and concentrates these nutrients as manure, which can be targeted to specific crops. Transhumerace, a nomadic livestock system that moves through field crop areas and trades residue grazing for transient manure deposition, was once one of the most common land use systems in the world.

Selecting and Testing Promising Nutrient Management Practices

Using this iterative process, a collaborative team decides what management strategies are worthy of testing and research. There is no single process that should be used in making these decisions; however, if a number of competing practices are identified, a simple method for comparing and contrasting these practices is to list the strengths and weaknesses of each. Also, the relationships between practices should be considered. Once you have agreements from farmers and other stakeholders about which practices they are most interested in, you can design research trials to evaluate and optimize these practices. This research should be conducted in farmers' fields as much as possible using experimental designs such as the mother-baby scheme (Snapp *et al.*, 2002). To succeed, research aimed at supporting ecological nutrient management must be conducting within a systems context and must apply participatory methodologies.

REFERENCES

Abawi, G. S., and Widmer, T. L. (2000). Impact of soil health management practices on soilborne pathogens, nematodes and root diseases of vegetable crops. *App. Soil Ecol.* **15,** 37–47.

Aber, J. D., Nadelhoffer, K. J., Steudler, P., and Melillo, J. M. (1989). Nitrogen saturation in northern forest ecosystems. *BioScience* **39,** 378–386.

Alves, B. J. R., Boddey, R. M., and Urquiaga, S. (2003). The success of BNF in soybean in Brazil. *Plant Soil* **252,** 1–9.

Ames, R. N., Reid, C. P. P., Porter, L. K., and Cambardella, C. (1983). Hyphal uptake and transport of Nitrogen from 2 N-15-labeled sources by Glomus-Mosseae, a vesicular arbuscular mycorrhizal fungus. *New Phytol.* **95,** 381–396.

Anderson, T. H., and Domsch, K. H. (1990). Application of eco-physiological quotients (qCO2 and qD) on microbial biomasses from soils of different cropping histories. *Soil Bio. Biochem.* **22,** 251–255.

Angers, D. A., and Mehuys, G. R. (1989). Effects of cropping on carbohydrate content and water-stable aggregation of a clay soil. *Can. J. Soil Sci.* **69,** 373–380.

Antos, J. A., and Halpern, C. B. (1997). Root system differences among species: Implications for early successional changes in forests of Western Oregon. *Am. Midl. Nat.* **138,** 97–108.

Aoyama, M., Angers, D. A., N'Dayegamiye, A., and Bissonnette, N. (2000). Metabolism of 13C-labeled glucose in aggregates from soils with manure application. *Soil Bio. Biochem.* **32,** 295–300.

Azam, F., Malik, K. A., and Sajjad, M. I. (1985). Transformations in soil and availability to plants of 15N applied as inorganic fertilizer and legume residues. *Plant Soil* **86,** 3–13.

Bah, A. R., Zaharah, A. R., and Hussin, A. (2006). Phosphorus uptake from green manures and phosphate fertilizers applied in an acid tropical soil. *Soil Sci. Plant Analysis* **37,** 2077–2093.

Bais, H. P., Weir, T. L., Perry, L. G., Gilroy, S., and Vivanco, J. M. (2006). The role of root exudates in rhizosphere interactions with plants and other organisms. *Annu. Rev. Plant. Bio.* **57,** 233–266.

Bringhurst, R. M., Cardon, Z. G., and Gage, D. J. (2001). Galactosides in the rhizosphere: Utilization by Sinorhizobium meliloti and development of a biosensor. *Proc. Natl. Acad. Sci. Unit. States Am.* **98,** 4540–4545.

Blake, L., Mercik, S., Koerschens, M., Moskal, S., Poulton, P. R., Goulding, K. W. T., Weigel, A., and Powlson, D. S. (2000). Phosphorus content in soil, uptake by plants and balance in three European long-term field experiments. *Nutrient Cycling Agroecosys.* **56,** 263–275.

Boddey, R. M., and Dobereiner, J. (1995). Nitrogen fixation associated with grasses and cereals: Recent progress and perspectives for the future. *Fert. Res.* **42,** 241–250.

Booth, M. S., Stark, J. M., and Rasstetter, E. (2005). Controls on gross nitrogen cycling rates in terrestrial ecosystems: A synthesis and analysis of the data from the literature. *Ecol. Monogr.* **75,** 139–157.

Bremer, E., Janzen, H. H., and Gilbertson, C. (1995). Evidence against associative N_2 fixation as a significant N source in long-term wheat plots. *Plant Soil* **175,** 13–19.

Buckley, D. H., Huangyutitham, V., Hsu, S. F., and Nelson, T. A. (2007). Stable isotope probing with $^{15}N_2$ reveals novel noncultivated diazotrophs in soil. *Appl. Environ. Microbiol.* **73,** 3196–3204.

Bundy, L. G., Andraski, T. W., and Powell, J. M. (2001). Management practices effects on phosphorus losses in runoff in corn production systems. *J. Environ. Qual.* **30,** 1822–1828.

Burger, M., and Jackson, L. E. (2003). Microbial immobilization of ammonium and nitrate in relation to ammonification and nitrification rates in organic and conventional cropping systems. *Soil Bio. Biochem.* **35,** 29–36.

Carberry, P. S., Probert, M. E., Dimes, J. P., Keating, B. A., and McCown, R. L. (2002). Role of modelling in improving nutrient efficiency in cropping systems. *Plant Soil* **245,** 193–203.

Cavagnaro, T. R., Jackson, L. E., Six, J., Ferris, H., Goyal, S., Asami, D., and Scow, K. M. (2006). Arbuscular mycorrhizas, microbial communities, nutrient availability, and soil aggregates in organic tomato production. *Plant Soil* **282,** 209–225.

Cavagnaro, T. R., Smith, F. A., Smith, S. E., and Jakobsen, I. (2005). Functional diversity in arbuscular mycorrhizas: Exploitation of soil patches with different phosphate enrichment differs among fungal species. *Plant Cell Environ.* **28,** 642–650.

Chen, J., and Ferris, H. (1999). The effects of nematode grazing on nitrogen mineralization during fungal decomposition of organic matter. *Soil Bio. Biochem.* **31,** 1265–1279.

Cheng, W., Zhang, Q., Coleman, D. C., Carroll, C. R., and Hoffman, C. A. (1996). Is available carbon limiting microbial respiration in the rhizosphere? *Soil Bio. Biochem.* **28,** 1283–1288.

Cheng, W. X., Johnson, D. W., and Fu, S. L. (2003). Rhizosphere effects on decomposition: Controls of plant species, phenology, and fertilization. *Soil Sci. Soc. Am. J.* **67,** 1418–1427.

Clapp, C. E., Allmaras, R. R., Layese, M. F., Linden, D. R., and Dowdy, R. H. (2000). Soil organic carbon and 13C abundance as related to tillage, crop residue, and nitrogen fertilization under continuous corn management in Minnesota. *Soil and Tillage Res.* **55,** 127–142.

Clarholm, M. (1985). Interactions of bacteria, protozoa and plants leading to mineralization of soil-nitrogen. *Soil Bio. Biochem.* **17,** 181–187.

Cross, A. F., and Schlesinger, W. H. (1995). A literature review and evaluation of the Hedley fractionation: Applications to the biogeochemical cycle of soil phosphorus in natural ecosystems. *Geoderma* **64,** 197–214.

Defoer, T. (2002). Learning about methodology development for integrated soil fertility management. *Agr. Syst.* **73,** 57–81.

Drew, E. A., Murray, R. S., Smith, S. E., and Jakobsen, I. (2003). Beyond the rhizosphere: Growth and function of arbuscular mycorrhizal external hyphae in sands of varying pore sizes. *Plant Soil* **251,** 105–114.

Drinkwater, L. E., and Snapp, S. (2007a). Understanding and managing the rhizosphere in agroecosystems. *In* "The Rhizosphere: An Ecological Perspective" (Z. G. Cardon and J. Whitbeck, eds.), pp. 127–154. Elsevier Press.

Drinkwater, L. E., and Snapp, S. S. (2007b). Nutrients in agroecosystems: Rethinking the management paradigm. *Adv. Agron.* **92,** 163–186.

Drinkwater, L. E., Wagoner, P., and Sarrantonio, M. (1998). Legume-based cropping systems have reduced carbon and nitrogen losses. *Nature London* **396,** 262–265.

Elser, J., and Urabe, J. (1999). The stoichiometry of consumer-driven nutrient recycling: Theory, observations, and consequences. *Ecology Washington DC* **80,** 735–751.

Erich, M. S., Fitzgerald, C. B., and Porter, G. A. (2002). The effect of organic amendments on phosphorus chemistry in a potato cropping system. *Agr. Ecosyst. Environ.* **88,** 79–88.

Eviner, V. T., and Chapin, F. S. (2001). Plant species provide vital ecosystem functions for sustainable agriculture, rangeland management and restoration. *Calif. Agr.* **55,** 54–59.

Ferris, H., Venette, R. C., Van, D. M. H. R., and Lau, S. S. (1998). Nitrogen mineralization by bacteria-feeding nematodes: Verification and measurement. *Plant Soil* **203,** 159–171.

Fierer, N., Schimel, J. P., Cates, R. G., and Zou, J. (2001). Influence of balsam poplar tannin fractions on carbon and nitrogen dynamics in Alaskan taiga floodplain soils. *Soil Bio. Biochem.* **33,** 1827–1839.

Fliessbach, A., and Mader, P. (2000). Microbial biomass and size-density fractions differ between soils of organic and conventional agricultural systems. *Soil Bio. Biochem.* **32,** 757–768.

Fliessbach, A., Mader, P., and Niggli, U. (2000). Mineralization and microbial assimilation of 14C-labeled straw in soils of organic and conventional agricultural systems. *Soil Bio. Biochem.* **32,** 1131–1139.

Foster, R. C., Rovira, A. D., and Cook, T. W. (1983). *Ultrastructure of the Root-Soil Interface,* p. 157. American Phytopathological Society, St. Paul, MN.

Gallandt, E. R., Liebman, M., and Huggins, D. R. (1999). Improving soil quality: Implications for weed management. *J. Crop. Prod.* **2,** 95–121.

Galloway, J. N., Aber, J. D., Erisman, J. W., Seitzinger, S. P., Howarth, R. W., Cowling, E. B., and Cosby, B. J. (2003). The nitrogen cascade. *BioScience* **53,** 341–356.

Gavito, M. E., and Olsson, P. A. (2003). Allocation of plant carbon to foraging and storage in arbuscular mycorrhizal fungi. *FEMS Microbiol. Ecol.* **45,** 181–187.

Ghosh, P. K., Bandyopadhyay, K. K., Wanjari, R. H., Manna, M. C., Misra, A. K., Mohanty, M., and Rao, A. S. (2007). Legume effect for enhancing productivity and nutrient use-efficiency in major cropping systems: An Indian perspective: A review. *J. Sustain Agr.* **30,** 59–86.

Giller, K. E. (2001). *Nitrogen Fixation in Tropical Cropping Systems.* CABI Publishing, New York, NY.

Godde, M., and Conrad, R. (2000). Influence of soil properties on the turnover of nitric oxide and nitrous oxide by nitrification and denitrification at constant temperature and moisture. *Biol. Fertil. Soils* **32,** 120–128.

Graham, P. H., and Vance, C. P. (2003). Legumes: Importance and constraints to greater use. *Plant Physiol.* **131,** 872–877.

Hamilton, E. W. III, and Frank, D. A. (2001). Can plants stimulate soil microbes and their own nutrient supply? Evidence from a grazing tolerant grass. *Ecology Washington DC* **82,** 2397–2402.

Harrison, K. A., Bol, R., and Bardgett, R. D. (2007). Preferences for different nitrogen forms by coexisting plant species and soil microbes. *Ecology* **88,** 989–999.

Haynes, R. J., and Beare, M. H. (1997). Influence of six crop species on aggregate stability and some labile organic matter fractions. *Soil Bio. Biochem.* **29,** 1647–1653.

Helal, H., and Dressler, A. (1989). Mobilization and turnover of soil phosphorus in the rhizosphere. *Zeitschrift für Pflanzenernährung und Bodenkunde* **152,** 175–180.

Hetrick, B. A. D. (1991). Mycorrhizas and root architecture. *Experientia* **47,** 355–362.

Hodge, A. (2004). The plastic plant: Root responses to heterogeneous supplies of nutrients. *New Phytol.* **162,** 9–24.

Hodge, A., Stewart, J., Robinson, D., Griffiths, B. S., and Fitter, A. H. (2000). Competition between roots and soil micro-organisms for nutrients from nitrogen-rich patches of varying complexity. *J. Ecol.* **88,** 150–164.

Holland, E. A., and Coleman, D. C. (1987). Litter placement effects on microbial and organic matter dynamics in an agroecosystem. *Ecology* **62,** 425–433.

Hooper, D. U., and Vitousek, P. M. (1997). The effects of plant composition and diversity on ecosystem processes. *Science* **277,** 1302–1305.

Hungria, M., and Vargas, M. A. T. (2000). Environmental factors affecting N-2 fixation in grain legumes in the tropics, with an emphasis on Brazil. *Field Crop. Res.* **65,** 151–164.

Illmer, P., Barbato, A., and Schinner, F. (1995). Solubilization of hardly-soluble AlPO-4 with P-solubilizing microorganisms. *Soil Bio. Biochem.* **27,** 265–270.

Jackson, L. E., Strauss, R. B., Firestone, M. K., and Bartolome, J. W. (1988). Plant and soil nitrogen dynamics in California annual grassland. *Plant Soil* **110,** 9–17.

Johnson, N. C., Graham, J. H., and Smith, F. A. (1997). Functioning of mycorrhizal associations along the mutualism-parasitism continuum. *New Phytol.* **135,** 575–586.

Kamh, M., Horst, W. J., Amer, F., Mostafa, H., and Maier, P. (1999). Mobilization of soil and fertilizer phosphate by cover crops. *Plant Soil* **211,** 19–27.

Kennedy, I. R., and Islam, N. (2001). The current and potential contribution of asymbiotic nitrogen fixation to nitrogen requirements on farms: A review. *Aust. J. Exp. Agr.* **41,** 447–457.

Kertesz, M. A., and Mirleau, P. (2004). The role of soil microbes in plant sulphur nutrition. *J. Exp. Bot.* **55,** 1939–1945.

Kirchner, M. J., Wollum, A. G. I., and King, L. D. (1993). Soil microbial populations and activities in reduced chemical input agroecosystems. *Soil Sci. Soc. Am. J.* **57,** 1289–1295.

Koch, B., Worm, J., Jensen, L. E., Hojberg, O., and Nybroe, O. (2001). Carbon limitation induces sigmaS-dependent gene expression in Pseudomonas fluorescens in soil. *Appl. Environ. Microbiol.* **67,** 3363–3370.

Kouno, K., Wu, J., and Brookes, P. C. (2002). Turnover of biomass C and P in soil following incorporation of glucose or ryegrass. *Soil Bio. Biochem.* **34,** 617–622.

Kramer, S. B., Reganold, J. P., Glover, J. D., Bohannan, B. J. M., and Mooney, H. A. (2006). Reduced nitrate leaching and enhanced denitrifier activity and efficiency in organically fertilized soils. *Proc. Natl. Acad. Sci. Unit. States Am.* **103,** 4522–4527.

Laboski, C. A. M., and Lamb, J. A. (2003). Changes in soil test phosphorus concentration after application of manure or fertilizer. *Soil Sci. Soc. Am. J.* **67,** 544–554.

Ladd, J. N., and Amato, M. (1986). The fate of nitrogen from legume and fertilizer sources in soils successively cropped with wheat under field conditions. *Soil Bio. Biochem.* **18,** 417–425.

Ladha, J. K., and Reddy, P. M. (2003). Nitrogen fixation in rice systems: State of knowledge and future prospects. *Plant Soil* **252,** 151–167.

Lewis, W. J., vanLenteren, J. C., Phatak, S. C., and Tumlinson, J. H. (1997). A total system approach to sustainable pest management. *Proc. Natl. Acad. Sci. Unit. States Am.* **94,** 12243–12248.

Liebhardt, W. C., Andrews, R. W., Culik, M. N., Harwood, R. R., Janke, R. R., Radke, J. K., and Rieger-Schwartz, S. L. (1989). Crop production during conversion from conventional to low-input methods. *Agron. J.* **81,** 150–159.

Liljeroth, E., Kuikman, P., and Van Veen, J. A. (1994). Carbon translocation to the rhizosphere of maize and wheat and influence on the turnover of native soil organic matter at different soil nitrogen levels. *Plant Soil* **161,** 233–240.

Lundquist, E. J., Jackson, L. E., Scow, K. M., and Hsu, C. (1999). Changes in microbial biomass and community composition, and soil carbon and nitrogen pools after incorporation of rye into three California agricultural soils. *Soil Bio. Biochem.* **31,** 221–236.

Lupwayi, N. Z., Haque, I., Saka, A. R., and Siaw, D. E. K. A. (1999). Leucaena hedgerow intercropping and cattle manure application in the Ethiopian highlands. III. Nutrient balance. *Biology and Fertility of Soils* **28,** 196–203.

Marschner, H., and Dell, B. (1994). Nutrient-uptake in mycorrhizal symbiosis. *Plant Soil* **159,** 89–102.

McCracken, D.V., Smith, M. S., Grove, J. H., MacKown, C. T., and Blevins, R. L. (1994). Nitrate leaching as influenced by cover cropping and nitrogen source. *Soil Sci. Soc. Am. J.* **58,** 1476–1483.

McGill, W. B., and Cole, C.V. (1981). Comparative aspects of cycling of organic carbon nitrogen sulfur and phosphorus through soil organic matter. *Geoderma* **26,** 267–286.

Menge, J. A. (1983). Utilization of vesicular-arbuscular mycorrhizal fungi in agriculture. *Can. J. Bot./ Revue Canadienne De Botanique* **61,** 1015–1024.

Neff, J. C., Townsend, A. R., Gleixner, G., Lehman, S. J., Turnbull, J., and Bowman, W. D. (2002). Variable effects of nitrogen additions on the stability and turnover of soil carbon. *Nature* **419,** 915–917.

Neumann, G., and Roemheld, V. (1999). Root excretion of carboxylic acids and protons in phosphorus-deficient plants. *Plant Soil* **211,** 121–130.

Nichols, P. D., Smith, G. A., Antworth, C. P., Hanson, R. S., and White, D. C. (1985). Phospholipid and lipopolysaccharide normal and hydroxy fatty-acids as potential signatures for methane-oxidizing bacteria. *FEMS Microbiol. Ecol.* **31,** 327–335.

Oberson, A., Friesen, D. K., Rao, I. M., Buhler, S., and Frossard, E. (2001). Phosphorus transformations in an oxisol under contrasting land-use systems: The role of the soil microbial biomass. *Plant Soil* **237,** 197–210.

Oberson, A., Friesen, D. K., Tiessen, H., Morel, C., and Stahel, W. (1999). Phosphorus status and cycling in native savanna and improved pastures on an acid low-P Colombian oxisol. *Nutrient Cycling Agroecosys.* **55,** 77–88.

O'Hara, G. W. (2001). Nutritional constraints on root nodule bacteria affecting symbiotic nitrogen fixation: A review. *Aust. J. Exp. Agr.* **41,** 417–433.

Ojiem, J. O., Vanlauwe, B., de Ridder, N., and Giller, K. E. (2007). Niche-based assessment of contributions of legumes to the nitrogen economy of Western Kenya smallholder farms. *Plant Soil* **292,** 119–135.

Paul, E. A., and Clark, F. E. (1996). *Soil Microbiology and Biochemistry*. Academic Press, San Diego, CA. p. 340.

Peoples, M. B., Giller, K. E., Herridge, D., and Vessey, J. (2001). Limitations to biological nitrogen fixation as a renewable source of nitrogen for agriculture. *In* "Nitrogen Fixation: Global Perspectives" (T. Finan, M. O'Brain, D. Layzell, J. Vessay, and W. Newton, eds.), p. 356–360. CABI Publishing, New York.

Powell, J. M., Ikpe, F. N., and Somda, Z. C. (1999). Crop yield and the fate of nitrogen and phosphorus following application of plant material and feces to soil. *Nutrient Cycling Agroecosys.* **54,** 215–226.

Puget, P., Chenu, C., and Balesdent, J. (2000). Dynamics of soil organic matter associated with particle-size fractions of water-stable aggregates. *Eur. J. Soil Sci.* **51,** 595–605.

Reilley, K. A., Banks, M. K., and Schwab, A. P. (1996). Organic chemicals in the environment: Dissipation of polycyclic aromatic hydrocarbons in the rhizosphere. *J. Environ. Qual.* **25,** 212–219.

Schimel, J-P., and Bennett, J. (2004). Nitrogen mineralization: Challenges of a changing paradigm. *Ecology Washington DC* **85,** 591–602.

Schlesinger, W. H. (2005). Biogeochemistry. Elsevier, Amsterdam; Boston. p. 702.

Schulten, H. R., and Hempfling, R. (1992). Influence of agricultural soil-management on humus composition and dynamics: Classical and modern analytical techniques. *Plant Soil* **142,** 259–271.

Shi, W., and Norton, J. M. (2000). Microbial control of nitrate concentrations in an agricultural soil treated with dairy waste compost or ammonium fertilizer. *Soil Bio. Biochem.* **32,** 1453–1457.

Siciliano, S. D., Germida, J. J., Banks, K., and Greer, C. W. (2003). Changes in microbial community composition and function during a polyaromatic hydrocarbon phytoremediation field trial. *Appl. Environ. Microbiol.* **69,** 483–489.

Silver, W. L., Herman, D. J., and Firestone, M. K. (2001). Dissimilatory nitrate reduction to ammonium in upland tropical forest soils. *Ecology* **82,** 2410–2416.

Snapp, S. S., Jones, R. B., Minja, E. M., Rusike, J., and Silim, S. N. (2003). Pigeon pea for Africa: A versatile vegetable—And more. *Hortscience* **38,** 1073–1079.

Snapp, S. S., and Silim, S. N. (2002). Farmer preferences and legume intensification for low nutrient environments. *Plant Soil* **245,** 181–192.

Snapp, S., Kanyama-Phiri, G., Kamanga, B., Gilbert, R., and Wellard, K. (2002). Farmer and researcher partnerships in Malawi: Developing soil fertility technologies for the near-term and far-term. *Exp. Agr.* **38,** 411–431.

Tan, K. H. (2003). *Humic Matter in Soil and the Environment: Principles and Controversies*. Marcel Dekker, New York. p. 386.

Tanaka, Y., and Yano, K. (2005). Nitrogen delivery to maize via mycorrhizal hyphae depends on the form of N supplied. *Plant Cell Environ.* **28,** 1247–1254.

Tate III, R. L., Parmelee, R. W., Ehrenfeld, J. G., and O'Reilly, L. (1991). Nitrogen mineralization: Root and microbial interactions in pitch pine microcosms. *Soil Sci. Soc. Am. J.* **55,** 1004–1008.

Tibbett, M. (2000). Roots, foraging and the exploitation of soil nutrient patches: The role of mycorrhizal symbiosis. *Funct. Ecol.* **14,** 397–399.

Tisdall, J. M., and Oades, J. M. (1979). Stabilization of soil aggregates by the root systems of rye grass Lolium-Perenne. *Aust. J. Soil Res.* **17,** 429–441.

Treseder, K. K. (2004). A meta-analysis of mycorrhizal responses to nitrogen, phosphorus, and atmospheric CO_2 in field studies. *New Phytol.* **164,** 347–355.

Vance, C. P., Uhde-Stone, C., and Allan, D. L. (2003). Phosphorus acquisition and use: Critical adaptations by plants for securing a nonrenewable resource. *New Phytol.* **157,** 423–447.

Vitousek, P. M., Aber, J. D., Howarth, R. W., Likens, G. E., Matson, P. A., Schindler, D. W., Schlesinger, W. H., and Tilman, D. G. (1997). Human alteration of the global nitrogen cycle: Sources and consequences. *Ecol. Appl.* **7,** 737–750.

von Lutzow, M., Kogel-Knabner, I., Ekschmitt, K., Matzner, E., Guggenberger, G., Marschner, B., and Flessa, H. (2006). Stabilization of organic matter in temperate soils: Mechanisms and their relevance under different soil conditions: A review. *Eur. J. Soil Sci.* **57,** 426–445.

Vong, P-C., Dedourge, O., Lasserre-Joulin, F., and Guckert, A. (2003). Immobilized-S, microbial biomass-S and soil arylsulfatase activity in the rhizosphere soil of rape and barley as affected by labile substrate C and N additions. *Soil Bio. Biochem.* **35,** 1651–1661.

Walley, F., Fu, G. M., van Groenigen, J. W., and van Kessel, C. (2001). Short-range spatial variability of nitrogen fixation by field-grown chickpea. *Soil Sci. Soc. Am. J.* **65,** 1717–1722.

Wander, M. M., and Traina, S. J. (1996). Organic matter fractions from organically and conventionally managed soils: II. Characterization of composition. *Soil Sci. Soc. Am. J.* **60,** 1087–1094.

Wander, M. M., Traina, S. J., Stinner, B. R., and Peters, S. E. (1994). Organic and conventional management effects on biologically active soil organic matter pools. *Soil Sci. Soc. Am. J.* **58,** 1130–1139.

Watanabe, I., Barraquio, W. L., Deguzman, M. R., and Cabrera, D. A. (1979). Nitrogen-fixing (acetylene-reduction) activity and population of aerobic heterotrophic nitrogen-fixing bacteria associated with wetland rice. *Appl. Environ. Microbiol.* **37,** 813–819.

Wedin, D. A., and Tilman, D. (1990). Species effects on nitrogencycling a test with perennial grasses. *Oecologia* **84,** 433–441.

Whitelaw, M. A., Harden, T. J., and Helyar, K. R. (1999). Phosphate solubilisation in solution culture by the soil fungus Penicillium radicum. *Soil Bio. Biochem.* **31,** 655–665.

Whitman, W. B., Coleman, D. C., and Wiebe, W. J. (1998). Prokaryotes: The unseen majority. *Proc. Natl. Acad. Sci. Unit. States Am.* **95,** 6578–6583.

Yin, S-X., Chen, D., Chen, L-M., and Edis, R. (2002). Dissimilatory nitrate reduction to ammonium and responsible microorganisms in two Chinese and Australian paddy soils. *Soil Bio. Biochem.* **34,** 1131–1137.

Yoneyama, T., Ladha, J. K., and Watanabe, I. (1987). Nodule bacteroids and Anabaena: Natural N-15 enrichment in the legume-Rhizobium and Azolla-Anabaena symbiotic systems. *J. Plant Physiol.* **127,** 251–259.

CHAPTER 7

Participatory Plant Breeding: Developing Improved and Relevant Crop Varieties with Farmers

Eva Weltzien and Anja Christinck

Agricultural Systems: Agroecology and Rural Innovation for Development

Summary

This chapter exposes the concept of participatory plant breeding (PPB) and relates it to the theory of plant breeding. Essential elements of PPB are new forms of cooperation between farmers and researchers, a detailed assessment of agroecological conditions and farmers' needs, the use of local germplasm, and a decentralized organization of selection, variety testing, and seed production. This chapter outlines how PPB can practically be organized in the different stages of a plant breeding program, starting from setting objectives, generating or assembling variability, selecting, variety testing and evaluation, and ending with production and diffusion of seed. The last section describes the types of impacts that can be achieved through PPB: food security is improved through rational use of resources and better adaptation of varieties to environmental conditions, and farmers are being provided with new options while eco-friendly traditional practices are strengthened and maintained.

INTRODUCTION

Wild and domesticated plants and animals are part of agroecological systems; their genetic and phenotypic properties are intimately related with the natural environment. However, plants and animals have a special position within the agroecological system; directly or indirectly, they serve the livelihood needs of people. Thus, since the very beginning of agriculture, people have tried to alter plants and animals in such a way that they are better adapted to their felt needs. Adapting plants and animals to human needs could be described as the most general goal of breeding.

Until quite recently in our history, breeding was done only by farmers. In many parts of the world, farmer-selected and farmer-produced seed continues to be the primary source of seed. Breeding activities of farmers usually form part of their general agricultural activities and include practices such as mixing, exchanging, selecting, and storing seed. Selection by farmers is usually based on their observation and understanding of environmental adaptation and quality, and is thus closely related to local knowledge and cultural traditions. Since plant breeding emerged as a scientific discipline, a new system of variety development, testing, and release was established. This "formal" system coexists with the "informal" farmer system of crop management and enhancement. Most farmers in the world currently use both formal and informal systems for sourcing seed of crop varieties. However, many local crops, even if economically important, as well as locally adapted landraces of staple food crops, continue to be mainly available from the informal system.

The rich diversity of adapted crops and varieties for specific agroecological "niches" has allowed people to settle and survive in diverse environments, and even under harsh climatic conditions, on steep slopes, poor soils, or under conditions such as recurrent droughts, floods, or storms. Many such farming systems are presently undergoing rapid change, also resulting in a demand for new crop varieties. For example, the sizes of landholdings are continuously declining in some areas, resulting in a reduction of fallow periods and a need for intensification. Furthermore, new options may emerge for marketing, processing, innovative products, fair trade, and so on, which also require new varieties. However, many traditional crops or crop varieties that were ideally adapted to certain farming practices and site-specific conditions tend to disappear because of technological change, economic pressure, changed food habits, or loss of traditional knowledge. Plant breeding could make an important contribution to safeguarding such varieties by readapting them to present conditions, technologies, and needs.

Formal plant breeding programs have clearly made major contributions to cropping system productivity around the world. However, the farmers' adoption of varieties from the formal system remains limited under certain conditions, particularly under marginal agroecological conditions or if access to resources is limited. Under such circumstances, "high-yielding" varieties may not be superior to traditional varieties. Thus, awareness is rising that plant breeding is not a user-neutral technology: different groups of farmers may need different types of varieties. However, even in better-off regions, there is rising awareness that formal plant breeding does not always address the farmers' preferences and needs and that the diversity and stability of farming systems could be better supported by applying different breeding strategies.

In recent years the application of concepts from participatory research and development to plant breeding has evolved rapidly. This has opened new options for project and program design, especially in view of addressing a wider range of development goals. Many international and national plant breeding institutions oriented their programs toward U.N. Millenium Development Goals and goals of the International Treaty for Plant Genetic Resources. Thus, plant breeding has been targeted to the needs of specific users, particularly poor farmers, and also addresses the conservation of agrobiodiversity through the sustainable use of crop genetic resources. One example is the adoption of the "Integrated Genetic and Natural Resource Management" (IGNRM) framework by ICRISAT and some of its partner institutions. IGNRM aims at developing agricultural technologies that can target specific, diverse agricultural systems and strengthen or empower users, with special focus on improving food security and livelihoods of poor farmers (http://www.icrisat.org/Vision/chapter3.htm#CI).

THEORY OF PARTICIPATORY PLANT BREEDING

Definitions and Terminology

Participatory plant breeding (PPB) includes all approaches of close farmer–researcher collaboration to bring about plant genetic improvement within a species. The basic idea is that farmers and researchers have different knowledge and practical skills, as well as different approaches to problem diagnosis and solving (Weltzien *et al.*, 2003). The strengths and weaknesses of both groups tend to be complementary so that better research results can be achieved through cooperation.

All the different phases or stages of a plant breeding program are concerned, and options for farmer participation exist for all of them: setting objectives, creating variability, and selecting experimental varieties and testing them, as well as producing and diffusing seed of new varieties (Fig. 7.1).

Collaboration between farmers and scientists can take many forms, and roles and responsibilities can be shared in many diverse ways. Some researchers have tried

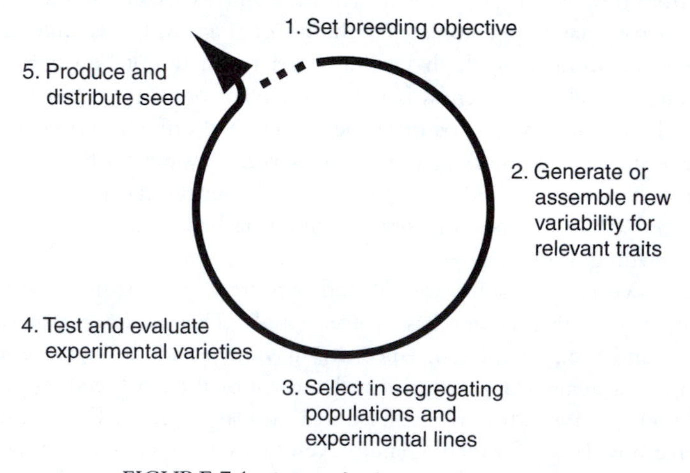

FIGURE 7.1 Stages of a plant breeding program.

to classify PPB approaches according to the form of collaboration or the locus of decision making (Farnworth and Jiggins, 2003; Lilja and Ashby, 1999). In any PPB program, farmers contribute knowledge and information to the joint program, in some cases also genetic material. For example, farmers can contribute their own check or control varieties to trials, and farmer varieties are also used as breeding parents in crossing programs. In addition, farmers may be directly involved in the breeding process by conducting and managing trials on their own land and making selection decisions in various ways. Thus, in addition to knowledge and genetic material, other major contributions of farmers in PPB programs are labor and practical skills regarding the evaluation and selection of test entries.

As with any developing field of research, the terminology for participatory plant breeding has not been fully standardized and is used differently by different groups of researchers. Some of the more commonly used terms are defined in the following paragraphs to assist the reader with interpretation of the growing PPB literature.

As in this chapter, *participatory plant breeding* is used as an overarching term that includes all approaches to plant breeding with close collaboration between farmers and researchers (Weltzien *et al.*, 2003).

However, some authors focus on the stage of the breeding program in which the collaboration takes place and on the status of the germplasm under consideration. In this context, one of the most commonly used terms is *participatory variety selection* (PVS). It is used to describe farmer participation in the process of evaluating finished, stable varieties. Accordingly, the term *participatory plant breeding* is used only when the project involves farmers in the earlier phase of variety development, that is, making crosses or selections in the early (segregating) generations. It is thus important to verify in which way the term PPB is being used in specific publications. Publications that use PPB in a "narrow" way tend to use *farmer participatory crop improvement* or *collaborative plant breeding* as overarching terms, which include then both PVS and PPB (Cleveland and Soleri, 2002; Witcombe *et al.*, 1996). These terms are, however, used only rarely.

The term *decentralized plant breeding* puts emphasis on the importance of selection in the target environment, that is, farmers' fields, based on considerations regarding the interaction between plant genotypes and the environment. This approach may imply farmer participation in the selection and diffusion of varieties (http://www.idrc.ca/fr/ev-85068-201-1-DO_TOPIC.html).

The term *client-oriented plant breeding* has been proposed as an overarching term, with the aim to avoid an artificial dichotomy between "participatory" and "nonparticipatory" breeding approaches (Witcombe *et al.*, 2005). The essential strength of participatory methods is seen in improving the client orientation of formal breeding programs, with productivity gains and research efficiency as the main goals.

In view of this rather confusing terminology, this chapter uses PPB in its most generalized meaning and attempts to describe the range of goals pursued by PPB programs and the various ways for achieving them.

Typical Elements of PPB Approaches

Even though various groups of researchers emphasize different aspects and potentials of PPB, all the approaches developed under the aforementioned terms have some essentials in common:

- New forms of cooperation
- Focus on detailed assessment of agroecological conditions and farmers' needs
- Use of local germplasm
- Decentralized organization and selection in target environments
- Innovative strategies for seed production and distribution.

In addition, there may be other elements that are important for some but not all PPB projects, such as strengthening indigenous knowledge and culture, conserving traditional crop varieties, or empowerment of farmers.

We will find certain theoretical considerations behind these elements, which have led various groups of researchers to depart from the ways that things used to be done "normally" in formal plant breeding programs. These considerations are outlined next.

New Forms of Cooperation and the Process of Setting Priorities in PPB Programs

A variety of goals of very different natures tend to be addressed by PPB programs, some of which may be quite general goals, which cannot be met by plant breeding alone, such as poverty alleviation or empowerment of farmers. However, PPB could be an important building block for addressing such goals, particularly if it would become part of a more far-reaching development strategy. In such a setting, PPB cannot be planned by one institution alone; very typically, PPB projects rely on various partners, including national or international research institutions, farmer organizations, nongovernmental organizations (NGOs), and local state authorities. In some cases, the private sector is also involved, for example, if industrial food processing and marketing are part of the followed strategy.

The process of setting priorities is extremely important for any plant breeding program, but even more for a PPB program: a shared vision about the goals needs to be achieved among all partners involved (Fig. 7.2). It is, therefore, important that discussions about the goals are held regularly to assure that the goals remain relevant and that they are evident and important to all partners involved in the program. These identified goals are then the guiding principles for priority setting and should be formulated in a way that facilitates regular adjustments and refinements as the program and the partnership evolve. Furthermore, indicators that could help monitor the progress should be identified, and a process of monitoring and evaluation installed (Germann *et al.*, 1996).

The priority setting process requires detailed information on a variety of key issues (see Box 7.1). This information is seldom "available" before the project starts,

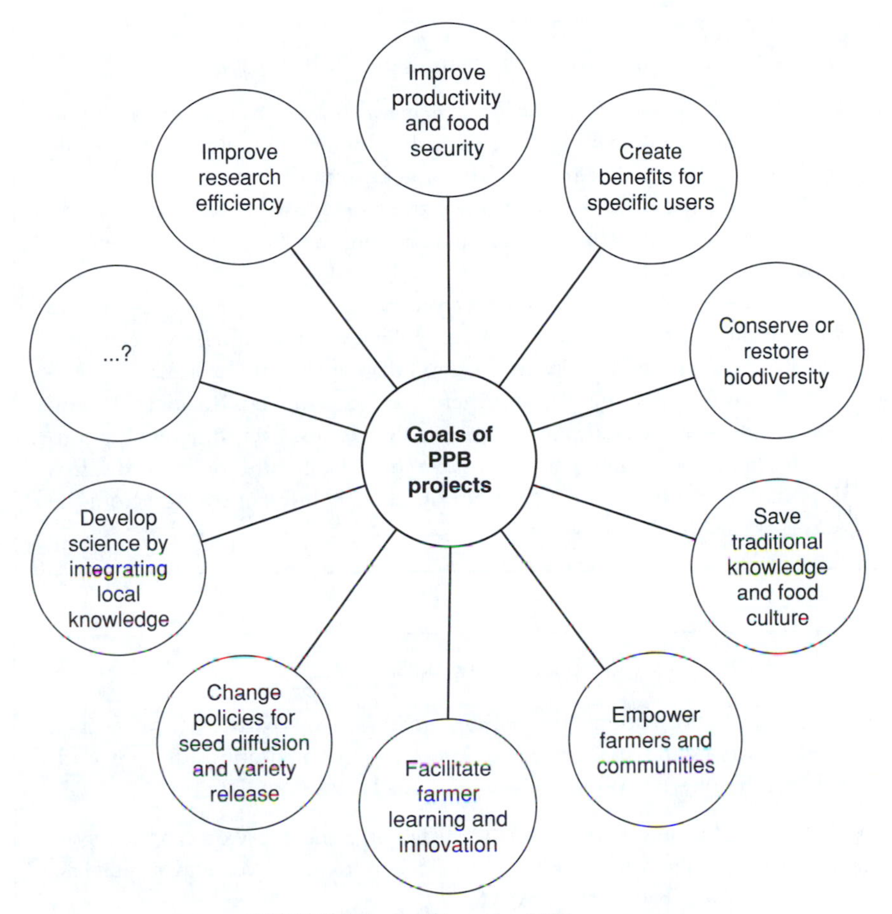

FIGURE 7.2 Possible goals of PPB programs.

and studies should be designed in such a way that the points of view of each partner group (particularly the farmers) will possibly be expressed and documented for the further planning process. Experience gained in a number of PPB projects has shown that participatory communication tools, such as semistructured or informal interviews, focus group discussions, wealth ranking, transect walks, time lines, mapping, classification, and ranking exercises, can be extremely useful for reaching a good base for further planning. The particular strength of such communication tools is that they facilitate direct dialogue between farmers and researchers and can help develop a common understanding of the situation, as well as of main constraints and needs. Practical guidelines for conducting such a situation analysis, particularly for plant breeding projects, have been suggested by Christinck *et al.* (2005c). Furthermore, many inspirations can be extracted from general guides and publications on participatory research (see Box 7.2).

BOX 7.1 Key Issues for Priority Setting in a PPB Program

1. Define the *target group and the environment*, that is, production conditions in which the newly identified varieties should perform better than existing cultivars, and the specific needs of the target group of farmers.

2. Closely linked to this are priority traits to be used as *selection criteria*.

3. To achieve good progress from selection, the *germplasm base* must be chosen appropriately.

4. It is also important to discuss what *type of variety* might be the most appropriate for achieving the goals, for example, open pollinated, rather diverse varieties may better achieve goals of diversity conservation than single-cross hybrids.

5. An issue that is often left until activities are planned is the identification of key *roles and responsibilities of partners*. Since, however, different options for sharing responsibilities have a major impact on some of the goals, it is important to consider roles and responsibilities of different partners from the outset of the breeding program.

BOX 7.2 Participatory Research Methods

Sources of information and training materials on participatory research methods are listed here. We concentrate on those publications that are available via the Internet, often for free download.

1. The Web sites of the FAO (www.fao.org) and the World Bank (www.worldbank.org) contain sections on publications for download and/or purchase (search for participation or PRA).

2. Further publications may be found via the online bookshop of the United Nations Environment Program (www.earthprint.com) in the section on participation and training.

3. Several guides can be downloaded from the GTZ homepage (some available in various languages, e.g., Spanish/French) (www2.gtz.de/participation/English/tools.htm).

4. Participatory Learning and Action (Methods and Approaches) is a series accessible through the IIED homepage (International Institute for Environment and Development, London, UK) (http://www.iied.org/sarl/planotes/index.html).

5. The Program for Participatory research and Gender Analysis has a Web site with a series of publications and resources, including a listing of cases for participatory plant breeding (www.prgaprogram.org).

6. The Reading University maintains a Web site with training materials and resources focusing on the analysis of participatory research results (www.reading.ac.uk/ssc/workareas/development/dfid-gga.html).

Detailed Assessment of Agroecological Conditions and Farmers' Needs

In general, it is impossible to successfully develop a highly specialized and adapted technology if the conditions under which it is going to be operated are only vaguely known. This was the situation of many formal breeding programs in developing countries; often, it was simply presupposed that farmers would need a particular variety type or that increasing the yield potential of certain major crops would per se be attractive for farmers; however, low adoption rates and no or insignificant yield increases under marginal conditions teach us that these assumptions were not generally true.

Under marginal conditions, farming is usually part of a complex livelihood strategy. However, it may interact with many other activities pursued by the members of a farm household, including animal husbandry, handicrafts, food processing and marketing, labor work, and seasonal migration. Therefore, understanding the general production goals and the importance of farming and certain crops within the people's livelihood strategy, as well as cultivation practices, uses, and the main constraints to yield increase or income generation, will be general preconditions for successfully developing plant varieties that meet the requirements of the farmers.

A crop can fulfill many different functions in the farming system. Farmers often use different products from one plant, and the value of these "by-products" can exceed that of the main product or be of great use in certain situations. The importance of multipurpose uses can vary largely from crop to crop and for different groups of farmers. Figure 7.3 summarizes the various functions of a crop for rural

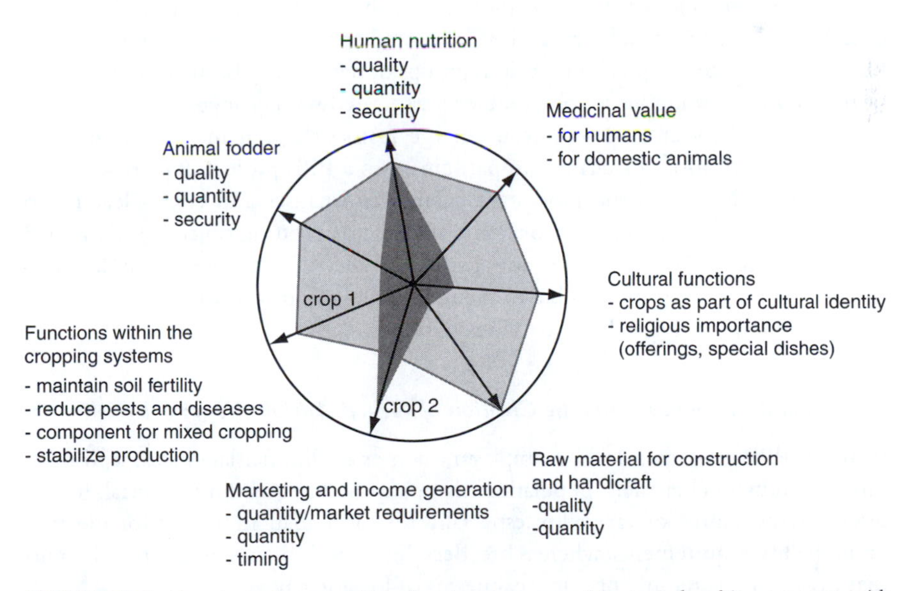

FIGURE 7.3 Functions of crops within a farming system; crop 1 is a typical multipurpose crop with high importance for most functions (except marketing), whereas crop 2 is a food crop important for nutrition and marketing, but not for other functions mentioned.

people's livelihoods and shows how different the situation can be for different crops or crop varieties.

However, a situation analysis should also focus on the typical constraints of the system, in view of both agroecological and socioeconomic considerations. This could also be an opportunity to recheck whether PPB is really the solution to the constraints and problems identified together with the farmers before starting the actual breeding work. Other options, such as reintroduction of landraces, improved seed production, or market development and training, should be weighed and considered carefully as alternatives or complementary options. Quite often, PPB may be only a part of the solution; strategic partnerships with NGOs or private enterprises, starting from the early stages of the project, can help increase the impact and sustainability of the PPB work.

Farming systems in many parts of the world are presently undergoing processes of gradual or rapid change. It is usually complex in nature because agroecological, sociocultural, political, and economic factors are combined, causing very specific sets of conditions at the local level.

Change can be an important motivation for farmers to search for new varieties. Generally, we should be aware that farming systems have not been static in the past and that adapting to variable conditions and allocating resources accordingly are key capacities of farmer families all over the world. However, the ability to adapt to change depends much on the natural, technological, economic, and human/social resources that are available to individuals or groups of people and whether these resources are useful to tackle the new situation. Change can be a slow and constant process that develops over decades, or it can be associated with catastrophic events, such as wars, economic crises, floods, and earthquakes; accordingly, people may or may not have useful resources at their disposal, and various groups of people may be affected differently or need different solutions to solve problems associated with change.

Interest in new crop varieties can emerge from such dynamics, thus being an important motivation for farmers to participate in a PPB project. That is why taking a deeper look into the process of change and causes and effects leading to change on the local level is a major point of interest in setting objectives for a PPB program (Table 7.1). Even if it is not always possible to anticipate future developments, it could at least be possible to identify major trends and their effects on the farming systems and people's livelihood strategies.

Use of Local Germplasm and the Creation of Genetic Variability for Selection

In many PPB programs, local germplasm, such as traditional landraces, seed mixtures, or individual farmers' populations, is being used as breeding material. It represents a key source of variability, especially for local adaptation, and for meeting grain quality requirements, whereas breeders' lines tend to contribute specific traits that may help overcome specific weaknesses of local germplasm.

Choosing and creating genetic variability for selection are of key importance in plant breeding programs. All the desired traits need to be present in the breeding

TABLE 7.1 Changes of the Production Environment and Possible Effects on Cropping System and Variety Use

Type of change	Possible effect on cropping system and variety use
Agroecological Conditions	
Reduced soil fertility due to erosion, reduced fallow periods, lower number of farm animals (manure)	Lower yield, increased intensity or greater variety of pests, diseases, or parasitic weeds
Amount of rainfall reduced or increased, onset of rainy season earlier or delayed, different rainfall distribution patterns	Reduced yield stability, higher risk of crop failures, higher or different pest and disease incidence, shift to other crops
Increased temperatures	Crops negatively affected by drought and high temperature
Higher frequency of adverse conditions, such as frost, thunderstorms, sandstorms etc.	Higher risk of crop failure (due to damage)
Newly introduced pests and diseases	Lower yield, risk of crop failure
Access to irrigation or changes in the quality/availability of irrigation water	Shift to crops or varieties with higher yield potential or specific adaptation (to water lodging, salinity, etc.)
Socioeconomic Conditions	
New marketing opportunities due to food processing factories, new infrastructure, export etc.	Crops/varieties with special characteristics required (to meet market demand)
Introduction of new farming technology (such as animal traction, tractor plowing, harvesting machines, etc.)	Need for adapted crops/varieties
Reduced availability of farm labor due to other economic activities	Crops/varieties which require less labor are preferred
Access to agricultural inputs (such as seed, fertilizer, pesticides)	New options to grow crops/varieties with higher yield potential
Failure of agricultural input supply (due to crises, wars, or disasters)	Low input varieties required
Culture and Knowledge	
Erosion of traditional knowledge and skills	Traditional varieties are abandoned or may need to be adapted to present knowledge and technologies
Different food preferences	Traditional crops/varieties are less preferred (or more preferred) than before

material, and the genetic variation with regard to important traits determines the level of improvement that can be achieved during the selection phase. Identifying and creating variability for selection can involve direct selection in suitable landrace populations, or identifying appropriate parents for crossing or creating new base populations. Various possibilities exist to ensure that important adaptive traits will be present in appropriate frequency and variability in the base populations (see Box 7.3).

BOX 7.3 Possibilities for Improving Chances of Success by Creating Variability

- All traits required for a successful variety need to be present: good local adaptation, grain quality for primary uses, and resistances to common pests and diseases
- Genetic diversity for the traits under improvement to assure rapid progress from selection
- Provide for ample recombination between different parents used in crossing, for example, large populations for biparental crosses, several random matings, and large population sizes while building base populations
- Use large parent population sizes when creating new population crosses or new bulks to avoid genetic drift and inbreeding
- Conduct evaluation of parents and base populations under target conditions to avoid loss of key adaptation traits
- Increase frequency of well-adapted genotypes in the base population(s)

In cross-pollinated crops (such as maize or pearl millet), crosses between different varieties and populations occur naturally, if the parental material grows in close vicinity and flowers at the same time. Farmers can thus easily produce crosses between local and exotic germplasm by growing them in the same field, possibly as a mixture. In situations where it is a normal practice of farmers to select seed, providing interesting, useful new germplasm for crossing may be a key input from researchers who wish to strengthen farmers' capacity to create their own new varieties.

Farmers can also learn to make targeted crosses, both in cross-pollinating and self-pollinating crops, and some farmer-managed PPB projects have asked researchers for this type of training. However, in most PPB projects, targeted crossing has been done by professional plant breeders because of efficiency considerations.

Decentralized Organization and Selection in Target Environments

Selection in Early Generations

The most general aim of selecting early generations of breeding populations in the target environment is to ensure adaptation to specific (often marginal) agroecological conditions, which cannot be "simulated" in research stations. These agroecological conditions include, in addition to location-specific factors (such as soil or water quality, climate), management practices that may affect the performance of a variety, for example, mixed cropping or local practices for soil preparation, sowing, weeding, or harvesting.

The issue that certain plant types or varieties may perform differently in different environments is called "genotype by environment interaction" by plant breeders. In general, most plant breeders tend to give preference to those populations that perform well under a wide range of conditions; "broad adaptation" is an important feature of new varieties, particularly with regard to breeding economics (size of the potential market). However, these varieties may fail under the conditions of poor farmers working with limited resources and under marginal agroclimatic conditions. Ceccarelli *et al.* (1996, 2000) have shown theoretically and practically that interactions between genotype and environment can be exploited positively if the selection is done in the target environment, for example, farmers' fields. Furthermore, narrow adaptation to specific site conditions, leading to the selection of many different cultivars for various conditions and purposes, is often regarded as an advantage of the PPB approach: it enhances the level of agrobiodiversity in farmers' fields (Joshi and Witcombe, 2001; Sperling *et al.*, 1993).

The involvement of farmers in selecting among early generation breeding material has various other advantages beyond ensuring environmental adaptation. Witcombe *et al.* (2006) listed the following circumstances under which farmer collaboration in the early selection phase is particularly beneficial.

1. Empowerment of farmers is a main objective of the breeding program, so strengthening their knowledge and skills regarding crop improvement and utilization of genetic diversity are important for reaching this goal.
2. The farmers' preferred traits and selection criteria are not (yet) well known, and/or joint learning is one of the objectives of the PPB program.
3. Consumer preferences of grain qualities are important and complex so that it is difficult to do selection for such traits on the basis of laboratory tests.
4. Farmers have complex selection criteria where they trade off traits among each other.
5. It is economically reasonable to do the selection on farm (rather than on research stations) because farmers are able and willing to contribute important resources and skills to the breeding program without receiving payment.
6. The size of the market is too small for private-sector investment (i.e., because of very specific site conditions or market preferences), or the needs of the farmers are not considered in public breeding programs because of lack of communication, lack of knowledge, or other priorities set.

Crops differ with regard to biological characteristics (self-pollinating, cross-pollinating, clonally propagated), and the breeding and selection methodology applied at this stage of a breeding program also depends on the type of variety targeted (hybrid varieties, open-pollinating varieties). Accordingly, the length of this phase varies enormously.

Along with methodological diversity, options for farmer participation become extremely diverse as well. Many farmers have strong skills and profound experiences in selecting individual plants from larger populations. In such cases, it is feasible to give them population bulks of diverse material that harbors all the necessary traits for improvement. Some interested farmers can manage the entire population for a number of generations using the selection technique he/she is familiar with, but applying it to material more diverse than what is normally available locally. There are cases were farmers have developed new open-pollinated varieties using this procedure. These varieties can later on be tested in a multilocation testing scheme to compare their yielding ability and range of adaptation to other varieties. Various factors for success are listed in Box 7.4.

Working with relatively few, but highly interested, farmers ("expert farmers") appears to be most effective for cooperation during this stage of a breeding program (Witcombe *et al.*, 2006)—if the goal is just developing one or several new varieties. Otherwise, if training and empowerment of farmers or biodiversity considerations are the focus of the PPB activities, more farmers may be involved. In some projects, farmers visited research stations and selected from early generation material for further testing and evaluation on their farms (Sperling *et al.*, 1993). Also, early generation trials were entirely grown on farm and the farmers of the village made their own selections from such trials (Ceccarelli *et al.*, 2001). Such practices can lead to a great number of selected and further propagated "varieties" and to the rapid spread of innovative material with preferred traits, but not necessarily to the development of stable varieties for release and diffusion through the formal seed system.

Farmer Participation during the Variety Testing Phase

The variety testing phase is the final stage of variety development. It requires that a number of experimental varieties exist, which are stable and reproducible; furthermore, sufficient quantity of seed should be available for testing on a larger scale.

BOX 7.4 Improving Chances of Success during Selection Phase in Early Generations

- High selection intensity, that is, selection of only a few clearly superior plants or lines out of a large number of individuals
- Homogeneous field conditions so that differences among plants are not masked by differences in soil conditions
- Selection criteria that can be evaluated on single plants/rows with reasonable heritability
- Good knowledge of parental material, inheritance of selection criteria, and methodological options

Chances for success during this phase of a breeding program are determined by the diversity and the trait combinations expressed in the new materials, as well as by the quality of testing (Box 7.5).

One key advantage of farmer participation during this phase is the evaluation of new varieties in farmers' own fields, under their own management. Thus, the varieties are tested directly in the target environment (see earlier discussion). Participatory evaluation trials are usually grown by a larger number of farmers, thus covering a wide range of possible growing conditions in the target environment. This gives farmers and breeders a chance to observe the varieties' responses to different, locally prevalent stress factors. Farmer-managed trials are often exposed to severe types of stress, for example, poor soil fertility, delayed weeding, or temporary flooding. In such cases, particular adaptation characteristics or weaknesses of the new varieties may be discovered. On research stations, such extreme stress conditions occur very rarely, and their relevance may not be judged easily by farmers.

The form of cooperation between farmers and plant breeders varies; sometimes farmers and breeders evaluate separately so that mainly the results of it "count"; in other cases, there is more intensive dialogue on relevant criteria and the underlying concepts. Learning from each other and integrating farmers' and "scientific" knowledge would then be part of the project outcomes.

Farmers can profit from participatory variety evaluation because they get exposed to a larger number of new options to choose from; in this way, they often identify varieties for different types of conditions, uses, and market opportunities. PPB practitioners regularly observe that different farmers prefer different varieties, for very specific reasons. Therefore, the participating farmers should ideally represent various groups and conditions. For example, subsistence farmers or people selling on local markets often prefer different varieties than farmers who grow the same crop for national or international markets. Also, if animal husbandry is an important part of the agricultural activities,

BOX 7.5 Improving Chances of Success during the Variety Testing Phase

- Genetic diversity expressed among experimental varieties for key selection criteria
- Experimental varieties do not have major weaknesses that hamper acceptability
- High selection intensity
- Testing environment reflects well the conditions in the target environment for the new varieties
- Trial management that maximizes heritability for key selection criteria, for example, sufficient replication, appropriate trial design, and sufficient number of locations for testing

a variety with a higher total biomass yield or better fodder quality may be preferred by the farmers. Furthermore, access to resources and agricultural inputs, such as irrigation water or fertilizer, availability of labor, or technical equipment, as well as cultural traditions and individual preferences, influence the varietal preferences of farmers.

A further advantage of farmer participation in the advanced stage of variety development is that seed of preferred varieties can be harvested and multiplied immediately. Very often, breeders observe that "adoption" of really promising varieties happens long before they are officially released. Neighbors and relatives can see the new varieties under "real" conditions and may ask the owner for seed. Therefore, the benefits from newly developed varieties reach the farmers much earlier than through the formal system (even though on a limited scale), and such informal seed diffusion in the variety testing phase should be observed and monitored carefully. This information could give important indications regarding the size of the future market and the potential customers of the new varieties.

Innovative Strategies for Seed Production and Distribution

For a plant breeder, the "normal" way of organizing the diffusion of seed of a newly developed variety is through the formal seed system. Private or parastatal companies sell the seed ("certified seed") to farmers, and depending on the legislation of the country, the seed has to fulfill certain legal requirements. Farmers, however, would usually use their own informal networks for seed distribution, such as direct distribution to neighbors, friends or relatives, or selling in local markets or fairs. Figure 7.4 shows schematically formal and informal seed systems and how they may interact.

Both formal and informal seed systems have their specific strengths and weaknesses (see Table 7.2); this is why many PPB projects have tried out innovative strategies, such as decentralized on-farm multiplication schemes (McGuire *et al.*, 2003), or special communication strategies using mass media (Joshi and Witcombe, personal communication, documented in Sperling and Christinck, 2005).

Table 7.2 shows that the informal seed system has various strong advantages regarding the diffusion of PPB varieties: as seed production forms part of the normal crop production, and the diffusion takes places mainly along the normal social networks (among relatives, friends and neighbors, at local markets, or through local traders), the cost for production and distribution is generally much lower compared to the formal system. This is particularly important if poor farmers are potential customers of the PPB variety. Also, narrowly adapted varieties can be multiplied and distributed easily on a limited scale, for example, within one or several villages, and even very small quantities can be distributed or exchanged. However, varieties from PPB programs are not necessarily narrowly adapted; they

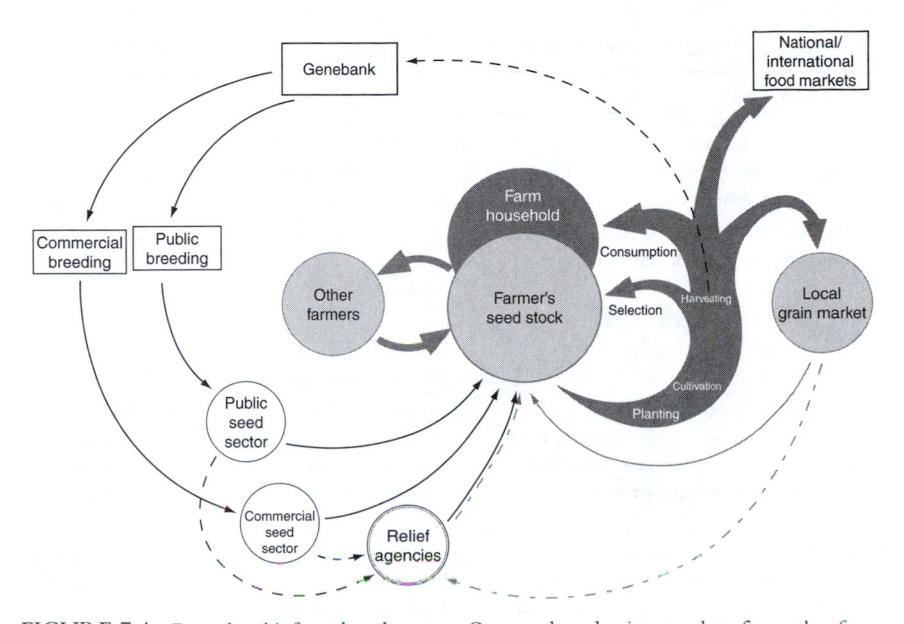

FIGURE 7.4 Formal and informal seed systems. Own seed production, purchase from other farmers, or local markets are informal channels (gray), whereas public or private seed outlets and relief supplies constitute formal channels (black).

can be equally relevant for farmers in other regions or even other countries (Joshi *et al.*, 2001, 2007). In such cases, particularly if the formal system is well developed, official variety release and distribution could also be considered for the diffusion of PPB varieties.

In any case, the distribution of seed from PPB programs needs to be planned carefully and strategically according to the situation of the seed market in the project area and the overall project goals (see Box 7.6). For example, PPB programs that have components targeting farmer empowerment may aim at creating local institutions or organizations that can sustain these activities possibly without project support, for example, farmer seed cooperatives or local seed enterprises. However, to achieve rapid adoption and widespread impact from use of the newly identified varieties, it is also common to partner with existing NGOs, community-based organizations, or extension services to rapidly diffuse small packs of seed widely among the target group of farmers. In all cases where seed is mainly distributed along informal channels, care should be taken to ensure the sustainability of seed supply: there should be possibilities to get new seed of the variety in cases of involuntary seed loss. Frameworks for planning seed outreach strategies in PPB projects have been summarized by Sperling and Christinck (2005).

TABLE 7.2 Strengths and Weaknesses of the Formal and the Informal Seed System

Formal seed system	Informal seed system
Makes varieties available that tend to be widely adapted and are often suited to favorable environments.	Makes all kinds of varieties available that can be reproduced on-farm: locally adapted modern varieties, traditional landraces, and farmer varieties.
Seed has to fulfill certain legal requirements regarding purity, germination rate, etc.	No "official" quality testing, but certain standards of reliability are being assured through social relationships.
Only varieties which fulfill the official requirements can be distributed.	Varieties, mixtures, or unstable populations (for further selection) can be distributed along informal channels.
Official recognition of ownership and intellectual property rights; however, it may be difficult to get farmer or community recognition inserted in formal release channels.	Ownership and intellectual property rights are usually not considered in the informal seed system: traditionally, seed is multiplied, sold or exchanged by all farmers, irrespective of their varietal identity.
Requires a certain "minimum market size" in order to make the investment in multiplication, certification, and distribution economically feasible.	Because seed production is embedded in the normal agricultural production, even small quantities of seed can be produced and distributed.
The seed price is often much higher compared to locally produced grain; this can hinder the access of poor farmers to certified seed.	The seed price is usually much lower, often similar to normal food grain price or slightly more.
Seed has to be paid with cash or credits/ loans.	Flexible modes of payment, depending on social relationships; seed may even be given for free in certain situations.
In some regions, the formal seed market is generally weakly developed, particularly in remote areas.	The informal system is locally organized, which ensures access to seed even in remote areas.
Large quantities of seed can be handled and distributed even in geographically distant areas and to different social groups.	The seed flow through the informal system is limited by quantity and geographical as well as social distance.
Formal seed distribution implies investment in communication strategies; however, the information may not always be relevant to farmers for taking informed decisions.	Communication is informal, from farmer to farmer; it is thus slower, but the information is highly relevant to other farmers of the region.

PRACTICAL ELEMENTS OF PPB IN VIEW OF AGROECOLOGICAL ISSUES

This section shows how agroecological issues can be practically addressed through PPB in the different stages of a breeding program (as shown in Fig. 7.1).

BOX 7.6 Building Blocks of Successful PPB Seed Distribution Strategies

- Identify and develop highly superior varieties that are attractive for farmers
- Estimate the market volume
- Develop a solid understanding of seed channels that farmers use and the key actors
- Consider the specific legal, political, and socioeconomic framework and conditions
- Find innovative solutions to overcome key weaknesses and gaps in the seed systems
- Include information exchange aspects along with the seed diffusion targets
- Collaborate with NGOs, farmer organizations, or the private sector for distributing seed on a larger scale

Develop Breeding Objectives Based on Deeper Understanding of Agroecological Conditions and Typical Constraints

Classical objectives of breeding programs, such as yield improvement or disease resistance, may be too general to develop varieties that will have to fulfill various functions in complex farming systems. Therefore, PPB projects have been particularly successful if they invested in the development of breeding goals based on a deeper understanding of agroecological conditions, farmers' needs, and the typical constraints of the farming systems.

This type of assessment can start from several points. One possibility would be to identify problems and constraints by means of a general assessment of the farming system, including soil qualities, crops and varieties grown, resources, and technologies used. Participatory communication tools can be used in order to make sure that the assessment is centered on the farmers' perceptions and their definitions of problems and needs. One important point is to carefully select the participating farmers; for example, various wealth or ethnic groups should be considered. It is also advisable to use a gender-sensitive study design. For this purpose it may be necessary to work with women and men in separate study groups. The methodologies could range from individual or group interviews to more specific tools such as transect walks, mapping, or modeling exercises. These communication tools have been described in the PRA literature (see Box 7.2) or, more specifically in view of breeding programs, by Christinck *et al.* (2005a).

However, not only are farmers' needs and constraints that relate directly to farming important for developing innovative breeding objectives, consumer and processing qualities are also an important point. New varieties are only acceptable for farmers if they can be processed with local technologies and meet a certain threshold level of important quality traits. Food habits and preferences of rural people are parts of their cultural identity; they may also be related to agroecological conditions, including, for example, seasonal fluctuations in production or problems of storability of certain products. If farm households are mainly the consumers of their own produce, interviews with farmers on preferred quality traits and testing of culinary quality could be organized. As women are mainly responsible for food storage, processing, and cooking in most cultures, their expertise should be searched through gender-sensitive participatory study designs. In those cases where crops are marketed and processed on an industrial level, it is advisable to consult key persons from relevant private or public enterprises with regard to required processing qualities.

Starting with a general assessment of problems and constraints is a good option if there has been little contact between farmers and researchers previously; PRA exercises and interviews can help initiate dialogue, identify potential research partners, and find a "common language."

Another possibility in developing breeding objectives would be to start by envisioning an end product with the required qualities. In this approach, farmers are exposed to a range of already existing varieties, including materials that may be unknown to them. The farmers are then asked to select those materials that appear most appropriate in view of the conditions of their own farming system. This activity is often done in combination with interviews or group discussions regarding the reasons for selecting certain varieties or discarding others. It is also possible that farmers evaluate one or several of the selected varieties on their farms over a longer period, that is, several growing seasons. This type of variety assessment has become known as participatory variety selection and is particularly applicable if it is assumed that varieties with the required qualities exist already but have not been accessible for farmers for various reasons, such as weaknesses of the seed distribution system and information flow. Furthermore, PVS is a good option wherever rather complex traits or trait combinations are under discussion that cannot be dealt with theoretically; real objects (plants, seeds, harvest products, etc.) can help to better focus such discussions.

In practice, both approaches (problem analysis and assessment of a desired end product) are often combined. For example, results of a problem analysis can be particularly useful for identifying appropriate material for PVS. However, PVS can help depart from discussions on problems and needs toward finding practical solutions.

Example: Improving a Local Maize Variety in Nepal

Maize is an important crop for the upland areas of Nepal. While 59% of the total maize planting area in Nepal is sown with modern varieties, the situation is completely different in the western hills of Nepal. Only during part of the year are these parts of the country accessible by roads, and farmers have very limited access to improved seed and information on new varieties. The existing crop research system has so far not addressed location-specific problems, resulting in very little impact of formal research and breeding institutions in the area. Thus, starting from 1998, a participatory maize breeding project was initiated jointly by LI-BIRD[1] and farming communities in the Gulmi district of Nepal, in collaboration with the National Maize Research Program.

Around 90% of the farmers in the study area depend on maize production for their livelihood. Five major cultivars are grown by the farmers, out of which the three most important ones are landraces. The widest distributed landrace is called *Thulo Pinyalo* and accounts for 75 to 80% of the planting area. After identifying two project sites, a village workshop was organized. During this workshop, farmers and researchers assessed the existing diversity of maize, analyzed problems and needs, identified preferred and undesirable traits, and set breeding goals. The methodology of this workshop was based on the PRA approach (see Box 7.2 for more information).

The local variety *Thulo Pinyalo* was appreciated by the farmers because of its high yield potential, good culinary and processing qualities, resistance to storage pests and diseases, fodder quantity and quality, and adaptation to local management practices. A major problem of the variety is its susceptibility to lodging. Furthermore, women and poor farmers particularly expressed a need for a maize variety that could be grown in mixture with legumes. Another important issue was that a new variety should be useful for roasting the immature cobs, as this is an important food supplement for poor farmers in the "hungry season" (before the main harvesting season).

The farmers selected a farmer research committee and a larger group of farmers, men, and women who would participate in the breeding activities. It was decided to introduce exotic germplasm for broadening the genetic base of the landrace by using a PVS approach. Elite germplasm provided by CIMMYT and the National Maize Research Program, as well as further exotic lines identified by the researchers based on the analysis of farmers' preferences for traits, was tested in farmers' fields. Results were evaluated through farm walks and a traveling seminar at the maturity stage. Farmers also conducted focus group discussions and preference ranking on their own. These activities helped farmers (1) evaluate the exotic material for performance under local conditions and identify new traits that could complement the traits of the landrace and (2) finally select breeding parents for improving the local maize variety *Thulo Pinyalo*.

After crossing the landrace with several exotic cultivars followed by several cycles of selection, the resulting populations were again tested extensively by

[1]Local Initiatives for Biodiversity Research and Development, Pokhara, Nepal; email: info@libird.org

farmers. The three most promising ones were less tall, less affected by lodging, and had shorter duration while still giving nearly the same grain and fodder yield as the original landrace. However, developing maize varieties for a marginal production environment was not the only outcome of this approach; with the help of LI-BIRD, farmers organized themselves collectively and established new linkages to government units, the National Maize Research Program, private companies, and NGOs. The project activities were an effective entry point for training activities and empowered farmers in their problem-solving capacities. Breeders from formal institutions appreciated the cooperation and changed their perspectives regarding their perception of varietal characteristics and the benefits of cooperating with farmers and informal institutions (project description based on Sunwar *et al.*, 2006).

Generate or Assemble Variability Based on Farmers' Knowledge and Local Germplasm

As outlined earlier, local crop varieties are often used as breeding parents in PPB projects, mainly because of their specific adaptation to agroecological conditions, as well as culinary quality traits. Farmers' traditional knowledge is often the main source for relevant information on local crop germplasm, thus being an important contribution that can be made by the farmers in the initial phase of a breeding program.

However, not all villagers may have the same level of knowledge in this regard. It is likely that local experts can be identified, often experienced farmers. Collections and inventories of local varieties can serve to assemble seed and planting material for the breeding program, while at the same time documenting the associated knowledge on adaptation and specific traits. Particularly in cases where breeding programs aim at safeguarding or increasing local biodiversity, such activities can also be an excellent means of raising awareness and increasing the information flow within or among village communities with regard to traditional crop varieties and their specific qualities.

Various participatory forms of action have been described for the purpose of sharing and documenting knowledge on traditional crop varieties (Christinck *et al.*, 2005b; Rana *et al.*, 2000; Rijal *et al.*, 2000). Some put more emphasis on informal communication and knowledge sharing, either among villagers or between farmers and researchers; examples are transect walks, diversity fairs, community biodiversity registers, or rural poetry or song festivals related to biodiversity. However, tools such as biodiversity mapping, four-square (or four-cell) analysis, or matrix ranking aim at systematic documentation of distribution, specific traits, and uses of a range of varieties. Various tools can be combined or brought into action in different phases of the research process.

In addition to selecting breeding parents, farmers can also be involved in generating new variability through crossing. In practice, this is often done in the case of cross-pollinating crops such as maize, where sophisticated technical operations are not required; this was the case, for example, in the maize breeding project described earlier. However, in some projects, farmers have also been trained to make crosses in self-pollinating crops (McGuire *et al.*, 2003).

Example: Selection of Breeding Parents for Participatory Rice Breeding in Nepal

In Nepal, a participatory plant breeding program was launched in 1998 by the national Agricultural Research Council together with LI-BIRD and other local institutions. This PPB program had a strong focus on testing and developing methodologies and aimed particularly at conserving local biodiversity, while at the same time providing benefits to the farming communities.

The main actors were aware that on-farm conservation of agrobiodiversity could only be achieved with a high level of community participation and that the capacities of local farmers to search, select, and exchange germplasm would be key to reaching the goals of the project. Furthermore, the setting of breeding goals and the selection of landrace parents for the PPB program were regarded as closely related activities for which community participation had to be organized.

As a first step, relevant agroecosystems and interested communities were identified. The study sites comprised mid- to high-altitude regions (Begnas village, 600–1400 m above sea level) with a high diversity of rice landraces, as well as sites with higher production potential in the Indo-Gangetic plains (Korchowa, 54–100 m above sea level).

In the initial phase of the program, "diversity fairs" were organized at each site with the goal of raising community awareness of local crop genetic resources, promoting the value of landrace diversity, and locating and collecting material. The local varieties presented at the fairs were then grown in farmers' fields as "diversity blocks," which were used to assess the performance and analyze preferred and undesired traits of local varieties with the participation of various groups of farmers. Furthermore, the diversity blocks were used as a seed source for crossing programs. In a next step, the communities were encouraged to maintain "community biodiversity registers" (CBRs), which means a systematic documentation of varieties held by the farmers, including information on special characteristics and uses. The CBRs allow monitoring dynamic changes in local crop diversity over time. Another tool used to prepare the selection of breeding parents was "four-cell analysis," through which the diversity of local varieties is grouped according to two criteria: the number of households that grow a certain variety and the area on which it is grown. This tool allows identifying those landraces that are in severe danger of being lost (grown by few households on small areas), just as those that are widely used (many households, large areas), or those that may be somehow specialized for certain conditions or uses (few households, large areas or many households, small areas) (Rana *et al.*, 2005).

The project then conducted focus group discussions in order to identify the breeding parents. The aim was to select at least one landrace from each of the four cells, and the project used a preference ranking exercise (Guerrero *et al.*, 1993) for identifying preferred traits and those traits that needed improvement. At least one landrace (the best one) from each cell was identified. The other parent (exotic parent) was then found by looking at the traits that needed improvement, new desirable traits, and adaptability to local conditions and germplasm.

Using this procedure, the research team finalized cross combinations for each study site. The resulting materials were then tested in farmers' fields and assessed during "farm walks" by farmers and researchers. Several new populations, which combined yield potential of introduced varieties with adaptive and quality traits of local landraces, were selected from this material and spread through farmer-to-farmer networks (project description based on Sthapit *et al.*, 2002).

Farmer Participation in the Selection Phase of a PPB Program

There are two types of contributions farmers can make in the selection phase of a breeding program: (1) identify the selection criteria and (2) perform selection in the breeding populations.

In many parts of the world, farmers invest considerable time in the production, selection, and storage of their own seed. They may harvest the seed grain from particular fields or field patches or have a preference for certain plant types, shapes or colors of leaves, stems, harvestable organs, or grains. Thus, farmers who select their own seed generally have knowledge as well as practical skills that could be important for reaching the goals of a breeding program.

Understanding selection criteria of farmers can help identify important adaptive or quality traits. Criteria used by the farmers may be directly or indirectly associated with environmental adaptation and/or quality issues. This knowledge is often embedded in traditional practices and ways of doing something that may not be communicated easily: it is implicit or "tacit knowledge." Communicating tacit knowledge generally requires methodology, which is based on action, demonstration, and observation rather than on questioning. Thus, understanding farmers' own selection criteria may be one reason why farmers are practically involved in the selection of diverse populations in many PPB projects.

However, many farmers also have the practical observation skills needed for performing selection. They are able to identify single plants that show unique properties or a trait combination that may be of particular use for certain purposes or conditions. This ability of farmers is in fact the origin of many successful farmer varieties.

Given the fact that selection in the target environment is generally an essential element of most PPB projects, farmer participation in selection is a logical consequence. In some cases, farmers and scientists made their selections independently from each other in the same material for later comparison of the results (Ceccarelli *et al.*, 1996).

Example: Learning from Farmers' Selection Criteria in Rajasthan, India

In Rajasthan, a semiarid state in northwest India, pearl millet is the staple food crop. Adoption of modern varieties has been very limited, particularly in the driest western parts of the state where low and unpredictable rainfall is the main problem faced by the farmers. However, landraces are commonly grown, and many farmers also grow mixtures of traditional and exotic varieties. With pearl millet being a cross-pollinating crop, these mixtures result in highly variable populations from which many farmers select seed for the coming season. However, also in traditional landrace material, many farmers select seed very keenly, most often from harvested panicles on the threshing floor and sometimes in the standing crop.

When scientists from ICRISAT started applying the PPB approach for developing new, drought-resistant varieties for western Rajasthan in the year 1991, they assumed that farmers' own selection criteria were probably based on their traditional knowledge of environmental adaptation and that understanding more of this knowledge could guide a way toward developing new, better adapted varieties.

Consequently, the seed selection practices of farmers in Rajasthan were assessed by applying a participatory research methodology that comprised observation of farmers' practices, simulation exercises, interviews, and workshops (in villages as well as on research stations).

One workshop was conducted in 1997, in the course of which farmers from various villages and researchers evaluated 15 demonstration plots grown at a research station near Jodhpur. Populations grown on the demonstration plots included traditional landraces from various parts of Rajasthan, farmer-selected varieties, and improved landrace-based materials selected by plant breeders, as well as some hybrid varieties bred from "exotic" germplasm. Farmers and researchers evaluated (separately) the 15 demonstration plots. Evaluation criteria used by farmers and scientists, respectively, were documented and compared.

The workshop revealed that the farmers used far more criteria (42) for their evaluation than the scientists (24). Grain and fodder yield were considered by both groups, as were time to maturity, panicle characteristics, number and productivity of tillers, stem diameter, disease incidence, and performance under drought. However, most quality-related criteria were not considered by the scientists, such as labor requirements and market price. The main differences were found, however, regarding many traits related to specific aspects of environmental adaptation, which were only considered by the farmer participants.

Many more interviews and PRA exercises, particularly direct observation and classification exercises using pearl millet panicles showing different traits and properties, revealed how farmers in Rajasthan relate panicle and plant characteristics to performance, environmental adaptation, and quality issues (Fig. 7.5).

Formal multilocational field trials in which farmers' evaluation and selection criteria were applied systematically (by scientists) showed that the farmers' way of

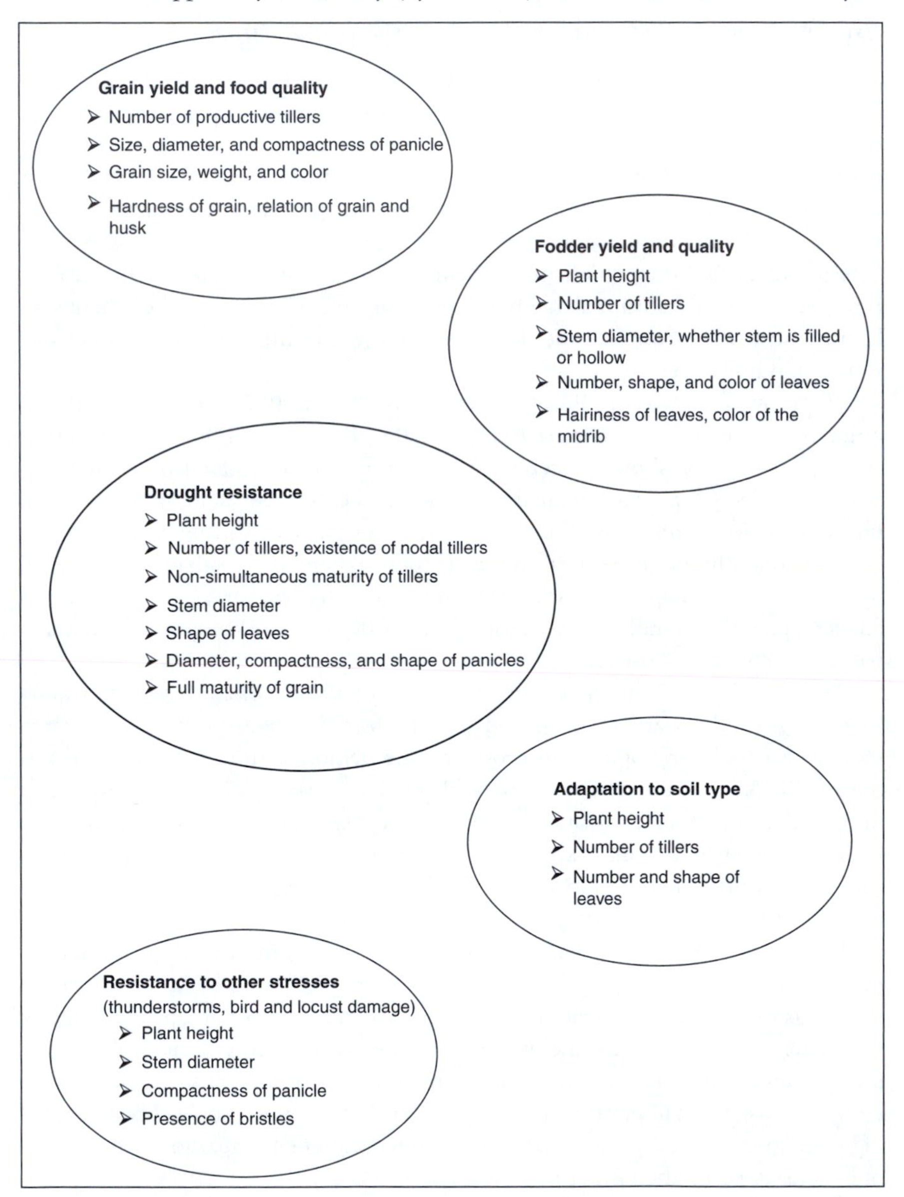

FIGURE 7.5 How farmers in Rajasthan, India, relate traits of pearl millet plants to relevant aspects of performance, stress resistance, and quality.

associating visual traits, such as stem and panicle diameter, tillering ability, or grain characteristics with drought resistance, is in fact supported by yield data gained in drought environments (Christinck *et al.*, 2002). This outcome allows two conclusions: (1) selection and evaluation criteria used by the farmers are based on a deep understanding of plants and the environment, and (2) farmers' knowledge and skills could contribute considerably to the selection of varieties adapted to the drought-prone environments and complex farming systems of western and central Rajasthan.

Participatory Methods for Variety Testing and Evaluation

As outlined earlier, variety testing and evaluation in the target environment are core issues of PPB. First, testing in the target environment reveals the degree of environmental adaptation of the experimental variety; second, it allows farmers to evaluate the overall "usefulness" of the new cultivar in the context of their own farm and family over the whole production cycle, including growing phase, harvesting, storage, consumption, and yield stability (if evaluated over several seasons).

Farmers generally tend to evaluate varieties in all the different stages of their development and spend considerable time in the fields while doing their normal field work. Furthermore, it is also a general custom to discuss the value of new materials with family members, neighbors, and colleagues in an informal way. Therefore, participatory variety evaluation is an activity that is relatively close to the farmers' reality.

Methods and tools for organizing farmer-managed trials for variety evaluation differ very widely and are usually determined by seed availability, commonly used field size, minimum plot size for reliable yield evaluations, and the number of farmers who are keen to conduct trials in a village. Furthermore, the tools used for assessing individual varieties differ with the level of literacy of the participating farmers, the types of observations required, and the number of varieties being evaluated (Weltzien *et al.*, 2005).

Generally, it can be distinguished between rather informal evaluation methods (i.e., farmers grow one or several test varieties and share their experience in interviews or village workshops) or more formal methods (evaluation trials with replications, documentation of observations, yield measurements, etc.). The decision about such issues depends much on the institutional background of a PPB program, the resources available, and the overall goals of the program. A higher level of "precision" may be required by scientists for their work, but does not necessarily lead to better results compared to informal methods, which require much fewer resources (Joshi and Witcombe, 2002; Witcombe *et al.*, 2005). A relatively widespread approach for decentralized participatory variety testing is the "mother–baby trial" design, which allows farmers to test subsets ("baby trials") of a general trial ("mother trial") on their farms. The "mother trial" can be grown either on a research station or in a village, and includes several within-site replications of

FIGURE 7.6 Women farmers scoring sorghum varieties in Mali (photograph by S. Siart).

treatments. The "baby trials" are grown in different farmers' fields, often with larger plot size and under farmers' normal management. They are considered as additional replications (Snapp, 2002). In this way, it is also possible to relate qualitative observations made by the farmers to yield and other quantitative data gained from the baby trial in which the farmer made his/her observations. Furthermore, ranking and scoring exercises are widely used for participatory variety evaluation, particularly during on-station workshops or for evaluating larger on-farm trials (Weltzien *et al.*, 2005)(Fig. 7.6).

Example: Participatory Variety and Clone Evaluation with Potato Farmers in Peru

In 1997, the International Potato Center (CIP) and CARE-Peru started a collaborative project on the integrated management of late blight, an economically important potato disease. Part of this participatory technology development process was to facilitate farmers' access to new, resistant genotypes in order to reduce the use of fungicides.

As a first step, farmers evaluated 12 varieties with different degrees of resistance to late blight. Based on the information gained through this activity, the breeding program of CIP provided 54 promising clones, which were divided into clusters and evaluated by 13 different farmer groups. In the following year, the farmers continued evaluating 25 selected clones from the previous season.

Taking into consideration that many farmers of the study area were illiterate, a simple evaluation method was used to classify the potato clones in three categories: good, moderate, and bad. The farmers received paper cards with smiling, serious, or "sad" faces drawn on them and were asked to put a respective card into a paper bag located near the genotypes for evaluation. At the end, all participants should have given their judgment on each genotype.

After finishing this activity, the cards were counted and the results were written on a board. Each genotype had a certain number of positive, medium, and negative judgments so it became clear which genotypes were preferred by the farmers. These results led to a discussion on the reasons and underlying criteria. The method is also useful for distinguishing the preferences of different groups of farmers (i.e., women and men) if done separately or with different colors of paper (project description based on Ortiz, 2002).

Seed Production and Diffusion Strategies in View of Agroecological Issues

Ideally, new crop varieties that are adapted to local agroecological conditions and farmers' needs are the final outcome of PPB programs. However, a positive impact can be realized only if the farmers have access to the seed of such varieties at the time needed on a sufficient scale and at a reasonable price. Here, PPB projects often face serious problems because efficient seed distribution in the longer term and on a large scale, on the one hand, and offering a range of diverse varieties for special agroecological conditions, particularly for poor farmers, on the other hand, are potentially conflicting goals.

As seen earlier, the informal seed system offers various potential advantages for seed diffusion, particularly in those cases where the amount of seed required is small, that is, varieties grown only locally and on small areas. However, the maintenance and distribution of seed require a high degree of long-term commitment and motivation from the involved actors (mainly farmers) if the seed supply should be sustainable. Traditional values and ways of sharing seed do not always continue to function. Therefore, awareness raising and building of organizational structures based on the local traditions may be required for ensuring seed supply from farmer to farmer in the longer term. Many PPB programs seek cooperation with NGOs or farmer organizations for this purpose.

It may be useful to effectively link "grass root" seed production to formal institutions (such as gene banks, private or public seed companies) to ensure a sufficient seed supply of agroecologically relevant varieties even in times of crisis, drought, or other events that may disturb the functioning of local seed supply.

However, it should be considered that the seed supply of varieties grown by few farmers and on small areas is generally vulnerable in the longer term. A solution could be to test the materials developed through PPB in other, agroecologically similar regions through participatory variety evaluation and selection schemes. Thereby, the demand for such varieties could increase.

A potential obstacle to linking informal and formal seed supplies could be seed legislation in some countries. Seed spread from farmer to farmer is often tolerated, as long as the varieties sold are not registered by any private company or breeder. However, as soon as the formal sector gets involved, official registration is inevitable in many situations. Some countries, such as India, have recently revised their seed legislation in order to allow for the registration of varieties also under the name of an individual farmer or farmer groups. The realization of "farmers' rights" in the near future (ensuring farmers' access to seed and the right to use it in multiple ways) is compulsory for all countries that have signed the International Treaty on Plant Genetic Resources for Food and Agriculture; however, uncertainty remains on how this can be done (www.fni.no/news/230606.html).

Example: Seed Fairs

Seed fairs provide an opportunity to facilitate farmer-to-farmer seed distribution, particularly of traditional or locally important varieties. Seed fairs can reach a large number of people, particularly if they are organized as a side event to other culturally important happenings, such as religious festivals, which attract people from larger areas. Furthermore, seed fairs can be a mean to facilitate seed exchange between people who do not interact otherwise because of geographical or social distances. Not only seed will be exchanged, but also there will be much discussion and sharing of information, for example, on cultivation practices, uses, and food preparation.

Seed fairs are a tradition in some countries, for example, in the Andean region of South America (Tapia and de la Torre, 1998; Tapia and Rosas, 1993). However, they have been organized successfully in many other countries, often by NGOs or other locally based institutions (Almekinders, 2003). The focus can be more on the diffusion of traditional varieties, or also on new varieties, such as PPB varieties (Weltzien *et al.*, 2006).

The farmers are usually invited to offer seed of their own production to other farmers. If required, some form of quality control may be installed prior to opening the fair. Competitions, prizes, awards, and cultural events can serve as further incentives to increase the attractiveness of the fair. Once established, seed fairs often catch the attention of more and more participants year after year. In addition to promoting the farmer-to-farmer exchange of varieties, they offer a range of possibilities for links to other biodiversity-related activities. For example, the diversity displayed can be monitored regularly, thus providing indications for the

FIGURE 7.7 Seed fair in Mali (photograph by S. Siart).

loss or revival of varieties. Furthermore, potential collaborators for planned *in situ* conservation or PPB activities could be identified among the participating farmers and personal relationships can be established (Almekinders, 2003) (Fig. 7.7).

EXAMPLES OF SUCCESSFUL PARTICIPATORY PLANT BREEDING FOR MARGINAL ENVIRONMENTS

The examples mentioned earlier merely focused on illustrating methodological aspects of participatory plant breeding. This section provides examples of how PPB successfully addressed specific agroecological problems or constraints, particularly in marginal environments. Such problems can be relatively straightforward to describe, such as adaptation to soil acidity or tolerance to a specific pathogen. However, we would like to emphasize that the breeding objectives of farmers are often multifaceted and include use-related parameters as well. Furthermore, the general wish to increase yield potential and/or yield stability of existing varieties can only be realized if the complex relationship between various plant traits and environmental adaptation is well understood.

Participatory Maize Breeding for Low-Fertility Soils in Brazil

In Brazil, as in other tropical countries, low soil fertility and either insufficient or excessive water supply are major problems in large areas of the country. Soil acidity associated with toxic levels of aluminum content and phosphorus and nitrogen deficiencies are the main soil parameters limiting agricultural production in the community of Sol da Manhã in the state of Rio de Janeiro. Maize, the economically most important crop in this area, is particularly affected by nitrogen deficiencies. However, chemical fertilizer is expensive for the farmers and is usually not applied. Therefore, the yield level of the maize crop used to be very low (roughly 1000 kg/ha).

Given the multiple difficulties faced by the farming community, some of their representatives sought technical support from the University of Rio de Janeiro and EMBRAPA.[2] A participatory program was set up, with the main objective to develop maize varieties with improved nitrogen use efficiency, using a participatory approach.

As a first step, a participatory variety evaluation of 16 maize varieties with good tolerance to low soil nitrogen levels was organized. Farmers selected one variety that they considered best, which was called *Sol da Manhã*, after their community. *Sol da Manhã* is a variety with a wide genetic base consisting of 35 populations from the Caribbean and South America. The sole introduction of *Sol da Manhã* in the community increased the average production level by 100%. The farmers reproduced and selected their own seed from it over several years, leading to various versions of the variety.

Simultaneously, a breeding program for improving and adapting *Sol da Manhã* for local conditions began in 1986. It was aimed at conserving the genetic variation and general crop characteristics of the variety while improving its nitrogen use efficiency and productivity under the conditions of the farming community. After six cycles of selection done by plant breeders, the "EMBRAPA version" of *Sol da Manhã* was introduced to the community in 1992 and was subjected to six further cycles of mass selection performed by the farmers. The selection was done at five community locations and at the EMBRAPA research station, and the locations showed different levels of nitrogen supply. The selected seeds were bulked at the end of each season to avoid loss of genetic variability due to severe stress conditions and to maintain an effective population size. To select for high nitrogen use efficiency, plants with more accentuated dark green coloration were marked at the flowering stage. After harvest, the farmers selected again for other plant characteristics, such as grain yield, plant and ear characteristics, and resistance to lodging. Introduction of the EMBRAPA version of *Sol da Manhã*, along with farmers' selection, again increased the yield level in the community by another 100%, from 2000 to 4000 kg/ha.

In 1994, after six cycles of EMBRAPA improvement and two cycles of community improvement, a further evaluation trial in the community proved the high yield potential of *Sol da Manhã* compared to other varieties and under low as well as

[2] Brazilian Agricultural Research Corporation.

higher levels of soil nitrogen. It was found to be in the group with the highest nitrogen use efficiency of the trial (Machado and Fernandes, 2001).

Participatory Barley Breeding for Drought-Affected Regions (Syria)

The barley breeding program of ICARDA (Syria) adopted the participatory plant breeding approach in the mid-1990s. Not only was the wish to better target the needs of farmers in marginal, drought-prone environments the background for this decision, but also theoretical considerations on plant breeding for drought environments.

Drought is one of the major factors limiting crop production worldwide, particularly in the rain-fed agricultural systems of the semiarid tropics and subtropics. One of the main characteristics of drought is its unpredictability in occurrence, timing, severity, and duration. In addition, the effects of drought on crop production are often further aggravated by high temperatures, reduced soil nutrient availability, or pest and disease incidence, as these abiotic and biotic stresses are closely interrelated with the occurrence of drought.

From a theoretical point of view, the variation between different target environments (i.e., test sites for trials) is much higher in dry areas compared to areas with high and reliable rainfall. Furthermore, the variation between years (for the same location) also tends to be very high in drought environments. Therefore, decentralized participatory plant breeding can address the complexity of dry areas more efficiently and effectively than a centralized plant breeding program.

The model for decentralized, participatory barley breeding that has now been brought into practice in Syria and various neighboring countries foresees that scientists make controlled crosses and select (on station) within the F1 and F2 generations. From the F3 generation onward, there is usually enough seed available for multisite testing, which is done during 3 more years in farmers' fields. Initially, a large number of entries (60 to 165) are grown in formal trials (unreplicated). Different sites may receive different germplasm from the beginning of the breeding cycle so that the total number of entries tested over all sites is even higher. Advanced trials in the following 2 years are grown with two replications, a reduced number of entries, and with larger plot size, as well as under farmers' own management strategies. The selection is done each year by farmers and scientists. After 3 years of testing, farmers of a given village usually select one to three entries for large-scale testing and seed multiplication. These selected test varieties also enter the next breeding cycle (crossing on station).

Practical results of this breeding strategy are higher yielding and highly drought-resistant barley lines with an average yield advantage between 7 and 47% compared to local varieties grown previously by farmers in areas where centralized breeding programs have not resulted in any alternatives to local landraces in the last 25 years (Ceccarelli *et al.*, 2006).

New Rice Varieties for West and Central Africa

In West and Central Africa, the demand for rice has been growing at a rate of 6% per annum in the last 3 decades. This demand has been met partly by the import of rice and partly by increasing domestic production, particularly in upland areas and under rain-fed conditions in lowland agriculture.

Rice cultivation has a history of 3500 years in Africa. The African rice *Oriza glaberrima* is well adapted to pests, diseases, low soil fertility, and other prevalent stress factors, but also has some undesirable traits, particularly lodging, grain shedding, and a low-yield potential. The Asian rice, *Oriza sativa*, which was brought to Africa about 500 years ago, has a higher yield potential and replaced the original species on a large scale. However, it is less adapted to typical stress factors prevalent in West and Central Africa.

There have been previous attempts to produce interspecific hybrids from both rice species, but they failed due to widespread sterility of the resulting hybrids. However, in 1991, WARDA launched a new effort to combine the potentials of both species by combining conventional breeding and tissue cultures in order to overcome sterility. By the end of the 1990s, a range of interspecific lines, showing radically new plant types, had been developed and were being tested and evaluated in a range of environments. This new rice was called NERICA (New Rice for Africa). The main characteristics of NERICA are as follows:

- Early maturity
- Strong stems that support heavy heads of grain without lodging
- More tillers with grain-bearing panicles than either parent and nonshattering grains
- Drought tolerance
- Tolerance to acidic soils
- Resistance or tolerance to important pests and diseases.

At this stage, WARDA adopted PVS for the further process of variety development. In a 3-year program, "rice gardens" with up to 60 different varieties (including *O. sativa, O. glaberrima*, NERICA, and local checks) were planted in the vicinity of study villages. Farmers were allowed to visit the gardens as often as possible; however, at three key stages (tillering, maturity, and postharvest), farmer groups were formally invited to participate in the evaluation. The farmers' evaluation criteria and selections were reported, and from the second year onward the participants received up to six varieties of his/her choice for further evaluation on the farm. The farmers' observations and the performance were recorded by NGO technicians, extension agents, or breeders. From the third year onward, farmers could buy the seed of desired varieties for sowing on a larger scale. Some of the most preferred varieties were then tested in official multilocation trials to generate data required for official release. NERICA varieties have now been introduced using similar approaches in 17 countries of West and Central Africa. The farmers' selection of varieties varies among the countries, reflecting a combination of differing varietal adaptation to the wide range of farm (micro-) environments and diverging consumer preferences (Gridley *et al.*, 2002).

Developing Sorghum Varieties for Changing Agroecological and Socioeconomic Conditions in Mali

Sorghum is the typical staple crop in the 700- to 1200-mm rainfall zones of southern Mali, where the soils are not too sandy. The duration of the rainy season is 4 to 5 months in this area (from May/June to September/October). It is important to note that agroecological conditions have changed markedly since the mid-1970s. The length of the rainy season has decreased, and the mean annual rainfall in the Sahel is now between 20 and 49% lower compared to the period between 1931 and 1960 (IPCC, 2001).

The status of soil fertility is also changing: it is decreasing in certain areas because of decreasing fallow periods and increasing in cotton-growing areas because of fertilizer use in cotton production. Therefore, there are demands for sorghum varieties that could profit from residual fertilizer effects and for others that are adapted to low-input conditions.

Lack of labor is an important limiting factor to agricultural production. There is a general trend that people seek other sources of income (in addition to farming), for example, through part-time jobs or temporal migration, and that children and young people go to schools. This trend can only partly be compensated for by the mechanization of agricultural works; less than 50% of the farm households in the project area have oxen for animal traction, and tractors are not used at all. It is quite common that people have to sell their oxen in low rainfall years.

Despite this situation, the adoption of new varieties from formal breeding programs has been very low, partly because of the weak development of the formal seed sector, but also because the varieties developed by breeding programs did not fit into the local farming systems. In Mali, the *guinea* race dominates sorghum production, as it does in the main sorghum-growing areas of Sudan and north Guinea. Improved populations of the *guinea* race show slightly earlier maturity, but no major yield advantages over local landraces. The *caudatum* and *kafir* races, which make up the bulk of the breeding materials that have been advanced in other regions of the world, are not adapted to the farming system in Mali because of photoperiodic sensitivity, which results in delayed flowering.

An economic impact assessment conducted in 1996 resulted in a complete reorientation of the sorghum breeding program of ICRISAT in Mali and the adoption of PPB methodologies in order to increase the relevance and impact of the program.

Participatory diagnostic studies revealed that yield increase was in fact the main objective expressed by farmers, while culinary and processing quality of local varieties, as well as adaptation to local conditions (soil, climate, pest resistance), had to be maintained.

Thus, participatory variety evaluation was found to be the starting point for future breeding activities. Interracial crosses between *guinea* and *caudatum* race parents form the major part of the breeding material, also including several *guinea* race dwarf lines. Farmers have been involved in on-station selection of these materials in various stages.

From 2003 onward, the project organized participatory variety testing of 32 entries in 12 villages and on three research stations. The participating farmers selected up to four varieties from these test entries for large-scale participatory on-farm evaluation using their usual farming methods. The set of varieties tested varied among locations. Results of these activities are encouraging. Farmers identified various varieties that yielded more (up to 20%) than their local control varieties and that also reached high preference ranks; furthermore, on-farm testing resulted in increased demand for seed for several varieties.

Establishing semiformal structures for seed production and diffusion is a further activity of the program, meant as an answer to the weaknesses of both formal and informal seed supply systems. Seed fairs and local seed producing associations were organized in cooperation with local farmer organizations and an NGO in order to meet the existing demand for seed of the new varieties (Weltzien *et al.*, 2006).

TYPES OF IMPACT ACHIEVABLE THROUGH PPB FOR SUSTAINABLE DEVELOPMENT OF FARMING SYSTEMS AND FARMERS' LIVELIHOODS

We found that adapting plants to human needs could be referred to as the basic goal of plant breeding. Plant genetic resources are, more than anything else, the "bridge" between the site-specific set of natural resources and people's livelihoods; only where a sort of balance can be achieved, between the nature and availability of resources, on the one hand, and human needs, on the other hand, is sustainable agriculture possible.

This section summarizes the types of impacts that can be achieved through PPB for the sustainability of farming systems by referring to some of the cases and examples given in the previous sections. Last but not least, the progress of rural peoples' livelihoods that may result from these impacts is mentioned.

Increase Farmers' Options to Adapt to Variable Conditions and Needs

Changing conditions can be a reason why farmers need to adapt traditional farming practices, including the portfolio of crops and varieties grown. This was discussed earlier in the example of Mali, where (like in other parts of the Sahel) rainfall patterns have gradually changed over several decades. A further common type of change is the reduced availability of family labor as a consequence of migration for jobs or education. However, interesting new options may also result from emerging food industries, for example, dairy products, convenience food (snacks, noodles, chips, etc.), brewery, and so on, wherever urban markets for such products are developed. New varieties that are tailored specifically to such new use options, while still being adapted to local environmental conditions, can be developed from local and/or exotic genetic resources.

Make Best Use of Limited Resources

Marginal environments are characterized by the limitation of natural resources essential for crop production; reduced availabilities of soil water and soil nutrients are very common limiting factors. Plant genetic resources show strong variation regarding their efficiency for using limited resources; traditional varieties, as well as wild or semiwild crop relatives, often show specific adaptation to marginal conditions, which can be combined with other important traits through breeding. Examples of maize breeding for nitrogen use efficiency in Brazil, as well as barley breeding for drought tolerance in Syria, were exposed in the previous section. A further advantage of improved efficiency would be that negative impacts on the environment, for example, through excessive use of fertilizer (groundwater pollution) or inadequate irrigation (decreasing groundwater levels, salinity), could be reduced.

Maintain Useful and Eco-Friendly Traditional Practices Related to Certain Crops or Varieties

Mixed cropping, typically a mixture of legumes and cereals or tuber crops, is a common practice in marginal agroecological systems, which fulfills a variety of functions, including complementary use of growth factors, such as soil nutrients, light, and water, reduced pest and disease incidence, reduced soil erosion, more total biomass production, more yield stability, and more food security. Furthermore, the mixtures can be flexibly adjusted to conditions such as late or early onset of the rainy season or status of soil fertility in different fields.

Because formal breeding programs seldom take such farming practices into account, the resulting varieties have usually never been tested for their ability to function under such conditions. In PPB programs, it is common for farmers to test materials on their own farms, thereby identifying genotypes for specific "niches" of their farming systems, including mixed cropping.

Reduce Susceptibility to Pests and Diseases

The incidence of pests and diseases not only reduces crop yields, but attempts to control them, particularly the use of pesticides, can result in negative effects on the environment and on people. Examples are soil and groundwater pollution, reduced biodiversity (of insects or soil organisms that may be affected by the pesticide use), accidents while handling pesticides, and long-term effects on human health.

Plant breeding in general can help identify resistance genes and incorporate them in new varieties, thus reducing pest and disease incidence. One example given earlier was the breeding of NERICA rice from African and Asian rice types. PPB and farmer breeding tend to work with broad resistance based on genetic diversity, both among and within varieties. Thus, biodiversity-oriented PPB programs per se reduce the possibilities that pests and diseases occur on a devastating scale, which has not alwayss been the case with formal breeding programs using varieties with a very narrow genetic base.

Improve Livelihoods through Increased Food Security and Empowerment of Farmers

Food security can be improved through PPB in various ways. For example, the yield or yield stability may be increased, the "hungry season" may be reduced (through earlier maturing varieties), or the quality of the diet may be improved through selection for important quality traits (vitamin or protein content, etc.). Furthermore, additional income can be generated in such cases where the plant breeding activities go hand in hand with improving or developing marketing options.

Empowerment of farmers, whether addressed directly or indirectly as a goal in a PPB program, can take many forms, starting with improved communication between farmers and researchers to farmers' gaining influence on scientific institutions and their research agendas, including raising and distribution of research funds. However, the most important impact of PPB with regard to empowerment is probably the improved access to seed of improved varieties that are suited to farmers' needs and can be reproduced on farm.

Farming system stability and agricultural livelihood options depend intricately on the available crop varieties and their specific traits. In the context of improving farming system stability and farmers' livelihoods options, it is of key importance to recognize that individual crops and specific varieties can serve very specific but very different functions within farming systems and that plant breeding offers targeted methods and tools for balancing agroecological issues and human needs. The particular strength of participatory plant breeding is that farmers' knowledge, as well as their practical selection skills, can be united with scientific knowledge focusing on how to achieve breeding progress for specific traits or trait combinations in a targeted way.

REFERENCES AND RESOURCES

Almekinders, C. (2003). Markets make a come-back: Diversity displays and seed fairs. Issue paper "People and Biodiversity." GTZ sector project "People and Biodiversity in Rural Areas." GTZ, Eschborn, Germany (download from: www2.gtz.de/agrobiodiv/download/Themenblaetter/Saatgutmaerkte_engl_05.pdf).

Ceccarelli, S., Grando, S., and Baum, M. (2006). Participatory plant breeding in water-limited environments. Paper presented at the 2nd International Conference on Integrated Approaches to Sustain and Improve Plant Production under Drought Stress, September 24–28, 2005, Rome, Italy. Download from: www.plantstress.com/id2 (forthcoming) or www.prgaprogram.org/modules/DownloadsPlus/uploads/Participatory_Plant_Breeding/ INTERDROUGHT.pdf.

Ceccarelli, S., Grando, S., and Booth, R. H. (1996). International breeding programmes and resource-poor farmers: Crop improvement in difficult environments. *In* "Participatory Plant Breeding" (P. Eyzaguirre and M. Iwanaga, eds.), pp. 99–116. IPGRI, Rome, Italy.

Ceccarelli, S., Grando, S., Tutwiler, R., Baha, J., Martini, A. M., Salahieh, H., Goodchild, A., and Michael, M. (2000). A methodological study on participatory barley breeding. I. Selection phase. *Euphytica* **111**(2), 91–104.

Christinck, A. (2002). This seed is like ourselves: A case study from Rajasthan, India, on the social aspects of biodiversity and farmers' management of pearl millet seed. Margraf Verlag, Weikersheim, Germany.

Christinck, A., Dhamotharan, M., and Weltzien, E. (2005a). Characterizing the production system and its anticipated changes with farmers. *In* "Setting Breeding Objectives and Developing Seed Systems with Farmers" (A. Christinck, E. Weltzien, and V. Hoffmann, eds.), pp. 41–62. Margraf Verlag, Weikersheim, Germany, and CTA, Wageningen, The Netherlands.

Christinck, A., vom Brocke, K., and Weltzien, E. (2000). What is a variety? Investigating farmers' concepts as a base for participatory plant breeding in Rajasthan, India. *In* "International Agricultural Research: A Contribution to Crisis Prevention." Proceedings of Deutscher Tropentag, October 11–12, 2000, University of Hohenheim, Stuttgart (download from: www.prgaprogram.org).

Christinck, A., Weltzien, E., and Dhamotharan, M. (2005b). Understanding farmers' seed management strategies. *In* "Setting Breeding Objectives and Developing Seed Systems with Farmers" (A. Christinck, E. Weltzien, and V. Hoffmann, eds.), pp. 63–81. Margraf Verlag, Weikersheim, Germany, and CTA, Wageningen, The Netherlands.

Christinck, A., Weltzien, E., and Hoffmann, V. (eds.) (2005c). "Setting Breeding Objectives and Developing Seed Systems with Farmers: A Handbook for Practical Use in Participatory Plant Breeding Projects." Margraf Verlag, Weikersheim, Germany, and CTA, Wageningen, The Netherlands.

Cleveland, D. A., and Soleri, D. (2002). Farmers, scientists and plant breeding: Knowledge, practice and the possibilities for collaboration. *In* "Farmers, Scientists and Plant Breeding: Integrating Knowledge and Practice" (D. A. Cleveland and D. Soleri, eds.), pp. 1–18. CABI Publishing, Oxon, UK.

Farnworth, C. R., and Jiggins, J. (2003). "Participatory Plant Breeding and Gender Analysis." PRGA Program, and CIAT, Cali, Colombia.

Germann, D., Gohl, E., and Schwarz, B. (1996). Participatory impact monitoring. Four volumes: (1) Group based impact monitoring; (2) NGO-based impact monitoring; (3) Application examples; (4) The concept of participatory impact monitoring. GATE/GTZ, Braunschweig, Germany.

Gridley, H. E., Jones, M. P., and Wopereis-Pura, M. (2002). Development of New Rice for Africa (NERICA) and participatory variety selection. *In* "Breeding Rain-Fed Rice for Drought-Prone Environments: Integrating Conventional and Participatory Plant Breeding in South and South-east Asia" (J . R. Witcombe, L. B. Parr, and G. N. Atlin, eds.), pp. 23–28 DFID/CASZ/IRRI, Bangor, UK, and Manila, Philippines (download from: www.dfid.co.uk/publications/IRRI_2002/nerica.pdf).

Joshi, K. D., Musa, A. M., Johansen, C., Gyawali, S., Harris, D., and Witcombe, J. R. (2007). Highly client-oriented breeding, using local preferences and selection, produces widely adapted rice varieties. *Field Crops Res.* **100**(1), 107–116.

Joshi, K. D., Sthapit, B. R., and Witcombe, J. R. (2001). How narrowly adapted are the products of decentralised breeding? The spread of rice varieties from a participatory plant breeding program in Nepal. *Euphytica* **122,** 589–597.

Joshi, K. D., and Witcombe, J. R. (2001). Participatory varietal selection, food security and varietal diversity in a high-potential production system in Nepal. *In* "An Exchange of Experiences from South and South East Asia." Proceedings of the international symposium on participatory plant breeding and participatory plant genetic resource enhancement, Pokhara, Nepal, May 1–5, 2000, pp. 267–274. PRGA Program, and CIAT, Cali, Colombia.

Joshi, K. D., and Witcombe, J. R. (2002). Participatory variety selection in rice in Nepal in favourable agricultural environments: A comparison of two methods by farmers' selection and varietal adoption. *Euphytica* **127,** 445–458.

Lilja, N., and Ashby, J. A. (1999). "Types of Participatory Research Based on Locus of Decision Making." Working Document No. 6. PRGA Program. Cali, Colombia.

Machado, A. T., and Fernandes, M. S. (2001). Participatory maize breeding for low nitrogen tolerance. *Euphytica* **122**(3), 567–573.

McGuire, S., Manicad, G., and Sperling, L. (2003). "Technical and Institutional Issues in Participatory Plant Breeding Done from a Perspective of Farmer Plant Breeding." PPB Monograph No. 2. PRGA Program, Cali, Colombia.

Ortiz, O. (2002). Participatory variety and clone evaluation within farmers' field schools in San Miguel, Peru. *In* "Quantitative Analysis of Data from Participatory Methods in Plant Breeding" (M. R. Bellon and J. Reeves, eds.), pp. 138–139. CIMMYT, Mexico.

Rana, R., Shrestha, P., Rijal, D., Subedi, A., and Sthapit, B. (2000). Understanding farmers' knowledge system and decision-making: Participatory techniques for rapid biodiversity assessment and intensive data plot in Nepal. *In* "Participatory Approaches to the Conservation and Use of Plant Genetic Resources" (E. Friis-Hansen and B. Sthapit, eds.), pp. 117–126. IPGRI, Rome, Italy.

Rana, R. B., Sthapit, B. R., Garforth, C., Subedi, A., and Jarvis, D. I. (2005). Four-cell analysis as decision-making tool for conservation of agrobiodiversity on-farm. *In* "On-Farm Conservation of Agricultural Diversity in Nepal (B. R. Sthapit, M. P. Upadhyay, P. K. Shrestha, and D. I. Jarvis, eds.), Vol. I, pp. 15–24. IPGRI, Rome, Italy.

Rijal, D., Rana, R., Subedi, A., and Sthapit, B. (2000). Adding value to landraces: Community-based approaches for in situ conservation of plant genetic resources in Nepal. *In* "Participatory Approaches to the Conservation and Use of Plant Genetic Resources" (E. Friis-Hansen and B. Sthapit, eds.), pp. 166–172. IPGRI, Rome, Italy.

Snapp, S. (2002). Quantifying farmer evaluation of technologies: The mother and baby trial design. *In* "Quantitative Analysis of Data from Participatory Methods in Plant Breeding" (M. R. Bellon and J. Reeves, eds.), pp. 9–16. CIMMYT, Mexico.

Sperling, L., and Christinck, A. (2005). Developing strategies for seed production and distribution. *In* "Setting Breeding Objectives and Developing Seed Systems with Farmers" (A. Christinck, E. Weltzien, and V. Hoffmann, eds.), pp. 153–176. Margraf Verlag, Weikersheim, Germany, and CTA, Wageningen, The Netherlands.

Sperling, L., Loevinsohn, M., and Ntabomvura, B. (1993). Rethinking the farmers' role in plant breeding: Local bean experts and on-station selection in Rwanda. *Exp. Agric.* **29,** 509–519.

Sthapit, B., Joshi, K., Gyawali, S., Subedi, A., Shreshta, K., Chaudhary, P., Rana, R., Rijal, D., Upadhaya, M., and Jarvis, D. (2002). Participatory plant breeding: Setting breeding goals and choosing parents for on-farm conservation. *In* "Quantitative Analysis of Data from Participatory Methods in Plant Breeding" (M. R. Bellon and J. Reeves, eds.), pp. 104–112. CIMMYT, Mexico.

Sunwar, S., Basnet, L. K., Khatri, C., Subedi, M., Shreshta, P., Gyawali, S., Bhandari, B., Gautam, R., and Sthapit, B. (2006). Consolidating farmers' role in participatory maize breeding in Nepal. *In* "Bringing Farmers Back into Breeding: Experiences with Participatory Plant Breeding and Challenges for Institutionalization" (C. Almekinders and J. Haardon, eds.), pp. 70–79. Agromisa, Wageningen, The Netherlands.

Tapia, M. E., and de la Torre, A. (1998). "Women Farmers and Andean Seed." FAO and IPGRI, Rome, Italy.

Tapia, M. E., and Rosas, A. (1993). Seed fairs in the Andes: A traditional strategy for *in situ* conservation of phytogenetic resources. *In* "Cultivating Knowledge: Genetic Diversity, Farmer Participation and Crop Research" (W. De Boef, K. Amanor, and K. Wellard, eds.), pp. 111–118. Intermediate Technology Publications, London, UK.

Welzien, E., Christinck, A., Touré, A., Rattunde, F., Diarra, M., Sangaré, A., and Coulibaly, M. (2006). Enhancing farmers' access to sorghum varieties through scaling up PPB in Mali, West Africa. *In* "Bringing Farmers Back into Breeding: Experiences with Participatory Plant Breeding and Challenges for Institutionalization" (C. Almekinders and J. Haardon, eds.), pp. 65–76. Agromisa, Wageningen, The Netherlands.

Weltzien, E., Smith, M. E., Meitzner, L. S., and Sperling, L. (2003). "Technical and Institutional Issues in Participatory Plant Breeding: From the Perspective of Formal Plant Breeding. A Global Analysis of Issues, Results, and Current Experience." PPB Monograph No. 1. PRGA Program, Cali, Colombia.

Weltzien, E., vom Brocke, K., and Rattunde, H. F. W. (2005). Planning plant breeding activities with farmers. *In* "Setting Breeding Objectives and Developing Seed Systems with Farmers". (A. Christinck, E. Weltzien, and V. Haffmann, eds.), pp. 123–152. Margraf Verlag, Weikersheim, Germany, and CTA, Wageningen, The Netherlands.

Witcombe, J. R., Gyawali, S., Sunwar, S., Sthapit, B. R., and Joshi, K. D. (2006). Participatory plant breeding is better described as highly client-oriented breeding. II. Optional farmer collaboration in the segregating generations. *Exp.Agric.* **42,** 79–90.

Witcombe, J. R., Joshi, A., Joshi, K. D., and Sthapit, B. (1996). Farmer participatory crop improvement. I. Methods for varietal selection and breeding and their impact on biodiversity. *Exp. Agric.* **32,** 445–460.

Witcombe, J. R., Joshi, K. D., Gyawali, S., Musa, A. M., Johansen, C., Virk, D. S., and Sthapit, B. R. (2005). Participatory plant breeding is better described as highly client-oriented breeding. I. Four indicators of client-orientation in breeding. *Exp.Agric.* **41,** 299–319.

CHAPTER 8

Livestock, Livelihoods, and Innovation

Czech Conroy

Summary

Livestock make important contributions to the livelihoods of most people living in rural areas in developing countries. Livestock ownership patterns are strongly influenced by households' access to various kinds of assets, including natural (fodder, water) and financial ones. People involved in livestock production may be spontaneously involved in innovation, driven by various factors, including population pressures, increasing land scarcity, or improved access to markets. These innovations can take various forms, including changes in livestock species kept, new production technologies, new arrangements for obtaining input services, or innovations in the way they process or market livestock and livestock products. Nevertheless, improvements in the productivity of, and returns from, resource-poor people's livestock in less developed countries have been disappointing and have not benefited from the livestock revolution as much as resource-rich and corporate livestock producers. Appropriate pro-poor technologies can be developed when researchers work closely with livestock keepers in an iterative research process.

Agricultural Systems: Agroecology and Rural Innovation for Development

LIVESTOCK AND LIVELIHOODS

Contributions of Livestock, Production Systems, and Agroecology

Contributions to Livelihoods

Livestock are a component of the livelihood systems of most rural households in less developed countries (LDCs), both farming and landless households. They often play a number of roles, including being a source of cash (planned sale) or serving as liquid assets (emergency sale); providing inputs to crop production; spreading a farmer's risks by acting as a buffer to poor crop yields; a source of food; a means by which the poor can derive benefits from land owned by others; and cultural value. In mixed farming systems, livestock may provide manure, fuel, and draught power as inputs, while receiving crop residues from the crop production side of the farm. The relative importance of different roles is liable to vary by production system and over time, as internal (to the farm and household) and external factors change.

Women often play a lead role in livestock keeping, particularly in the case of small stock such as goats and scavenging poultry. They often have discretion as to how income from sale of livestock is used, and owning livestock, even just one goat, can raise a poor woman's status and self-esteem significantly.

In most rural households the contribution of livestock to livelihoods is seen as being less than that of crop production. However, in some situations, depending on the production system and the agroecology, livestock enterprises are the most important. Livestock tend to be particularly important in dryland regions, in which crop production is either not possible or is less remunerative and relatively risky.

Types of Production Systems

There are different ways of classifying livestock production systems. The Food and Agriculture Organization (FAO) has a typology that is a three-tiered one, in which the two main categories are "solely livestock" and "mixed farming systems," each of which has further subcategories (see Fig. 8.1). Another typology, not including industrial systems, has the following five categories: (1) commercial ranching; (2) smallholder mixed farming/agropastoral; (3) pastoralism (landless owner/herder); (4) sedentary landless—extensive; and (5) sedentary landless—intensive. Box 8.1 describes goat-keeping systems in India that roughly correspond to categories 2, 4, and 5, respectively.

Livestock and Agroecology

In extensive livestock systems, a large proportion of animal feed has traditionally come from grazing on common lands, in forests, and on fields of stubble after crops have been harvested. In the latter case, there is a symbiotic relationship between

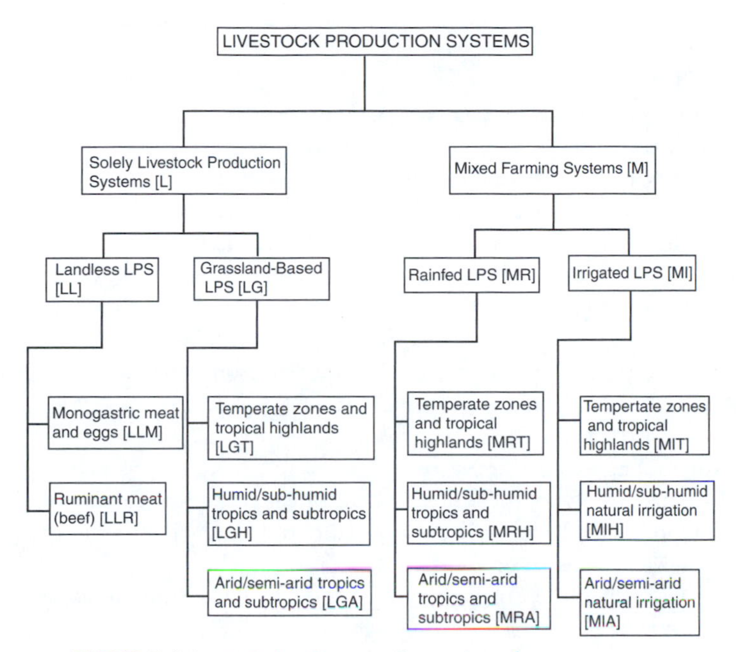

FIGURE 8.1 FAO classification of livestock production systems.

BOX 8.1 Livelihoods and Goat-Keeping Systems in India

Smallholder Agropastoral Systems

These systems can be subdivided into those of small ruminant specialists and nonspecialists. In south Rajasthan, there are many tribal people who are non-specialists, who sometimes live in the same village as specialist castes. What they have in common is that the feed resources include crop residues from their own land, forage from their private wasteland, and forage from common lands. Specialists have larger herds, comprising mainly sheep, with flocks of 50 to 150, which are kept for their meat and wool, but also some goats. They tend to be more self-sufficient in feed resources and the animals are herded by men. Nonspecialists keep goats rather than sheep: they have smaller landholdings and hence depend more on common grazing lands around the village, where the goats are usually herded by children, women, or old people. Adult nonspecialists are often involved in seasonal migration for wage labor, which may be a constraint on the numbers and types of animals that can be kept.

(*Continues*)

BOX 8.1 *(Continued)*

Marginal/Landless Livestock Specialist

In the Bhavnagar district, Gujarat, the Rabaris are the specialist livestock caste. They tend to be landless, or own only a small amount of land, which may be of poor quality: at certain times of the year some of them do wage labor. Their livestock include goats and large ruminants. Some of them spend most of their time herding their animals. Being landless, they graze them primarily on common lands and on other people's agricultural fields after harvesting. Nearly all of their income comes from livestock production: milk production is the major source of revenue, with some income also from the sale of manure and the occasional animal sale.

Landless Wage-Laborer System

In Bhavnagar, some of the scheduled caste goat keepers are landless and their livelihoods depend primarily on wage labor, usually by both adult males and females. Most wage labor is agricultural labor, and there are about 4 months in the year when there is little, if any, available. These households keep one to two breeding does. The feeding system is a combination of stall feeding, with forage collected in the fields where they work or on their way home, and some herding by old family members or children. Families who do not have any members available for herding tend to pay the livestock specialists, Rabaris, to do the herding for them. Milk for subsistence use is the main product, and people of higher castes would not want to purchase milk from them anyway (social stigma).

Source: Conroy, 2005.

livestock and crop production, with crop residues an input to the former and manure to the latter. Two common trends have been for human populations and livestock numbers (see Table 8.2) to increase over time, leading to shrinkage of the common land areas (as people privatize land) and degradation of the forage resource base as the pressures on remaining pastures intensify. These trends may eventually encourage people to ban or control grazing on common pastures, invest in the improvement of the pastures (by fencing them off, enrichment planting, etc.), and switch to cut-and-carry fodder systems, as has happened in numerous villages in Rajasthan (Conroy and Lobo, 2002). However, this is only likely to happen where there is good social cohesion and a history of collective action, and where the social group is confident in its rights to the land and its capacity to benefit exclusively from the investment in rehabilitating it.

In dryland regions that normally experience seasonal fodder scarcity, the cutting and storage of fodder can smooth out and extend its availability over a year (Conroy and Lobo, 2002). This is useful for farmers who want to become involved in dairying and who need to supply milk on a fairly continuous basis, although they will also need a continuous supply of green fodder. There usually needs to be a high degree of land or fodder scarcity before planted fodder, and cut-and-carry methods of feeding, become more attractive than open grazing (Tiffen *et al.*, 1994).

Intensive landless production systems are becoming increasingly important in some LDC regions. These rely on purchased feed rather than local feed resources, and hence the local agroecological linkages are severed in these systems.

Factors Influencing Livestock Ownership Patterns and Systems

Household Access to Assets

Livestock ownership patterns are strongly influenced by access to various kinds of assets or capitals (natural, financial, human, social, and physical). This is illustrated by data from a village in Rajasthan (see Table 8.1). Relatively speaking, members of the Rebari caste owned more of productive lands (*natural capital*) than members of the Bhil tribe and had greater access to irrigation facilities (*physical capital*). They were also relatively wealthy and had better access to credit (*financial capital*)

TABLE 8.1 Livestock Ownership by Ethnic Group in Barawa Village, Udaipur District, Rajasthan, 1999–2000

	Bhils (tribals)		Rebaris	
	Total	Mean/household	Total	Mean/household
Households	32	—	63	—
Beeds* (hectares)	10.1	0.33	111.25	1.78
Arable land (ha)	11.4	0.35	38.38	0.6
Buffaloes	2	0	61	0.97
Cows	22	0.7	60	0.95
Camels	0	—	102	1.6
Bullocks**	40	1.3	29	0.5
Goats	117	3.7	213	3.4

Source: Jindal (2000).

*Beed is poor quality private land that is used mainly as a source of fodder.

**The lower numbers of bullocks owned by *Rebaris* were due to most families having at least one male member employed in a permanent city job and not being available to plough the land.

and were better educated (*human capital*) with better connections to traders (*social capital*). Rebaris estimated that there had been a 90% decline in their camel ownership during the previous 50 years, partly because of reduced access to fodder from state forests (*natural capital*). However, relatively good access to all or most of these assets appears to have contributed to a rapid increase in buffalo ownership by the Rebaris.

Villagers estimated that the number of buffaloes in the village had more than doubled in the previous 10 years. Their milk was sold for a good price in nearby towns, as its very high fat content is much valued by consumers, but only the Rebari households had acquired buffaloes and taken advantage of this income-generating opportunity. Buffalo rearing was deemed as "risky" by tribals, as it involved allocating considerable cash for purchase of the animal and regular expenditure in terms of good feed, which is necessary for high milk yields. Only the Rebaris could take this risk and afford the cash outlays involved. Furthermore, their superior endowments of natural capital enabled them to obtain a large proportion of the necessary fodder (including green fodder) for most of the year.

Socioeconomic Status and Livestock Ownership: The "Livestock Ladder"

The types and numbers of livestock owned by households or individuals can vary substantially over time. The concept of a "livestock ladder" is sometimes used by people and organizations (such as the International Livestock Research Institute [ILRI]) working in livestock research and development to describe the relationship between a household's financial status and livestock ownership. The livestock ladder is seen as a series of steps from one livestock type to another as people become less and less poor (see Fig. 8.2). For example, ILRI states that "in some situations, the

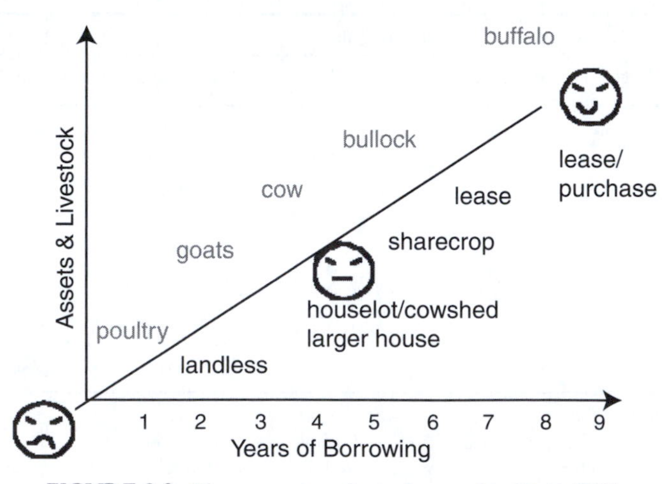

FIGURE 8.2 The progression of animal ownership (Todd, 1998).

'livestock ladder' may allow the poor to progress from modest livestock holdings, typically consisting of a few local poultry, and ... to acquire sheep and goats or pigs and perhaps, in some cases, also cattle or buffaloes" (ILRI Revised Strategy for 2003–2010). The sequence is assumed to be related to which species people can afford to buy, the market value of chickens being the lowest and that of large ruminants, especially milch animals, the highest. There is also an implicit assumption that larger animals give better financial returns. This kind of sequence has been described for Asian countries, based on survey information relating to female recipients of microcredit:

> Across projects and countries in Asia there is an ideal progression that even the poor aspire to. It starts with ducks and chickens; then a few goats are kept for milk or fattening and to slaughter for a day of sacrifice; next a milch cow; then a bullock for ploughing in cooperation with another one-buffalo family; then two bullocks, which can be used to plough the fields of others .... In India, one would add a milch buffalo at the apex of desirable animals on the farm (Todd, 1998).

A similar sequence has been described for Kenya, that is, chicken → sheep or goat → local cattle → dairy cattle (Kristjanson *et al.*, 2004); and in Zimbabwe, poultry have been described as forming "the first step on the ladder of livestock ownership for the poorest families" (Bird and Shepherd, 2003).

Critique of the Livestock Ladder Concept

There are a number of weaknesses in the way this concept is sometimes used. First, it has been argued that women *prefer* to start with small stock, such as chickens or goats, for various reasons, including large animals need large spaces, shelter in the form of cowsheds, security from theft, and large amounts of fodder, which a poor woman cannot provide; the risk of death and theft are probably equally high with small as with large animals, but the loss of a goat is less catastrophic than the loss of a cow (Todd, 1998). However, evidence shows that poor women do not always prefer to start with small stock and can sometimes provide the resources required to keep large ruminants. In a project in Bangladesh in which ultrapoor women were offered a choice of livestock types on a grant (not credit) basis, the vast majority preferred to start with cattle. The main reason they gave for this was that chickens and goats had much higher mortality rates and hence were a riskier investment (Bond, 2006; Conroy, 2006), which has been borne out by subsequent experience. The women have been able to feed the cattle adequately and theft has been minimal (Fig. 8.3).

Second, it is often assumed that larger and more valuable animals generate the best returns and that people will therefore automatically choose to move up the ladder to the next livestock step when they can afford to do so. While this assumption may generally be valid, differing conditions may change the profitability of different livestock species and enterprises. For example, one of the reasons why returns on scavenging poultry keeping tend to be low is that mortality rates tend

FIGURE 8.3 Yemeni woman with cow.

to be high (e.g., 40% in India; Conroy *et al.*, 2005) because of disease and/or predation (Conroy *et al.*, 2005). The availability of effective vaccination services could significantly enhance the profitability of scavenging poultry, in which case people might prefer to expand their poultry flock rather than move up the livestock latter. Third, it is important to recognize that villagers may sometimes prefer to move off the livestock ladder before reaching the top of it and to invest money in other productive resources, such as land for crop production.

Long-Term Trends in Livestock Ownership, Production, and Consumption

Livestock Ownership

Smallholder mixed farming systems have been strongly influenced by some important trends in many LDC regions. These include replacement of animal traction with tractors, reductions in farm sizes, and reductions in off-farm grazing and water resources. The latter two factors have made it increasingly difficult to make a living from agriculture and, together with "pull" factors (better wages), have encouraged members of a rapidly growing number of households to migrate for labor, either seasonally or long term, as with the Rebaris in the aforementioned example (see Table 8.1). However, there have also been some positive trends for livestock production, namely the development of stronger market linkages and a rapid growth in the demand for livestock products in many LDCs, particularly Asia.

These trends have brought about significant changes to livestock in mixed farming systems, including a reduction in the number of large ruminants per farm, a shift from cattle to small ruminants on farm, a shift from cattle for traction to cattle for milk (particularly in India), and intensification of livestock production, especially in periurban areas. Some of these trends are reflected in livestock census data for Rajasthan, India, that are summarized in Table 8.2. Decadal data are also shown in Fig. 8.4 (i.e., 1988 and 1997 data are not included).

Highly aggregated data like these may conceal various important differences among districts, ethnic groups, or production systems. Nevertheless, as shown over a period of 50 years, there have been major aggregate changes at the state level in the numbers and proportions of different types of livestock:

- numbers of all categories of livestock, except cattle, were far higher in 2003 than they were in 1951
- the cattle population peaked in 1983 and declined in 1988 and again in 2003
- the buffalo population has grown the most rapidly, more than trebling, and has increased in each successive census year

TABLE 8.2 Populations of Various Types of Livestock in Rajasthan, 1951–2003 (millions)[a]

Livestock category[b]	1951	1961	1972	1983	1988	1992	1997	2003
Cattle	10.8	13.1	12.5	13.5	11.0	11.7	12.1	10.9
Buffaloes	3.0	4.0	4.6	6.0	6.3	7.8	9.8	10.4
Sheep	5.4	7.4	8.6	13.4	9.9	12.5	14.6	10.1
Goats	5.6	8.1	12.2	15.4	12.6	15.3	16.9	16.8

[a]From Sagar and Ahuja (1993); livestock censuses.

[b]Other types of livestock found in Rajasthan, such as camels and poultry, are relatively unimportant at the state level: their combined total in 1997 was 1.2 million. However, they may be important in particular districts or blocks.

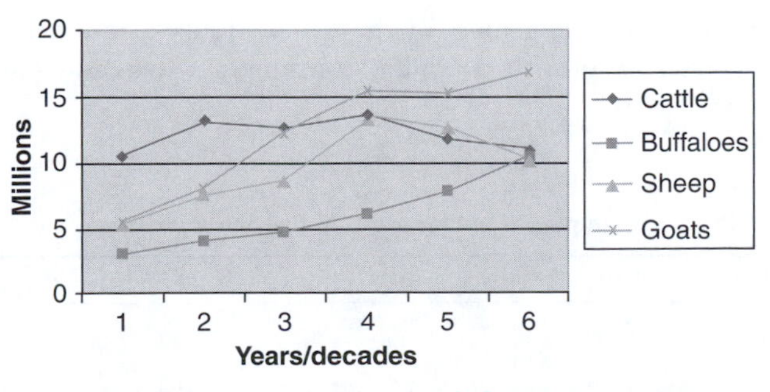

FIGURE 8.4 Livestock populations in Rajasthan (1951–2003).

- the sheep population nearly doubled over the whole period, but decreased in 1972, 1988, and 2003
- the goat population more than trebled over the whole period, but decreased in 1988 and stabilized in 2003.

The declines in numbers of some species in 1972, 1988, and 2003 were largely because of droughts. Growth in the buffalo population reflects a growth in the market for milk, as does growth in the numbers of crossbred cattle (not shown in Table 8.2), which more than doubled between 1997 and 2003, from 211,000 to 464,000.

The "Livestock Revolution"

Livestock consumption and production have been growing faster than any other agricultural subsectors, and it has been predicted that by 2020 livestock will account for more than half of total global agricultural output in value. This process has been termed the "livestock revolution." Most of the growth in consumption of livestock products is taking place in LDCs, as can be seen from Table 8.3, particularly in Asia and Latin America. Total meat consumption in developing countries overtook that of developed countries in the mid-1990s and is now substantially higher. The gap is projected to grow still wider by 2020.

The livestock revolution has been characterized by the following production trends:

- from resource driven (shaped by local feed availability) to demand driven
- from local demand to regional, national, and international demand, and from rural to urban
- from extensive (land-based) to intensive
- production expansion has mainly involved monogastrics (pigs, poultry) rather than ruminants
- geographical clustering of production units, either in a periurban belt around consumption centers or close to commercially produced feed resources.

Data on Indian meat consumption shown in Table 8.4 mirror the aforementioned bullet point regarding most of the production expansion involving pigs and poultry: percentage growth in poultry consumption was easily the most rapid, followed by pig.

TABLE 8.3 Meat Consumption (million tons), Actual and Projected

	1983	1993	2001	2020
Developing world	50	88	132	188
Developed world	88	97	102	115

TABLE 8.4 Meat Consumption in India (million tons)

Meat type	1983	1993	2001
Poultry	0.2	0.5	1.3
Bovine	1.9	2.5	2.6
Mutton and goat	0.5	0.6	0.7
Pig	0.3	0.5	0.6
Other	0.0	0.2	0.1
Total	2.9	4.3	5.3

Source: FAO statistics.

There are two schools of thought on the implications of the livestock revolution for poor producers: some observers see it mainly as an opportunity, whereas others believe it poses a serious threat. The optimists argue that it represents an opportunity for bringing about sustained and increased revenues for the poor and making a major contribution to achievement of the millennium development goals, whereas the pessimists are concerned that livestock production by the poor could be undermined by increased competition from larger production units, with their economies of scale, and by more stringent sanitary requirements with their high compliance costs.

It is likely that both schools are right in the sense that either of these outcomes could materialize, depending on the enabling environment (prevailing policies, laws, livestock services, and marketing systems), which may vary from country to country. This point can be illustrated by the example of India. Despite the large numbers of various types of livestock in India, the productivity of all species is low, with the exception of commercial poultry. This suggests that the enabling environment there, primarily livestock services (research, extension, veterinary services) and policies, leaves much to be desired.

Most of the growth in poultry consumption in India has been supplied by the commercial and industrial poultry subsector.

LIVESTOCK RESEARCH AND INNOVATION

Despite the important contribution commonly made by livestock to poor people's livelihoods in LDCs, the productivity of these animals tends to be well below what is possible due to a variety of problems that livestock keepers and their animals face. If these constraints could be overcome, the benefits to huge numbers of resource-poor people would increase significantly. There is a major need, therefore, for innovations that will enable poor livestock keepers to improve the productivity of their livestock, enhance the subsistence and income benefits that they derive from them, and benefit from the livestock revolution. Unfortunately, most traditional livestock research has tended not to benefit poor livestock keepers.

General Problems with Traditional Livestock Research

Animal science research in the south has been strongly influenced by that in the north. The latter has a history of being orientated toward meeting the needs of estates, and more recently "factory farms," and being geared to increasing the production of livestock and their products (Waters-Bayer and Bayer, 2002). Specialization and commercialization have been common themes. Another feature of this research has been manipulating the environment so that it contributes to maximum production or productivity; for example, feeding systems are based on the nutritional "demand" of the animals rather than on the availability of various feed resources at different times of the year (Bayer and Waters-Bayer, 1998).

The traditional "northern" or "western" paradigm of animal science research for developed countries has been dominant and pervasive. It was transferred directly to LDCs by researchers from the north who turned their attention to these countries, and indirectly by its influence on the education of animal scientists from LDCs. If scientists had been more sensitive to the needs and priorities of livestock keepers in LDCs; they might have reoriented their research so that it was more appropriate and relevant to them. However, they were not particularly sensitive or responsive; they often failed to understand the circumstances of small farmers (Roeleveld and van den Broek, 1996), and the old paradigm persisted. They were aware that traditional systems in LDCs were often substantially different from those in the textbooks, but they saw these traditional systems as backward and in need of change. Hence they did not take much effort to understand why these systems were different, and it did not occur to them that resource-poor livestock keepers might have different objectives from resource-rich, commercially oriented ones. They failed to take proper account of the fact that most livestock in LDCs belong to farmers and are an integral part of a mixed farming, crop–livestock system, providing inputs into crop production (in the form of draught power and manure) and receiving inputs (e.g., crop residues) from crop production.

Another reason why old attitudes, methods, and beliefs persisted was that researchers' contact with resource-poor livestock keepers was quite limited. They did most of their research in the research station because it was more convenient and also because it enabled them to exert more control over treatments and nonexperimental variables. This in turn meant that they could produce sound, scientifically valid results that were publishable in journals, which has been more important for scientists' promotion than has the usefulness of the results for farmers (Bayer and Waters-Bayer, 1998). (Reward systems in research organizations tend to be strongly dependent on the extent to which staff are able to publish articles in respected scientific journals. Such journals tend to be prejudiced against material based on on-farm trials, particularly participatory ones, because it may not satisfy conventional criteria for experimental design and statistical rigor [Chambers, 1997; Morton, 2001].) It is hardly surprising, therefore, that there has been a "lack of participation and interest among animal scientists" in on-farm animal research (Amir and Knipscheer, 1989).

Furthermore, scientists' accountability to resource-poor livestock keepers has been almost nonexistent, and there has been little pressure on them to work with these groups. Where research has been geared to livestock keepers' needs, it has been primarily addressing the needs of relatively resource-rich, commercially oriented groups because they have more influence and also because traditional research is more likely to be relevant to their needs anyway, as their production systems tend to be more similar to those in the north.

Case Studies of Innovation Related to Livestock

Despite the problems just described, the picture has not been entirely gloomy. This section describes a number of case studies of livestock-related innovations that have been developed in recent years, most of which have been suitable for, and popular with, poor livestock keepers.

Minor/Medium Innovations

The first three case studies describe the development of technologies that could be considered to be minor or medium in terms of the degree of change from existing practices that they involve. These are as follows:

1. Development of a low-cost egg-cooling technology (India)
2. Fodder innovations in southeast Asia
3. Low-cost treatment of mange in goats in semiarid Kenya.

Egg-cooling technology involves the use of locally available goods (bowls, jute bags) that people tend to have in their possession anyway or which they can make or obtain at low cost. Similarly, the mange treatment technology utilizes locally available and accessible plant materials, and the only cost is the labor required to harvest and process them. In the fodder innovations project, planting materials are supplied to farmers at no cost, and the main costs to farmers of planting and using fodder crops are the labor cost of harvesting the fodder and the opportunity cost of land taken up by the fodder crop. However, farmers were given complete discretion as to where they planted crops and tended to choose ones where the opportunity cost was minimal.

Major Innovations

Case studies 4 and 5 describe the development of technologies that could be considered to be major in terms of the degree of change from existing practices that they involve, namely

1. Development of the Kebkabiya donkey plow in Darfur, Sudan
2. The "Bangladesh model" of poultry production and marketing.

Case studies 1–4 are examples of technological innovations; case study 5 is a combination of technological innovations and innovations in marketing arrangements; and case study 6 is primarily about social and marketing innovations.

Development of the donkey plow represented a major innovation as it took place in an area with little previous history of plowing and with no experience of donkeys being used for plowing: at the start of the project, many farmers laughed at the idea of using donkeys for plowing. The Bangladesh poultry model is another major innovation, as it changes the whole poultry production and marketing system, including the birds used (from local to exotic), the type and source of feed (from scavenging and household scraps to commercial feed), the housing, the main product (from meat to eggs), and the marketing.

Projects described in the first four case studies can be considered to have been successful in that the technologies developed have been adopted and disseminated by farmers. The Bangladesh poultry model appears to be unsustainable, as the system tends to break down after donor funding has been withdrawn. Thus, although it has been extremely widely implemented, it has not been effective in the medium to longer term. In the four successful projects there was close collaboration between researchers and users and a high degree of user participation in the process; in most cases, technology development was iterative, with repeated modifications being made over time. In most cases, there were no other major stakeholders, but manufacturers played a key role in the donkey plow development.

In general terms, the case studies can be said to confirm the inappropriateness of the traditional transfer of technology model, with its conception of users as passive recipients of technologies, and the validity of the emphasis given by several more recent models to users and researchers working closely together on technology development. The case studies also show that the technology development process may go through several cycles, with technologies gradually becoming more and more suitable to farmers' needs.

CASE 8.1 Development of a Low-Cost Egg-Cooling Technology

A 5-year scavenging poultry research project in Rajasthan looked at issues of egg management. The project was managed by the Scottish Agricultural College's Avian Science Research Centre and implemented, in collaboration with the BAIF Development Research Foundation, an Indian nongovernmental organization (NGO), with inputs from a socioeconomist. Poultry keepers in Udaipur, Rajasthan, informed the research team that during the summer months (March–June), when temperatures can reach more than 40°C, the percentage of spoiled eggs increased. It is well known in poultry science that high temperatures (>27°C) can increase the incidence of abnormal

CASE 8.1 (*Continued*)

embryos and the percentage of embryos that die during incubation. Thus, the project team hypothesized that this was the cause of poor hatchability and suggested to poultry keepers in the project villages running an experiment to test a technology to address the issue based on the principle of evaporative cooling.

After discussions with the poultry keepers, a simple technology was identified based on locally available materials that had the potential to reduce and stabilize the temperature of the eggs. The technology involved the use of a half-moon-shaped bowl in which the eggs would be kept cool by evaporative cooling. The bowl was filled with an earth/sand mixture that was kept moistened with water. Then a piece of jute bag was placed on the sand to prevent the eggs coming into direct contact with water (which could facilitate contamination). The eggs were placed on the bag, and a cotton cloth or woven basket was placed over them. The bowl was placed either on a shelf or ledge or on the floor inside a family building. When the hen stopped laying, all the eggs were placed under her, as per existing traditional practice.

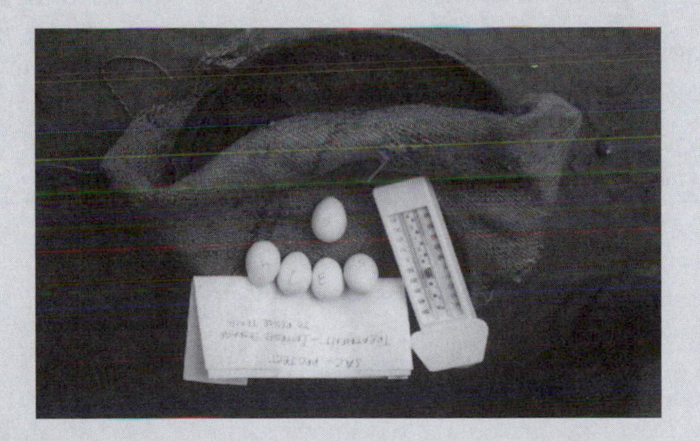

The project conducted a pilot trial in February–May 2003 with two groups of poultry keepers to test this technology in which all eggs were candled first to confirm fertility. The temperature in the vicinity of the eggs and in the egg store room (ambient) was recorded each morning (between 8:00 and 10:00). Numbers of eggs that hatched viable chicks, that contained dead-in-shell embryos, or that had spoiled (infertile or bacterial rot) were recorded. The first trial, held in 2003, showed promising results and hence was repeated on a larger scale, with more birds and eggs, in March–June 2004.

(*Continues*)

CASE 8.1 (*Continued*)

Of the fertile eggs available for hatching in the first trial (2003), the percentages of chicks that hatched were 97.0 and 69.0% for the modified storage and control groups, respectively. In the second trial (2004), the equivalent figures were 84.3 and 69.5%, respectively. The minimum room temperature during storage tended to exceed physiological zero and often the maximum temperature achieved was in excess of 32°C: the highest temperature recorded was 42°C. Results provide clear evidence that the modified storage of eggs did improve the overall hatchability of the eggs, and data were consistent with the hypothesis.

Development of the cooling technology went though an iterative process. Initially, clay pots were used, but because these had a tendency to crack, then locally available iron pots were used (e.g., in the 2004 trial). Although the latter proved to be effective, reed baskets lined with cloth have been used more recently. One advantage of these is that evaporation may also occur through the side of the container, leading to greater cooling than the iron pot technology: they may also be less expensive. The technology was adopted by a large proportion of the poultry keepers in the project villages and by many others in nearby villages who heard about it from people in the project villages. It has also been adopted in a few villages in the state of Tamil Nadu, where it was publicized through farmer poultry schools.

Source: Sparks *et al.* (2004).

CASE 8.2 Fodder Innovations in Southeast Asia

This case study was based on experiences of a project over 8 years, from 1995 to 2003, eventually operating in six countries. The Forages for Smallholders Project (FSP) was coordinated by the International Centre for Tropical Agriculture (CIAT) whose goal was to work with resource-poor upland farmers. In this project, "forages" mean grass and legume crops that are specifically cultivated to provide feed for animals. These are usually planted within a complex pattern of other food and cash crops, utilizing farm space and labor in a multiple and optimal way, such as in lines along contours on farm land; as cover or green manure crops in fruit trees, coffee, and tea; as live fences for demarcation of external and internal boundaries; and as pastures and fodder banks in backyards or under young palm oil or coconut plantations.

CASE 8.2 (*Continued*)

Forages are often of secondary importance to poor farmers, as food security is their main concern, so developing technologies of interest to them can be a major challenge. The project initially evaluated some 500 species and accessions of forages, and found 25 to 40 of them to be well adapted to climate, soils, and diseases: these were the ones recommended for evaluation by new farmers. In FSP, the process of farmer participatory research, in which farmers were involved in planning and carrying out the evaluation of new species and in adapting the management of them to their farming system, has been a major contributor to farmer adoption of forage technologies. A farmer was considered to have adopted a technology when (s)he experimented with a species or a forage technology and subsequently expanded the cultivated area with his/her own resources. About 25% of farmers dropped out of the evaluation process after 1 to 3 years. Farmers have developed some unique systems that they discovered to be more profitable, such as feeding cut fodder to carp instead of cattle.

More than 4000 farmers benefited from this project over a 3-year period (2000–2002).

One reason for the project's success has been its recognition that no two smallholder farms are the same, and that farmers need to experiment with and develop their own forage systems. Thus, the project aims to provide "building blocks" and not "finished products." In other words, the project shows the farmers the species and forage systems that have worked in other places, while at the same time allowing new farmers to evaluate a range of optional species and develop their forage systems within their overall farming system. Where feasible, new farmers were taken on cross visits to other farmers who had been working with the project for several years, as these were seen to be best placed to demonstrate how forage can make a positive contribution to livelihoods, livestock, and the environment.

Perhaps because it covers such an unusually large geographical area, the project identified important ways of enrolling in-country partners, both organizational and individual, into supporting and promoting the project. At the organizational level, it was found that building partnerships at local, provincial, and national levels is crucial to obtain broad support for the initiative. Therefore, the project makes a serious effort to invite key agricultural or political officials at district or provincial levels for various training workshops and courses.

At the individual level, the project seeks to identify enthusiastic farmers and extensionists. In every new community exposed to cross visits from participating farmers, new champion farmers emerged, whose enthusiasm and

(*Continues*)

CASE 8.2 (*Continued*)

experience is harnessed by the project. They in turn will become key farmers able to receive other farmers from new areas to show them their experience in forage evaluation and utilization. Promising field staff are often identified during training courses: apart from skills, attitudes are also an important selection criterion for staff. In very remote areas, where extension workers can be scarce, another option that has worked well is the use of experienced farmers as extension workers.

Source: Roothaert and Kerridge (2005).

CASE 8.3 Low-Cost Treatment of Mange in Goats in Semiarid Kenya

The Dryland Applied Research and Extension Project (DAREP) covered approximately 70,000 smallholder farming families in semiarid areas of Embu, Tharaka-Nithi, and Central Isiolo districts in Kenya. DAREP, which was operational in the mid-1990s, followed a process of farming systems characterization, diagnosis of priority constraints, trial planning, technology testing, evaluation, and extension/dissemination. A project survey found that mange was the second most important cause of mortality in goats in the districts. Mange occurs in most arid and semiarid lands of Kenya, including Kajiado, Kitui, Makueni, Machakos, Mwingi, and Narok. Some farmers had found commercial solutions to mange too expensive and had started looking for locally available alternatives.

A focused participatory rural appraisal, using group discussions and visits to a few local herbalists, came up with a list of about eight local concoctions that some farmers had been testing or using. This list was further screened through discussion with farmers, and a trial was designed comparing four local concoctions that both farmers and the researcher felt reasonably comfortable using, with two recommended commercial medicines. The two recommended commercial medicines were an organophosphorus acaricide, *Supa dip,* and ivermectin, *Ivomex*; these were used as experimental controls. A mange control trial was conducted on-farm with infected herds belonging to farmers. To maintain experimental standards for comparison, the local concoctions were supplied and prepared by the researcher.

Two of the treatments based on local technical knowledge proved to be significantly more effective than the other treatments, including the commercial ones. One of these two treatments was a paste made from a mixture

CASE 8.3 (*Continued*)

of tamarind fruit and crushed roasted castor oil seed in equal proportions (by weight), which had certain advantages over the other one. This technology does not require land or cash, unless the fruit is purchased. Tamarind (*Tamarindus indica)* is quite widespread in the villages where the trials were conducted and in much of semiarid Kenya. The fruit is also sold in local markets, where it is inexpensive in season. Castor (*Ricinus comunis*) is also widespread. Both ingredients are reasonably accessible (even to resource-poor people), particularly castor, which has become a weed in some places. The only resource required to utilize the technology is the labor required to harvest and process the raw materials. However, the commercial drug for treating mange, ivermectin, was quite expensive. Thus, this technology has proved to be popular with goat keepers, many of whom have been adopting it. The government funded the production of 1000 copies of a brochure about the technology, aimed at farmers, for distribution by extension agents; about 3500 farmers were informed about it through field days.

Other factors contributing to project success were the facts that the research was demand driven and that the farmers/livestock keepers were involved from diagnosis and prioritization, through to experimentation and dissemination. In addition, there was full involvement of livestock and veterinary extension officers throughout the project, which gave them ownership of and commitment to the new technology and made the veterinary officers less skeptical about local remedies.

Sources: Kang'ara (2005) and Conroy and Sutherland (2004).

CASE 8.4 Development of the Kebkabiya Donkey Plow in Darfur

The Kebkabiya smallholder project (KSP) was initiated by Oxfam after the 1984–1985 drought. KSP aimed to empower the communities, strengthen the position of the poor, and increase food security. This project focused on the development of an animal traction technology, something that was largely absent from the area at the outset, but which project staff saw as an important means of increasing food security. The only plowing was that provided to rich farmers by people hiring out camel-plowing services, but this practice had been declining for various reasons, including increased theft of camels. The only widely owned animal capable of plowing was the donkey. A careful needs assessment was undertaken, which proved to be a prerequisite

(*Continues*)

CASE 8.4 (*Continued*)

for success. It was decided to develop a donkey-drawn plow suitable for the local conditions that could address farmers' main constraints on crop production, which were weeding and excessive runoff. A participatory process was adopted, involving blacksmiths as producers and farmers as users.

The project developed and tested a number of plow designs over a period of several years. Initially, the traditional moldboard camel plow was adapted for use by donkeys, and a moldboard donkey plow was brought from the United Kingdom in 1987, which was found to be unsuitable when tested. A donkey-drawn seeder/weeder developed in another area for a different soil type was tested; it was found to facilitate quick sowing but was less effective at weeding, which was the main constraint in Kebkabiya. Local blacksmiths were then brought into the project by Oxfam to play a role in further modification of donkey plow designs. Because they had no previous experience of manufacturing plows, they were given some training first. In addition, a blacksmith from a nearby area, who did have experience of plow manufacture, was involved. Plows produced by blacksmiths, based on a modified design, were tested in community demonstration farms. A few farmers showed interest and borrowed the plow. It was too heavy and did not speed up agricultural operations significantly, but plow development continued and in 1988 ITDG was contracted to provide technical support.

ITDG decided to develop two different types of plows—a moldboard one and a chisel plow, such as the traditional ard—so that farmers would have choices. A prototype ard was brought from the United Kingdom, copied, and tested by local blacksmiths. Various problems were identified by the blacksmiths who produced modified ards to address these problems. At the same time, modifications to the moldboard design were being made by another blacksmith group. Both new designs were tested in the same demonstration farms; various problems were identified with the ard design and there was greater farmer acceptance of the moldboard design, although this had its own problems. Blacksmiths then produced seven moldboard plows, which were tested in 1990 in demonstration farms and by some farmers in their fields. The results were encouraging, so development of the design was continued. This was done in a participatory, iterative annual process involving blacksmiths as producers and farmers as users, plus a little technical advice from the project engineer. By 1994 the technical weaknesses had been addressed and a final and accepted donkey plow was developed. This reduced labor requirements and increased crop yields, planted area, and food security.

CASE 8.4 *(Continued)*

However, the poorest farmers faced two barriers to using the plow: lack of cash to buy plows and lack of donkeys; about one-third of the poorer households had none. These were addressed by developing a pay-by-installment system and by facilitating the sharing of a donkey and plow between two households, one owning the donkey and the other the plow. Now more than 3000 plows have been distributed. Capacity building of blacksmiths and farmers, through training and group formation, has been an essential part of the project, as has provision of credit. These activities were made possible because the technology development process was part of a larger development project. The development of the plow took about 10 years, longer than most research projects last.

Source: Suliman (2005).

CASE 8.5 The Bangladesh Poultry Model

The Bangladesh poultry model aimed to increase household income by up to 50% by offering combined packages of credit and technical training in poultry rearing to the beneficiaries, as well as supporting marketing of eggs. It was developed during the second half of the 1980s and was replicated from 1992 to 2003 through three livestock development projects funded by Danida.

The model was a holistic but complex one. It comprised seven cadres: a production chain, market outlets, and input service suppliers. NGOs were contracted to implement the village-based activities in collaboration with the GoB's Department of Livestock Services (see Fig. 8.5).

The main target group (95%) of project beneficiaries in the poultry model is the key rearers, who invest in 2-month-old crossbred chickens and rear them to egg layers, while at the same time keeping a small flock of local hens for egg hatching and chick rearing to pullet size. The other 5% of the project beneficiaries are service deliverers, who are linked to the key rearers in order to ensure that they receive the necessary input supplies (e.g., vaccination, chickens, and feed) and output-related services. The model breeders produce quality fertilized eggs from crossing Fayoumi hens and Rhode Island Red cocks; these are sold to minihatcheries (small, low-cost ones), which in turn produce day old chicks and sell them to NGOs to sell to key rearers. Support services

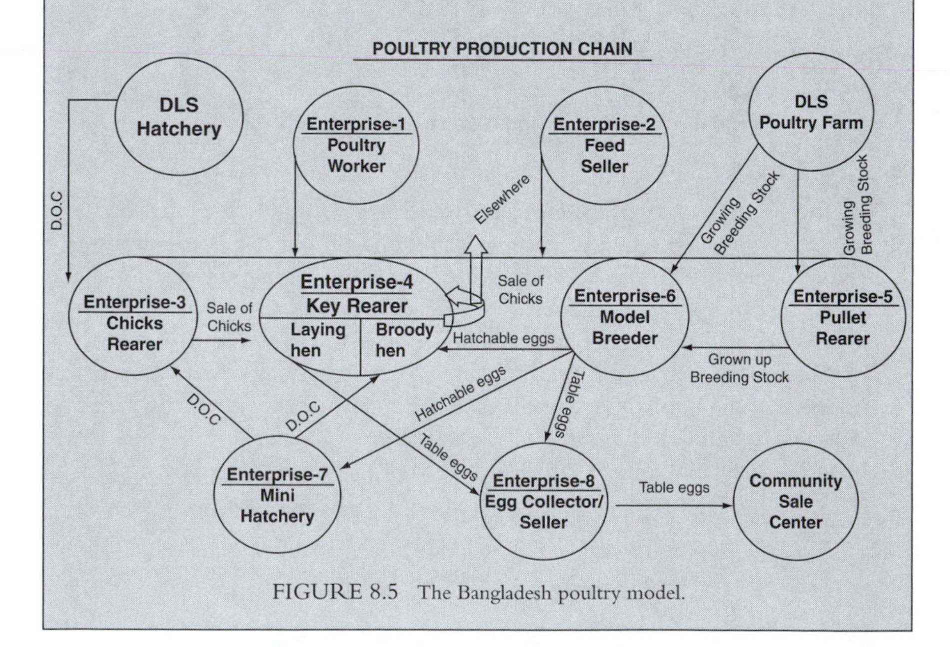

FIGURE 8.5 The Bangladesh poultry model.

CASE 8.5 *(Continued)*

for the key rearers are provided primarily by the poultry workers, who are local women trained and equipped to vaccinate poultry against the most common poultry diseases. In addition, feed sellers procure various feed ingredients available at the local market, or supplied by the supporting NGO, and sell compound feed or ingredients to the poultry keepers. The egg seller buys eggs from the producers and sells them at the market, and is also expected to transport fertile eggs from model breeders to the minihatcheries.

During the last few years it has become clear that the Bangladesh model only generates limited net increases in income and is unsustainable; when donor support stops, the largely supply-driven structure, involving linkages between the various component enterprises, falls apart (Riise *et al.*, 2005).

The provision of credit for poultry keeping (and a strong emphasis on this in the project monitoring) had a distorting effect in three ways. First, the implementing NGOs gave higher priority to credit provision to beneficiaries than to technical support; second, NGOs tended to exclude the main target group, the ultrapoor, because of doubts about their ability to repay loans; and third, 90% of beneficiaries enrolled in the project primarily to get credit, and only 10% because they were interested in improved poultry production.

The Bangladesh model was designed and implemented in a top-down manner—the "rigid model approach, with its fixed numbers per cadre and pre-set production outputs, resulted in ... NGOs having a less demand-driven approach [and] not involving farmers or villager organisations in decision-making (e.g., on size of loan or type of enterprise)" (Riise *et al.*, 2005); a "sufficiently demand driven, beneficiary participatory approach was not applied in designing the model or in introducing changes" (Islam and Jabbar, 2005).

Sources: Riise *et al.* (2005) and Islam and Jabbar (2005).

CASE 8.6 Cooperative Village Milk Collection Systems in India

One livestock innovation that has changed lives in many parts of India, including Rajasthan, is milk collection systems based on a cooperative model. There is a dairy in the village, people bring in milk, the dairy in charge places a sample on an instrument, checks the fat content, and prints a receipt that tells the seller the fat content and the price. Once a week, the milk seller encashes receipts. As most villages do not have electricity, instruments and computers work on diesel generators. Every day the cooperative's van arrives to take the milk for sale in the nearby town.

(*Continues*)

CASE 8.6 (*Continued*)

In Laporiya village, Rajasthan, young girls, women, and men stream into the dairy in the evening. Their milk is checked and they collect their receipts whose numbers are written in English. They do not know the language, but can read their receipts. One buffalo gives roughly 5 liters of milk each day, and milk sells at Rs 15 to Rs 25 per liter, depending on the fat content, so even the poorest one-buffalo owners earn Rs 75 per day, which could be more than they would earn from a day's labor.

Laporiya has seen a back-breaking drought since 1997. Meteorological data show that the last good monsoon was in 1997; it rained 700 mm. Since then, rainfall has varied from 300 to 400 mm and comes in a few cloudbursts. It is in this situation that animals become the mainstay of the economy. Animal care is much less risky than agriculture. The dairy is the vital link in adversity—it links people to the market. It helps them cope with scarcity.

This system is simple but not simplistic. It provides for the poorest and most marginalized by investing in improving the productivity of common grazing lands. This is critical in ensuring a continuous supply of fodder, particularly during scarcity periods, and hence in sustaining the production of milk. In 2006, after 9 years of persistent drought, it rained less than 300 mm, but the village of 300 households was still able to sell milk worth Rs 17.5 lakh.

Source: CSE Newsletter (2007). Editorial: "When markets do work" by Sunita Narain.

Supporting Livestock Innovation

This section pulls together implications of the preceding discussion for organizations intending to support innovation among poor livestock keepers. It has been shown that people, households, and communities involved in livestock production may be spontaneously involved in innovation, driven by various factors such as changes in their personal circumstances, population pressures, increasing land scarcity, or improved access to markets. These innovations can take various forms, including changes in livestock species kept, new production technologies, new arrangements for obtaining input services, or innovations in the way they process or market livestock and livestock products. Nevertheless, improvements in the productivity of, and returns from, resource-poor people's livestock in LDCs have been disappointing and have not benefited from the livestock revolution as much as resource-rich and corporate livestock producers. In this 21st century era of globalization, if small-scale livestock producers are to survive they must increase the efficiency of their operations and the productivity of their animals.

Technological innovations developed and promoted by the private sector, such as vaccines and other veterinary products, tend to be inaccessible or unaffordable, and technologies developed by public sector researchers are often inappropriate and unaffordable (often in terms of the opportunity costs of labor or land involved, as well as the cash expenditure required). The formal research system may even completely fail to address key constraints, such as high mortality rates as a consequence of predation in the scavenging poultry system (Conroy *et al.*, 2005). However, the case studies presented earlier, and others (see Conroy, 2005a), show that livestock research by government researchers and NGOs can be relevant and beneficial to the resource poor, provided various conditions are satisfied.

Social and institutional innovations can be as important as technical ones (as illustrated in case study 6) and may take two forms: (1) innovation among producers and (2) development of innovatory linkages/networks between producers and service providers. Social innovation among producers may be formal or informal and includes the development of cooperatives, farmer groups, and self-help groups. The formation of groups of farmers or livestock keepers can have a number of benefits, including

- making government research and extension services more client driven and efficient
- strengthening farmers' bargaining power with traders
- reducing transaction costs for input suppliers and output buyers
- economies of scale (e.g., from bulking up in output marketing or storage)
- facilitating savings and access to credit
- reducing public sector extension costs.

The Enabling Environment

A whole raft of measures are required to create an enabling environment that will boost the small-scale livestock producer sector in developing countries, including an *institutional revolution* on the scale of the "livestock revolution" itself. Livestock service organizations must be made more accountable and responsive to poor livestock keepers by facilitating the articulation of the latter's priorities and demands, and research organizations should give greater emphasis to species that are important to the poor (Conroy, 2005b).

Changing the public sector working environment so that it supports demand-led, pro-poor, participatory multistakeholder research rather than hinders it is a major challenge, particularly as government agencies' rules and norms may be determined outside the agency itself. Common constraints include lack of *incentives* (or even perceived disincentives) for this kind of work and lack of *resources*, including funds to cover the travel and subsistence costs of fieldwork. In Kenya, the

National Agricultural Research System has taken the following initiatives to address these constraints (Okothe *et al.*, 2002):

- *incentives*: changes in appraisal procedures so that staff are rewarded for undertaking participatory work instead of being penalized
- *resources*: the establishment of competitive research funds specifically for demand-led participatory research.

With a supportive enabling environment, the livestock revolution can become an opportunity for resource-poor livestock keepers in LDCs rather than a threat. Their ability to take advantage of the burgeoning demand for livestock and livestock products can make a significant contribution to poverty eradication and broader social and economic developments.

REFERENCES

Amir, P., and Knipscheer, H. (1989). "Conducting On-Farm Animal Research: Procedures and Economic Analysis." Winrock and IDRC.

Bayer, W., and Waters-Bayer, A. (1998). "Forage Husbandry." Macmillan Education Ltd., London and Basingstoke.

Binswanger, H. P., and McIntire, J. (1987). Behavioural and materials determinants of production relations in land-abundant tropical agriculture. *Econ. Dev. Cult. Change* **36**(1), 73–99.

Bird, K., and Shepherd, A. (2003). "Chronic Poverty in Semi-arid Zimbabwe." CPRC Working Paper No 18. Chronic Poverty Research Centre, Overseas Development Institute, London.

Bond, R. (2006). CLP Asset Transfer Programme Review. July 2006. Chars Livelihood Programme, Bogra, Bangladesh.

Chambers, R. (1997). Whose Reality Counts? Putting the first last. London: Intermediate Technology Publications.

Conroy, C. (2005a). "Participatory Livestock Research: A Guide." ITDG Publishing, Bourton-on-Dunsmore, UK.

Conroy, C. (2005b). "What Does It Take to Make Livestock Research Relevant for the Poor? Paper presented at international workshop "Does Poultry Reduce Poverty and Assure Food Security?" Copenhagen, Denmark August 30–31, 2005.

Conroy, C. (2006). Chars Livelihood Programme Pilot Asset Transfer Programme: Review and Recommendations, June 2006. Chatham, Kent: Natural Resources Institute.

Conroy, C., and Lobo, V. (2002). "Silvipasture Development and Management on Common Lands in Semi-arid Rajasthan." Pune: BAIF Development Research Foundation and Chatham: Natural Resources Institute.

Conroy, C. and Sutherland, S. (2004). Participatory Technology Development with Resource-Poor Farmers: Maximising Impact through the Use of Recommendation Domains. AgREN Network Paper No. 133, January 2004. London: Overseas Development Institute.

Hayami, Y., and Ruttan, V. W. (1985). "Agricultural Development: An International Perspective." Johns Hopkins University Press, Baltimore, MD.

ILRI Revised Strategy for 2003–2010. International Livestock Research Institute, Nairobi.

Jindal, R. (2000). Case study 1: Barawa village. *In* "Silvipasture Management Case Studies by Seva Mandir." BAIF/NRI Goat Research Project Report Number 5. Natural Resources Institute, Chatham, UK.

Kang'ara, J. N. (2005). Case Study B: Participatory Development of Mange Treatment Technology in Kenya. In "Participatory Livestock Research – a Guide" (C. Conroy, ed.), pp. 165–174. ITDG Publishing, Bourton-on-Dunsmore, UK.

Kristjanson, P., Krishna, A., Radeny, M., and Nindo, W. (2004). Pathways out of Poverty in Western Kenya and the Role of Livestock. Food and Agriculture Organization, Pro-Poor Livestock Policy Initiative Working Paper 14. FAO, Rome. Available at: www.fao.org/ag/againfo/projects/en/pplpi/project_docs.html.

Morton, J. (2001). "Participatory Livestock Production Research in Kenya and Tanzania: Experience and Issues." Natural Resources Institute, Chatham, UK.

Okothe, O. S., Kuloba, K., Emongor, R. A., Ngotho, R. N., Bukachi, S., Nyamwaro, S. O., Murila, G., and Wamwayi, H. M. (2002). "National Agricultural Research Systems Experiences in the Use of Participatory Approaches to Animal Health Research in Kenya." Paper presented at the international conference Primary Animal Health Care in the 21st Century: Shaping the Policies, Rules and Institutions, October 15–18, 2002, Mombasa. African Union's Interafrican Bureau for Animal Resources, Nairobi.

Riise, J. C., Kryger, K. N., Seeberg, D. S., and Chistensen, P. F. (2005). Impact of smalholder poultry production in Bangladesh-12 years experience with Danida supported projects in Bangladesh. Paper presented at international workshop "Does Poultry Reduce Poverty and Assure Food Security?" Copenhagen, Denmark. August 30–31, 2005.

Roeleveld, A., Broek A, van den, (eds.) (1996). "Focusing Livestock Systems Research." Royal Tropical Institute, Amsterdam.

Roothaert, R., and Kerridge, P. (2005). Case study G: Adoption and scaling out—Experiences of the Forages for Smallholders Project in South-east Asia. *In* "Participatory Livestock Research—A Guide" (C. Conroy, ed.), pp. 225–236. ITDG Publishing, Bourton-on-Dunsmore, UK.

Sagar, V., and Ahuja, K. (1993). "Economics of Goat Keeping in Rajasthan." Indo-Swiss Goat Development and Fodder Production Project, Jaipur.

Sparks, N., Acamovic, T., Conroy, C., Shindey, D. N., and Joshi, A. L. (2004). "Management of the Hatching Egg." Paper presented at XXII World Poultry Congress June 8–13, Istanbul, Turkey.

Suliman, M. S. (2005). Case study I: Development of the Kebkabiya donkey plough in Western Sudan. *In* "Participatory Livestock Research—A Guide" (C. Conroy, ed.) pp. 247–256. ITDG Publishing, Bourton-on-Dunsmore, UK.

Tiffen, M., Mortimore, M., and Gichuki, F. (1994). "More People, Less Erosion: Environmental Recovery in Kenya." Wiley, Chichester, England.

Todd, H. (1998). Women climbling out of poverty through credit; or what do cows have to do with it? *Livestock Res. Rural Dev.* **10**(3).

Waters-Bayer, A., and Bayer, W. (2002). Animal science research for poverty alleviation in the face of industrialisation of livestock production. *In* "Responding to the Increasing Global Demand for Animal Products: Programme and Summaries of an International Conference" organized by BSAS. ASAS, and MSAP, Merida. Mexico, November 2002.

World Bank (2006). "Enhancing Agricultural Innovation: How to Go Beyond the Strengthening of Research Systems." The World Bank, Washington, DC.

CHAPTER 9

Gender and Agrarian Inequality at the Local Scale

Rachel Bezner Kerr

Summary

This chapter focuses on gender dimensions of inequality in agriculture and rural development. A basic definition of gender is explained, and the ways in which gender inequality occurs in terms of access to productive resources are discussed. These productive resources include land, labor, inputs, and knowledge. This chapter explores the complex ramifications of how agricultural scientists conceptualize gender roles and households and engage with cooperation, competition, and conflict dynamics. A case study in northern Malawi outlines specific examples of gender and other inequalities (e.g., age, residency status, and ethnicity) and how agricultural research can try to address these inequalities in the approach and method of research.

INTRODUCTION

This chapter focuses on inequalities at community and household levels, giving special attention to the gender dimensions of inequality and associated struggles to insert attention to gender equity into rural development planning and agricultural innovations. A case study in Malawi helps highlight the need to consider gender and other social inequalities in agroecological approaches.

THE STUDY OF GENDER RELATIONS

Inequality is multidimensional and multiscaled, and one of its most important dimensions relates to gender, which can be understood within and across different scales. While the suggestion that there are inequalities between men and women is not likely to surprise many people, understanding gender and the study of gender relations are more complicated than they might appear at first.

To understand gender it is first necessary to distinguish it as a social category distinct from sex. A person's *sex* is his or her *biological attributes* as a man or a woman. A person's *gender* constitutes a multifaceted set of relations and characteristics that are related to his or her biological sex, but also involve social meanings, position, and relationships to others as a man or a woman. These are, in turn, constructed and interpreted through social interactions and vary across time, space, and culture, which is why gender is referred to as something that is *socially constructed*.

The study of gender relations explores the different and often highly uneven roles, responsibilities, access to resources, authority, decision-making patterns, and perceptions about gender held between men and women within societies. This often starts with fundamental questions about inequality, such as the following. What is the division of labor? Who determines the access to resources? Who benefits from the use of these resources? It also involves asking these questions across different scales, from within households, communities, and institutions, and up in scale to national and global levels. Understanding the implications of these differences supports finding ways to reduce inequalities between men and women.

Gender is not the only socially constructed category that influences a person's position or activity; other social differences, such as class, age, ethnicity, and occupation, also influence social outcomes and interact with gender in complex ways. Two examples illustrate how age and class intersect with gender roles to influence agricultural systems. First, in India, while rural farming women carry out much of the agricultural labor in poorer households, they often have only a limited role in agricultural decision making. At the same time, Indian rural women have a much heavier workload than men in terms of household work, including food preparation, child care, and the collection of fuel wood and water. However, these roles vary with class, as some women in wealthier households are able to "buy out" of agricultural labor by purchasing casual laborers. A second example comes from northern Malawi, where older women usually have more decision-making authority than younger women and have more ability to mobilize labor through social obligation. If older women have married sons living nearby, they are able to ask sons to assist with different agricultural tasks. A younger married woman, who has usually moved to her husband's home, is less able to draw on social and kin ties for mobilizing help in agricultural activities, and her children may be her only source of assistance (Fig. 9.1). Further, older women are facing escalating household burdens in the north. There, increases in HIV/AIDS rates and adult mortality are pushing greater responsibility upon older women for early child care, which is compounding the workload associated with farming responsibilities.

The study of gender relations is informed by a variety of theoretical approaches, and reviewing some of the major developments provides useful background and tools for addressing gender issues in agriculture. One of the most important pioneers in the study of gender relations was Ester Boserup, whose seminal book *Woman's Role in Economic Development* (1970) heralded a new attempt to conceptualize and analyze the role of women in agricultural systems. Boserup examined women's roles in different farming systems and how such cultural practices as dowries and bride price are related to women's economic status.

Around the same time, feminists were also working to raise the profile of women's issues within various international and national bodies. For instance, the Conference of the Status of Women began to raise the profile of what were framed as distinctively women's issues within the United Nations (UN) system, and the establishment of the Office of Women in Development within the U.S. Agency for International Development (USAID) helped raise the profile of development issues specific to women within the sphere of official development assistance. The 1975 UN Mexico City Conference on Women, coinciding with the first International Women's Year, highlighted the need for enhanced legal rights for women and for their economic empowerment. In terms of development policy and planning, the most recognizable outcome of the conference was the adoption of the women-in-development (WID) approach.

FIGURE 9.1 Children carrying out "hand hoe" crop management in Malawi (photograph by Sieglinde Snapp).

The WID approach focused on increasing women's access to training and resources, emphasizing women's individual legal rights to social, economic, and political advancements, and it became a fairly standard operating guideline for development agencies in the 1970s and 1980s. The WID approach did draw attention to the issue of gender equality (see Box 9.1) as well as monies to women's programming at a time when overseas development assistance made up a much greater proportion of total money flowing into the global South than it does today. However, by the end of the 1980s there were widespread concerns that the WID approach tended to marginalize women's concerns by confining them to a specifically "women's" office or program, at the same time as they continued to be ignored within the most significant development policies. Another criticism was that the emphasis on individual rights was too Western in approach and ignored structural economic inequalities.

Conservation tillage provides an example of the challenges posed by an isolated "women's office" approach. Protection of the environment is a laudable goal of conservation tillage; less well understood are the implications of different soil conservation techniques for women's labor and decision making. Weed management during the cropping season is a responsibility of rural women throughout much of sub-Saharan Africa. Initial soil preparation for crop production is often men's responsibility. There are profound implications for gendered labor requirements if conservation tillage markedly alters the labor required for weed management, for example, from two to six weeding operations. Conversely, novel types of conservation tillage show potential to suppress weeds (see reduced tillage and weed ecology discussion in Chapter 3). A women's bureau is unlikely to be engaged in policy or research discussions on which conservation tillage approaches to pursue.

BOX 9.1 Gender Equality

Gender equality essentially means that women and men should have equal conditions to realize their full human rights and equal potential to contribute to national development in all of its facets. Work toward gender equality must start from recognition that current social, economic, cultural, and political systems are gendered; that women's unequal status is systemic; that this pattern is further affected by race, class, ethnicity, and disability; and that it is necessary to incorporate women's specificity, priorities, and values into all major social institutions. Gender equality is essential for progress in human development and peace.

Source: Adapted from the Canadian International Development Agency's gender definitions www.acdi-cida.gc.ca/equality.

In the 1990s, a new approach emerged based on the basic argument that considering women's issues in a "silo" was, in many respects, counterproductive and that a fundamental change in gender relations required the integration of men's concerns and perspectives with those of women. An associated image is the tearing down of the silo where women's issues were separated and largely isolated. In contrast, the gender and development (GAD) approach encouraged what became known as gender *mainstreaming*—the attempted integration of gender concerns into all development programs. GAD also emphasizes the diversity of cultural perspectives on gender issues globally and the need to take a participatory, empowerment approach to addressing the needs of poor women from the global South.

GENDER AND UNEQUAL ACCESS TO PRODUCTIVE RESOURCES IN AGRICULTURAL SYSTEMS

Gender relations profoundly influence agricultural outcomes in a number of different ways. While women are critical cultivators throughout much of the world, they tend to have very unequal access to land, as well as to other productive resources in agriculture: labor, capital, inputs/technology, and the institutions that support agriculture (Ravazi, 2006; Whiteside and Kabeer, 2001). The marked gendered character of land inequality must first be understood in the context of the gendered division of agricultural labor. On a global scale, according to the Food and Agriculture Organization (FAO), women make up 51% of the total agricultural labor force, with particularly high levels in the world's poorest countries (see Table 9.1 for selected African examples). For instance, throughout most of sub-Saharan Africa, a region where agriculture also tends to be extremely labor intensive, women have long had the heavier share of labor in farming, contributing between 60 and 80% of the total (FAO, 1995), and these levels are expected to rise due to dynamics such as an increase in male migration and the HIV/AIDS pandemic (Garcia *et al.*, 2006). Related to this is the fact that agriculture is the dominant livelihood—over 70%—for women who are considered to be "economically active" in the world's least developed countries. The FAO (2003) expects this pattern to continue for the foreseeable future and notes that this "is even more significant given that data for the economically active population in agriculture tends to exclude the unpaid work by rural women in farm and family economies. If unpaid work were included, the figures for female employment in agriculture would be even higher."

In marked contrast to their contribution to agricultural labor across much of the global South, women are often treated as unequal partners within their households and, as a general rule, hold much less and lower quality land than men (FAO, 1995). There are complex and culturally diverse reasons for this, but it generally cannot be understood without reference to the colonial period. Boserup (1970) was one of the seminal authors outlining how the colonial promotion of male land ownership in Africa and southeast Asia and the associated privileging of male access to

TABLE 9.1 Role of Women in Agriculture: Selected African Examples[a]

Benin	Seventy percent of the female population lives in rural areas, where they carry out 60–80% of the agricultural work and furnish up to 44% of the work necessary for household subsistence.
Burkina Faso	Women constitute 48% of the laborers in the agricultural sector.
Congo	Women account for 73% of those economically active in agriculture and produce more than 80% of the food crops.
Mauritania	Despite data gaps, it is estimated that women cover 45% of the needs in rural areas (further details not specified).
Morocco	Approximately 57% of the female population participates in agricultural activities, with greater involvement in animal (68%) as opposed to vegetable production (46%). Studies have indicated that the proportion of agricultural work carried out by men, women, and children is 42, 45, and 14%, respectively.
Namibia	Data from the 1991 census reveal that women account for 59% of those engaged in skilled and subsistence agriculture work[b] and that women continue to shoulder the primary responsibility for food production and preparation.
Sudan	In the traditional sector, women constitute 80% of the farmers. Women farmers represent approximately 49% of the farmers in the irrigated sector and 57% in the traditional sector. Thirty percent of the food in the country is produced by women.
Tanzania	Ninety-eight percent of rural women defined as economically active are engaged in agriculture and produce a substantial share of the food crops for both household consumption and export.
Zimbabwe	Women constitute 61% of the farmers in communal areas and comprise at least 70% of the labor force in these areas.

[a]Modified from FAO (1995).
[b]Distribution by sex is statistically significant at 99% level of confidence.

technology and cash incomes during colonialism meant that it was not only small farmers who were disadvantaged in a general sense, but women specifically within this broad category, with the net result being worsened food production in these regions.

Women have fewer representatives in political power, and their interests are often marginalized in politics. Women's representation in decision-making positions within Ministries of Agriculture and other government bodies dealing with rural development is similarly low. At the local governmental level, few women hold decision-making positions and are very rarely involved in traditional authority structures. As these bodies are often responsible for the allocation of resources, women's lack of representation at this level has many negative implications in terms of how these get distributed (FAO, 1995). In addition, most financial institutions have policies and practices (e.g., collateral requirements) that systematically discriminate against the poor and women in particular. The unequal access of women to land, capital inputs, and other productive resources within different agricultural systems can mean that different land management strategies are used.

Access and control over land is a fundamental inequity facing women farmers. In some places, women simply have no or limited land property rights, or gain access to land through men. While women do have legal rights of access in many freehold land sectors, they generally lack the economic resources to acquire such land. In many parts of Africa, women are often unpaid laborers on their husbands' land while simultaneously cultivating separate plots of their own, which they may not have legal ownership to and thus risk losing upon the death of their spouse (FAO, 1995; following Bullock, 1993). African women who become "de facto" heads of households when males migrate for work also sometimes face threats to their access, as they are rarely endowed with stable property or user rights (FAO, 1995).

In the 1960s and 1970s there was considerable attention to land reform, but most of the land reform that took place was "gender blind," that is, it failed to recognize how gendered relations were embedded in different cultures and legal practices. This failure produced outcomes where women were either not empowered by land reforms or were made more vulnerable if the household dissolved (e.g., separation, divorce, or widowhood), in some instances producing outcomes where women had less "bargaining power" within the household in terms of workload and resource management.

The dominant approach to land reform has been through market forces from the 1990s onward. While some have professed hopes that this could improve gender equity in land access if accompanied by efforts to improve women's legal rights to land tenure, a study by the United Nations Research Institute for Social Development (UNRISD) suggests otherwise, having found that women's access to and control over land has not improved significantly on a global level. Although there are some contextually specific reasons for the failure of post-1990s land reforms to improve women's access to land (see Box 9.2), the report concludes that legal improvements alone are not likely to be a transformative force; rather, they are part of an array of changes needed to improve women's access to land, with the state, political parties, and social movements also named as having crucial rules in addressing rural women's needs in agriculture (Ravazi, 2006).

Countries in Latin America also showed little improvement or worsening land ownership for women in the last decade (Table 9.2). One very notable regression occurred in Mexico, where communal lands were opened for privatization in order to accede to the North American Free Trade Agreement, and the outcome was to reduce women's traditional access to land by granting formal title to only the household head, usually male (Deere and Leon, 2003). Similar concerns are arising now in land reform processes underway in southern Africa.

BOX 9.2 Why Have Land Reforms Failed to Improve Women's Access to Land in South Africa, Brazil, and Tanzania?

South Africa: Fall of apartheid and transition to democracy in 1994 promised improvements for women, but:

- institutional weaknesses and lack of political accountability for gender policy at high levels
- weak rural women's movements since 1994
- market-based land reform model tends to build on inequitable community structures and disadvantage women

Brazil: Constitutional guarantees (1988) and vibrant rural social movements promised hope for rural women to gain access to land:

- by the mid-1990s, Brazilian women were only 12.6% of land reform beneficiaries
- women's land rights have not been a priority within social movements; in some cases, they are seen as being "incompatible" with class issues for landless peasants

Tanzania: Market-based land reform processes (1991–1999) combined with gender advocates did not result in substantial change:

- divisions about how to change customary inheritance laws
- disagreement amongst gender advocates over whether land markets are best solution for land reform

Source: Ravazi, 2006.

Varying land tenure patterns in Asia pose different problems for women. In India, patrilineal inheritance laws have limited women's access to and use of land. In Vietnam, Laos, and China, where the state is the dominant landowner, land allocated through contract is often the highest quality and tends to go largely to men. Women have the right to own property in China, but married men often control the land in practice; further, if the husband dies, the land sometimes gets taken by the husband's kin (Agarwal, 1994).

There is strong evidence indicating that women have unequal access to and control of land throughout sub-Saharan Africa. State support for customary land tenure systems has often reinforced discriminatory practices toward women, while increased privatization of land tenure has generally worsened women's access to land (Khadiagala, 2001; Lastarria-Cornhiel, 1997; Whitehead and Tsikata, 2003). In southern Africa, the increased levels of HIV/AIDS infection have also made

TABLE 9.2 Forms of Acquisition of Land Ownership in Six Latin American Countries[a]

Form of land acquisition	Inheritance[b]	Community	State	Market	Other	Total	Total sample size
Brazil[c]							
Women	54.2	0	0.6	37.4	7.8	100	4,345
Men	22.0	0	1.0	73.1	3.9	100	34,593
Chile[d]							
Women	84.1	0	1.9	8.1	5.9	100	271
Men	65.4	0	2.7	25.1	6.8	100	411
Ecuador[e]							
Women	42.5	0	5.0	44.9	7.6	100	497
Men	34.5	0	6.5	43.3	15.6	100	1,593
Mexico[f]							
Women	81.1	1.8	5.3	8.1	3.7	100	512
Men	44.7	14.8	19.6	12.0	8.9	100	2,547
Nicaragua[g]							
Women	57.0	0	10.0	33.0	0	100	125
Men	32.0	0	16.0	52.0	0	100	656
Peru[h]							
Women	75.2	1.9	5.2	16.5	1.3	100	310
Men	48.7	6.3	12.4	26.6	6.0	100	1,512

[a]From Deere and Leon (2003).

[b]In areas of community ownership, distribution by the communal authority is one channel through which women access or acquire land.

[c]"Other" includes donations by private parties.

[d]For farms larger than 5000 m^2 only. "Other" includes imperfect donations by private parties and other responses.

[e]Based on total parcels acquired by 1586 individuals assuming principal agriculturalist is the owner. "Other" includes land held in usufruct, which is treated as private property.

[f]From a nationally representative sample of *ejidatarios* and *posesionarios*; based on total parcels titled to 1576 individuals. "Other" includes adjudications based on judicial actions.

[g]For individual land owners only.

[h]"Other" includes parcels held in co-ownership with family and nonfamily members of unspecified sex.

women's access to land more precarious, with the forced removal of widows and property seizures serious concerns in many countries.

Agricultural inputs such as fertilizer, seeds, and equipment are important resources for improving agricultural systems. Structural adjustment policies (SAPs) and neoliberal development policies have reduced state subsidies for inputs and have increased corporate control over the agro-input sector, which has also led to rising costs. In general, women typically face greater barriers than men in accessing agro-inputs for a variety of reasons, including

- having less access to credit and capital, as women typically have a harder time getting loans, particularly if credit is tied to evidence of surplus production of cash crops (Doss, 2001)
- having fewer assets to sell and having fewer and lower paid off-farm employment options than men (Whiteside and Kabeer, 2001)
- having less access to land to use as collateral to gain loans
- the fact that within married households, in many cultural contexts, control of money to purchase inputs is considered the male responsibility (Jewitt, 2002; Whiteside and Kabeer, 2001).

There is considerable empirical evidence that SAPs have had a disproportionate and largely negative impact on rural women and children. Under SAPs, large-scale farming and commercial crop production were promoted, and some productive resources got reallocated from subsistence production to the production of export crops. Women farmers, largely concentrated in the subsistence sector, had limited ability to move into export crops due to various constraints. The increased emphasis on export crops associated with neoliberal development policies can force women to reduce time spent on food production and move them into the export sector. With less access to capital, land, and inputs, women can be pushed out of food production and into marginal employment opportunities with low wages. Even if women are successful in participating in the production of export crops, they often do not control the marketing and sale of these products. Thus the push to increase export crops can exacerbate existing unequal gender and community relations (Garcia *et al.*, 2006). The gendered dimensions of this can be broadly understood in terms of

(a) Time—given the double burden of productive and reproductive tasks facing women
(b) Systemic discrimination—as women tend to have lower access to credit, technological packages, and marketing information
(c) Sociocultural traditions—in which women act as the primary caregivers responsible for feeding and taking care of the household.

Debt and adjustment also involved the reduction of government expenditures on social services such as education, health, and rural infrastructure (e.g., water and energy supplies), which levied further demands on women's time and energy to make up for shortfalls in these areas (FAO, 1995). The fact that these public expenditure cuts had an uneven gendered impact stems in large part from underappreciation of the all-encompassing nature of women's household work, which gets magnified yet largely unmeasured.

In general, women work longer hours than men, contributing significant agricultural labor in addition to a myriad of other types of work within the household that are gender specific (Doss, 2001) (see Box 9.2). Women are almost exclusively responsible for household food production and preparation, child care, cleaning, and the collection of water and fuel wood. The range of responsibilities women face can add up to a very significant amount of time, although much of this is not measured as "work." This is partly because many of the aforementioned gender-specific tasks take place outside the market and are therefore seen to have less value in a monetized system, whereas market-based activities (whether from selling agricultural produce or from waged earnings) are conceptually privileged by the fact that they bring money into the household. This conceptual privilege, part of the disciplinary bias of economics, can range from things such as elevated household status for the money earners to the fact that development policy historically tended to focus on "economically active" (earning an income) men and women. This focus on the economically active, in turn, led to development interventions that sometimes ignored those who had no measurable income, as well as those who were at once "economically active" while retaining heavy household responsibilities (what was described as the "dual burden"), with women heavily overrepresented in both groups.

Scholars such as Dixon-Mueller (1985) and Samarasinghe (1997) have drawn attention to this dual burden and reconceptualized the distribution of work within the household. A detailed time budget (which measures a broader conception of work by hours rather than by income) is one useful methodological tool for highlighting the disproportionate share of work that tends to fall to women. Empirical time budgets have shown that irrespective of how household monetary earnings were generated, women tended to be working more hours per day than men.

One of the implications for this unequal workload in agriculture is that it can be difficult for women to take on new agricultural tasks or technologies. So while the application of fertilizer might seem like an unmistakably beneficial technology, for example, it may also increase women's weeding tasks; further, if the increased production is not controlled by women and gets used for other purposes, the overall outcome may be negative. Decision-making processes and control over agricultural outputs are, therefore, important aspects in considering the implications of gender relations on agricultural systems, and conversely the implications of agricultural interventions on gender relations.

Another dimension of inequality in agricultural supports is the fact that research and extension efforts have typically focused on male farmers. One facet of this is the shortage of women in agricultural research: there are few women agricultural scientists; an absence of women farmers in demonstration plots and engaged in on-farm experimentation; and limited numbers of women extension workers (Doss, 1999). Further, women farmers' knowledge about seed varieties, crops, or land management is often ignored, unknown, or underreported by agricultural scientists and extension workers (Ferguson, 1994), while crops that women tend to have the greatest time invested in managing, such as legumes, hardier grains, and vegetables, have long been underresearched (Fig. 9.2). The absence of women in agricultural research and the lack of attention to gendered divisions of labor on the

FIGURE 9.2 South African student conducting research on a traditional woman's crop, Bambara groundnuts (*Vigna subterranean* L.) (photograph by Sieglinde Snapp).

farm and in households may mean that efforts to increase agricultural production inadvertently expand women's workloads, for example, through weeding, which can result in reduced time for other important household activities, such as child care or food processing (Karl *et al.*, 1997). Several studies have documented how women farmers are excluded from extension services (see, e.g., Due *et al.*, 1997; Saito and Weidemann, 1990). Often it is poor farmers in general who are excluded; for instance, one study in Zambia found that extension services only reached 25% of the nation's farmers, with this support concentrated among the wealthier farmers (Alwang and Siegel, 1994).

The net result of the gendered nature of agricultural science is that women farmers often lack substantial technical information that might assist them in farming, and their needs, preferences, and concerns are systematically excluded from agricultural research priorities. By primarily focusing agricultural research on male farmers' production issues, researchers have neglected areas, such as food processing, that could significantly reduce women's work burdens and improve the quality of life for women and children.

The neglect of women farmers in agricultural research and extension is linked to the ways in which households have been modeled by the sciences and economics. Systematic exclusion from agricultural research has had major implications for women's labor and, consequently, household food security and child nutrition. To better understand these linkages, we turn to an examination of household level studies in agriculture.

UNDERSTANDING GENDER DYNAMICS AT THE HOUSEHOLD SCALE IN AGRICULTURAL SYSTEMS

A fundamental starting point for approaching household gender relations in agricultural systems is to ask what is a *household* and what *household model* best approximates reality. This task can be partially framed with the questions: to what extent do households act as a unit, and to what extent are their interactions competitive or cooperative? These questions are important when considering the implications of policies, environmental changes, and other dynamics affecting poor rural households in the global South—from trade liberalization to climatic stresses to the challenges of coping with the HIV/AIDS pandemic.

At times, development policy makers, practitioners, and agricultural scientists have assumed that households everywhere follow the Western norm of nuclear families with a married couple and children. However, households can be defined in many ways, often by what they share (e.g., a home, a "common pot," economic resources), and many different types of households exist between different cultures. For instance, a basic household unit might include grandparents and grandchildren, several generations of married families, same-sex couples, or polygamous couples with separate households for each wife. An extension of this is that a married couple might not meet the definition of a household in some cultures, and assuming some level of coordinated household effort in agricultural practices can be very misleading, as husbands and wives may maintain separate fields, harvests, and act largely independently of one another.

Despite the variety of forms that households take and the array of dynamics within them, many economic and agricultural studies still assume that households have a set of common preferences and act as a unit when making decisions and allocating resources. However, a number of empirical studies have challenged this assumption (see, e.g., David, 1998; Dwyer and Bruce, 1988), demonstrating that households are not always sites of sharing and equity and that, by contrast:

- men and women can have different preferences for household resource use that can lead to overt or subtle conflicts
- resources are not always "pooled" within households, with incomes sometimes kept separately and spent for individual gains
- men and women sometimes farm different plots of land and manage crops separately

- different forms of conflict and cooperation play out over household resource and labor use
- domestic violence and other abuses of power have an important negative role in household relations in myriad ways.

Some theorists argue that a *bargaining model* in which women and men negotiate for different resources is the most appropriate theoretical tool for examining households (Agarwal, 1997). This model explicitly integrates issues of *power* within the household into the framework and encourages the consideration of a number of factors. For instance, what different elements of cooperation and conflict are evident? How much "bargaining power" do different members possess (and what sort of outside options do they have if cooperation fails)? How do different members frame the benefits and disadvantages of cooperation? What are the general decision-making patterns and how do these relate to the division of household labor? How does this intrahousehold bargaining connect to the gender relations beyond the household, such as those evident in the market, community, or state? Women's access to land and other resources outside the household, in particular, has been found to have strong effects on their bargaining position within households. Women may also have relatively greater decision-making power within a marriage if they have been married longer or if they earn more money through small businesses (Doss, 2001).

Household conflict should not be assumed either, as other theorists have emphasized that joint interests and cooperation can be the overarching dynamic even when women and younger males are systematically disadvantaged (Whiteside and Kabeer, 2001). A central point here is that the different priorities, preferences, tasks, and control over resources that occur within households often have major implications for agricultural systems in terms of crop types, crop use, land management (e.g., soil and water conservation strategies, intercropping, the use of tree crops), and other farming decisions, and conversely that agricultural development interventions can have critical implications for household relations. The bargaining power and amount of cooperation within households will influence agricultural systems and therefore need to be understood if agricultural scientists want to appreciate how their technological recommendations will be utilized by farmers.

An example from the Gambia illustrates how the introduction of a new agricultural technology can change household decision-making patterns and power dynamics. Introduction of a centralized pump irrigation system was expected to benefit women farmers, along with the granting of legal land title to women for rice plots. Increased rice production subsequently shifted from being controlled by individual women to being under the authority of a male compound head. Women's labor in agricultural production went up because pump-irrigated labor required 61% more labor than swamp rice. While overall calorie production went up, consumption of nutritious upland crops (e.g., groundnuts) went down. Women had less control over many aspects of rice production and had increased labor requirements, which they often used hired labor for, thereby increasing costs (Von Braun and

Webb, 1989; Von Braun *et al.*, 1989). (The issue of decision making and introduction of new technologies will be explored further in the case study of Malawi.)

The influence that gender can have on crop types can be seen clearly in an example from Malawi where research has found that gender differences relate to whether there is a preference for growing a flint versus a dent maize variety. Flint maize varieties are easier to pound into flour and can be recycled, whereas hybrid dent varieties need to be purchased annually but have higher yields if grown with fertilizer. Women were found to be more likely to grow local flint maize varieties, in large measure because they tended to have less access to credit to buy seed and fertilizer but could access flint seed through social networks and because their responsibility for food processing made its ease of pounding and storability very desirable features (Smale and Heisey, 1995).

The management of harvested crops and the use of income generated from crop sales have also been found to differ considerably between men and women in many places, with evidence demonstrating that women tend to invest more in children's nutrition and health than men (Kennedy and Peters, 1993; Quisumbing and Maluccio, 2000). In a number of contexts, however, men have demonstrated a willingness to invest in children and households, highlighting the fact that these patterns and gender roles in decision making are not static but rather are dynamic and complex (Whitehead and Kabeer, 2001).

Agricultural labor is another arena in which gender relations need to be considered. In many parts of the world, agriculture continues to be a very labor-intensive activity (Fig. 9.3), with tillage, planting, weeding, and harvesting carried out mostly by hand, using simple tools such as hoes and machetes. The labor intensity of these agricultural systems makes the ability to mobilize labor within households and communities an important difference that affects production, management decisions, and agroecological methods.

In many parts of the world, the gendered divisions of agricultural labor get segregated by task, crop, and even by crop variety. Although it is often stated that women are more responsible for subsistence crops and men for cash crops, the reality is often much more complex. For instance, it may be that women do not have access to the inputs or information to grow cash crops, yet they may be carrying out much of the labor but not controlling the income from the sales of cash crops, or a crop may be both a cash crop and a food crop (Doss, 2001). As new opportunities with a crop or technology arise, the gender and other divisions of labor may change for that crop, and the control and management of the crop may also change (Doss, 2001; Due, 1988) (see Box 9.3 and Table 9.3).

In married households, men may be able to mobilize women's labor to a much greater extent than vice versa, as well as having a greater ability to hire labor outside the household, as a result of differences in status, income, access to credit, and other gender inequalities (Whitehead and Kabeer, 2001). In some places where there are shared or cooperative labor practices, men may still be better positioned to mobilize such arrangements than women. The seasonal nature of agricultural tasks adds another dimension to the importance of being able to mobilize labor, with labor

FIGURE 9.3 Jennie Mumba, woman farmer participating in a SFHC project in Zombwe, outside Ekwendeni, Malawi, incorporating crop residue in April 2004 (photograph by Carl Hiebert).

bottlenecks a major issue that often emerges during critical periods in planting and harvesting. Women and men may have specific tasks in agriculture and in other arenas that conflict at these critical periods, and these specific tasks may vary by class, ethnicity, marital position, and other social distinctions.

The challenge of mobilizing labor is often greater for female-headed households (FHH), which tend to have lower incomes, less land, and fewer adult people within them to carry out agricultural labor. They also may not be able to mobilize labor as easily outside the household due to inequalities in social status and access to cash (Doss, 2001). Understanding how a FHH is formed is necessary to know whether the headship is relevant. For example, if a FHH is one in which the husband has migrated to another region for work (called a FHH de facto) and if remittances are sent back, labor can be hired. In this case, the household may be female headed because there were few opportunities for agricultural production, causing men to migrate to another region, and the poverty of a household cannot be seen as having been *caused* by being female headed (Doss, 2001). Focusing too much attention on the headship of a household can simplify problems and mask other factors that affect poverty and low agricultural production.

The critical role that women play in child care and in food production and processing and the different ways in which women and men use household resources

BOX 9.3 Women and Irrigation: Do They Benefit?

Irrigation is vital to certain crops and crop types and may prove increasingly important with a changing climate in the increasing variability of rains, especially in savannah regions. It is also a technology that is typically dominated by men. For instance, one study in highland Peru noted that although women contribute labor in agricultural systems with irrigation, they had little decision-making power over irrigation systems at the household or state level. The result was the exclusion of women from irrigation access as farmers by a male-dominated irrigation bureaucracy, with women provided only marginal access based upon domestic needs (Lynch, 1997). Another study in southern India noted that irrigation rights were allocated based on land title, thereby excluding most women and reinforcing gender inequalities, as land tenure is granted primarily to men in this region (Ramamurthy, 1997; Ravazi, 2006). In areas where irrigation had been introduced, women's workloads in poorer agricultural households had increased significantly (see Table 9.3), and the increased input costs associated with irrigated agriculture (e.g., hybrid seeds, fertilizer) meant that poorer women had to also work as agricultural laborers to pay for these inputs. Wages for agricultural labor were lower for women because the work was considered "lighter" (Ramamurthy, 1997). The overall effect of irrigation for poor households was negative, increasing costs and workloads but not incomes.

TABLE 9.3 Comparison of Women's Labor Demands (Irrigated vs. Rain Fed) in Southeast Andhra Pradesh[a]

Crop	Labor demand (in woman-days/acre)
Rain-fed sorghum	25
Rain-fed tobacco	55
Rain-fed cotton	44
Irrigated cotton	112
Rain-fed groundnut	23
Irrigated groundnut	45
Irrigated paddy rice	53
Irrigated onion	125

[a]From Ramamurthy (1997).

also mean that changing agricultural activities or women's control over agricultural resources can have significant effects on child nutrition and household food security (Berti *et al.*, 2004). These linkages become very evident when considering the effect of HIV/AIDS infection rates in sub-Saharan Africa. In Malawi, for example, women carry out much of the agricultural work and also are the primary caregivers of family members who become ill. Increasing rates of adult morbidity from HIV/AIDS place a disproportionate burden on women as predominant caregivers for sick family members. Illness of a sick adult can mean the loss of two adults working in the fields and can mean that key activities (e.g., weeding) do not get done, leading to lower agricultural yields.

CASE STUDY: GENDER, INEQUALITY, AND AGROECOLOGICAL APPROACHES IN NORTHERN MALAWI

This case study examines how agricultural scientists need to understand the broader historical context, as well as the social inequalities of a place, and can develop progressive partnerships to work toward agroecological solutions. I begin by considering how multiple types of inequalities both created agricultural problems and affected efforts to improve smallholder farmers' food security and soil fertility. This is followed by an overview of how the Soils, Food and Healthy Communities (SFHC) project in northern Malawi has integrated an equity focus into its science, extension, and education outreach, with evidence given about how this has improved not only agricultural systems and nutrition, but also equality at a household and community scale.

Some brief context is first necessary. On a per capita level, Malawi is one of the poorest nations in sub-Saharan Africa, and recent estimates place 68% of the population below the poverty line (Ellis *et al.*, 2003). It is predominantly agrarian, and most smallholder farming families suffer from food shortages and declining agricultural productivity and soil fertility. In addition to problems of food insecurity and child malnutrition, women in Malawi face difficulties of high workloads, unequal decision making, and domestic violence. Malawi's experience with structural adjustment has compounded the problems facing most small farmers. Adjustment policies, such as currency devaluation, privatization of the National Seed Company of Malawi, and reduced funding for the Ministry of Agriculture, have resulted in dramatic increases in fertilizer prices, declining availability of legume seeds, and the reduction of extension services (Devereux, 2002; Peters, 1996).

In light of the challenges facing smallholder farmers in Malawi, a project was initiated in 2000 by community nurses at a hospital in the northern region in collaboration with researchers from both Malawi and Canada. The SFHC project aims to improve the health, food security, and soil fertility of smallholder farming families through participatory research using legume intercrops and pursues this by employing a holistic approach to understanding linkages between agriculture and health. The SFHC project is located near the town of Ekwendeni in the midaltitude region (1200 m) of northern Malawi (Fig. 9.4). The average landholding of farmers in the region is roughly 1.1 ha. The long-term average annual rainfall is 1300 mm,

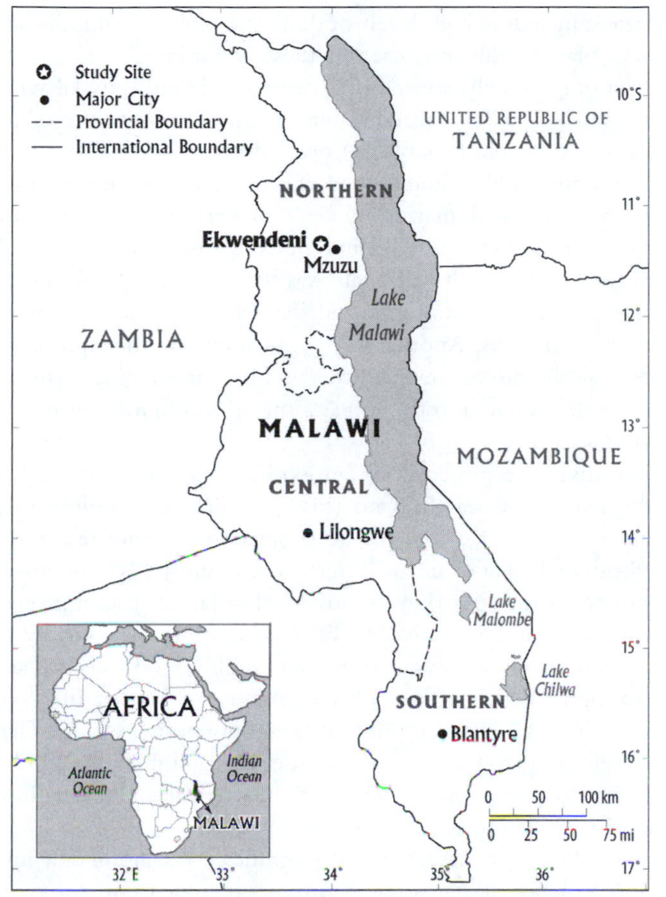

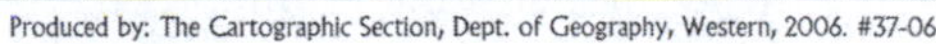

FIGURE 9.4 Site location of SFHC project, Ekwendeni, Malawi.

but it is highly seasonal with most (~85%) occurring from November to April. Smallholder farmers have limited access to irrigation and grow primarily maize (*Zea mays*), which is planted in an estimated 60% of smallholder land, alongside a wide range of other crops grown at low density (Snapp *et al.*, 2002). Prior to the onset of the SFHC project, legumes grown in the region, in order of decreasing frequency, were groundnut (*Arachis hypogaea)*, bean (*Phaseolus vulgaris*), cowpea [*Vigna unguiculata* (L.) Walp], soybean (*Glycine max*), Bambara groundnuts (*V. subterranean* L.), and pigeon pea [*Cajanus cajan* (L.) Millsp].

The SFHC began after hospital staff and researchers interviewed farmers who were highly food insecure with malnourished children and found that many had few options for improving soil fertility and food availability. Further, unequal gender relations, including high levels of domestic violence and misuse of household resources, played a role in worsening these conditions.

Earlier scientific research carried out in central and southern Malawi with smallholder farmers had identified several viable options for improving soil fertility and providing other household benefits: (1) groundnut and pigeon pea intercropped, (2) soybean and pigeon pea intercropped, (3) maize and pigeon pea intercropped, (4) *Mucuna* spp. rotated with maize, and (5) *Tephrosia voglii* relay intercropped with maize (Snapp *et al.*, 1998). These legume options were chosen by SFHC project for on-farm testing. Part of the rationale was that several of the legumes are well-known edible crops, and it was expected that increasing cultivation would help improve nutrition in diets. Another key motivation was to improve soil fertility, and one key way the project pursued this was by encouraging farmers to incorporate the legume residue into the soil as a means to improve nitrogen levels and organic material.

From the outset, the project took an explicitly participatory approach, centered on the Farmer Research Team (FRT), which is a volunteer, farmer-led organization formed at the start of the project to conduct research and share knowledge both on behalf of and within the community. FRT members are critical to the project's success, as they are involved in farmer training, seed distribution, data collection, and research. The FRT is composed of a variety of different social groups (e.g., widows, divorced women, highly food insecure and well-off farmers), and approximately 40% of FRT members are women. In 2000, the project began with 30 FRT in seven pilot villages, but because of high farmer interest, FRT membership had grown to include 120 members by 2007, while the project itself had grown to involve over 5000 participating farmers in more than 100 villages.

After 7 years, the project has facilitated a significant expansion of legume options and has found evidence of increasing legume residue incorporation by the majority of participating farmers (Bezner Kerr *et al.*, 2007). The outgrowth of this is that farmers are citing local indicators such as improved maize growth, soil color, and legume harvest as evidence that their soil fertility has improved (Fig. 9.5), as well as pointing to enhanced food security within their households.

Another important facet of the SFHC project has been its explicit attention to problems relating to gender and other inequalities. One of the key mechanisms for this has been for project members to conduct periodic research and participatory workshops with farmers to assess these issues to develop, in effect, an evolving baseline inventory on inequality, including how it is related to project interventions. Project staff has also contributed to this evolving inventory since the start of the project. Several key issues have been identified from this process, largely centered on decision making, division of labor, and project and village leadership.

Initially, women played a minimal role in decision making about crop sales, although they made important contributions to the labor involved in legume

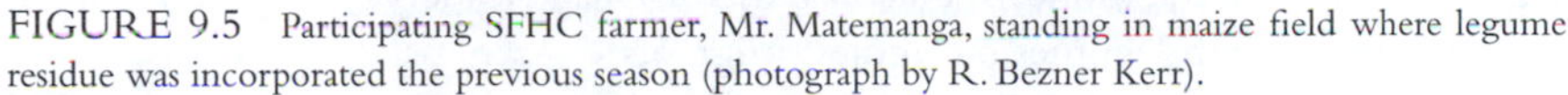

FIGURE 9.5 Participating SFHC farmer, Mr. Matemanga, standing in maize field where legume residue was incorporated the previous season (photograph by R. Bezner Kerr).

production. Women noted that men sometimes sold legume crops and used the money for nonhousehold use, such as alcohol. At the same time, women noted an increase in labor requirements for crop residue incorporation and a change in the gender division of labor with harvesting. In the past, crop residue was incorporated just before planting, usually by men, but the need for incorporating residue just after harvest meant that women often did this work, as women usually harvest the majority of legumes. Women also had difficulty feeding young children frequently in the rainy season because of high agricultural labor requirements.

The legume options and project approach had appealing qualities for women farmers. High numbers of women joined the project following a severe national food shortage in 2002, primarily because of the project focus on child nutrition and food security. Women tended to select the doubled-up edible legumes, whereas men were more likely to select the nonedible *mucuna* legume crop. The average area of expanded legume production, however, was lower for women than for men (Bezner Kerr *et al.*, 2007). Highly food insecure and/or HIV/AIDS-affected households had difficulty utilizing the legumes due to conflicts with other labor requirements (e.g., care for sick family members, labor on other peoples' farms to get seed or food).

Another complicating factor was that women had difficulty participating in the leadership of the FRT, despite making up a large proportion of SFHC participants, due to accusations of adultery if they visited other farms or had to stay overnight in other villages. Village leaders also played an important role in the successful use of legumes to improve food security for poor households. For example, some landless migrants who gained access to communal land and improved the land with legume intercrops and residue incorporation had their land seized by village leaders. Other village headmen let their cattle roam into pigeon pea fields of participating farmers to eat the pigeon pea and residue.

To help keep gender issues (e.g., household crop decision making) at the forefront of project planning, a flexible, participatory process of problem identification is used, which in turn gets integrated into various project activities with farmers, such as seed distribution events, training programs for new participants, and annual "field days." (Field days involve farmers, hospital staff, government, media, and other farmer organizations visiting selected fields of SFHC participants and learning what crops and cropping methods they are using, followed by a range of social activities.) Additionally, the FRT is involved in giving presentations, dramas, and talks that emphasize the links among improved gender relations, agricultural improvements, and improved child nutrition. For example, farmers perform dramas highlighting common household and community conflicts, such as men selling legumes and using the money for alcohol use or village leaders seizing land that has been improved with legumes by landless migrants.

Another important response to identified gender issues occurred when the FRT decided to make "family cooperation" a core project theme, emphasizing the need to work together within families and communities to improve food security. This decision led to the establishment of a nutrition research team, which was tasked with carrying out informal education during mobile clinics, recipe days, home visits, and in public spaces more generally on themes such as dietary diversity, family cooperation, and early child-feeding practices.

FRT members and project staff also raised the issue of legume residue incorporation, which then became a topic of intensive discussion within households and communities (see the example provided in Chapter 4). The FRT subsequently decided to organize legume residue promotion days in which the FRT visited villages and had a very public demonstration of legume residue incorporation in village plots in an effort to encourage men to participate in residue incorporation and address women's workload concerns. Public discussions followed thereafter, with the FRT again emphasizing the importance of family cooperation and male involvement in residue incorporation. The residue incorporation days have continued, and are now held annually, and they have contributed to the dramatic increases in legume residue incorporation that has been documented, from 15% of farmers in 2000 to over 70% in 2005 (Bezner Kerr *et al.*, 2007).

Another institutional innovation—the Agriculture and Nutritional Discussion Groups (ANDG)—was initiated in 2005. The ANDG seek to integrate agriculture, health, gender, and social relations in discussions, with the primary purpose of enhancing problem solving and conflict resolution at a household level in a way that will support the adaptation of equitable legume systems that enhance food and nutrition security. The village area groups comprise about 80 members each and have subgroups based on age and sex (i.e., older men, older women, younger men, and younger women). Each group meets monthly and carries out participatory, problem-solving activities on different agricultural and nutritional themes, first in the subgroups and then in the larger forum. For example, one month the groups might discuss ways that men and women can work to incorporate crop residue, and another month they might discuss how households can feed their children frequently during the "hungry season" (i.e., when food stores are at their lowest ebb) while ensuring that agricultural activities are carried out. The approach tried to build on previous work done on small group discussions and participatory educational methods by recognizing the inherent power dimensions at work within community households and by addressing these dynamics explicitly but with cultural sensitivity (Cornwall, 2003; Humphries *et al.*, 2000).

Already there is evidence of improved gender relations. One clear sign of this emerged from qualitative research done in 2005 and 2006, which indicates that women who participate in the SFHC project play an increased role in decision making regarding legume crops. In 20 focus groups with men and women of different ages in the project's "catchment," more than half the participants said that women are increasingly involved in the decision making with regards to legume crops. In 36 interviews in 2006 with SFHC participants, all but 2 talked about the ways that their relationship with their spouse had improved following participation in the ANDG. Respondents, particularly younger fathers, remarked that spousal interactions had improved markedly and spoke about the increased collaboration and cooperation with their partners that they felt resulted from these discussions. They also provided many examples of husbands making increased efforts to help their wives with such things as cooking, carrying materials home from the field, looking after their baby, and making a fire for cooking in the evening. There is also considerable evidence that crop residue is being incorporated with the active participation of men in many cases (Bezner Kerr *et al.*, 2007).

Project emphasis on "family cooperation" and the role of husbands in promoting good child nutrition through careful use of household resources seems to be having some effect on knowledge and practice, although documenting changes in gender relations is difficult to ascertain. While gender has been one prevailing concern, the SFHC project has also sought to identify other dimensions of social inequalities through various mechanisms such as

- project meetings with village leaders to address land seizures of landless migrants that improved land with legumes
- community meetings to discuss livestock management and to reduce cattle roaming on pigeon pea fields
- legume seed distribution targeting more food-insecure households and HIV/AIDS-affected households
- agricultural and social research to identify appropriate agricultural options for HIV/AIDS-affected households
- development of a community legume seed bank to try to improve legume seed access in the long term
- formation of a farmer organization to link with other smallholder farmers and advocate for better access to productive resources at national and international scales.

Qualitative research carried out in 2005–2006 also indicated that there were reduced problems with livestock eating pigeon pea in many villages and that land seizure from new migrants was rare. Farmers also indicated that they felt there was increased cooperation at the village level to work together to solve problems and that there was an increased pride and dignity and a feeling that they have their own resources to solve problems.

In sum, the SFHC project has used participatory methods, bridging social and agricultural sciences, to address a range of inequalities, including rising fertilizer costs, declining access to legume seeds, unequal women's workloads, and household conflict over use of legumes. Many of the inequalities identified are long-term problems that will require considerable social mobilization, but already clear strides have been made along a number of fronts. While agricultural development problems are always to some extent contextually specific, this case study nevertheless suggests a variety of ways in which agricultural research can partner with progressive rural groups and take flexible, participatory approaches as a means to addressing social inequalities in rural areas in the global South.

CONCLUSION: ADDRESSING GENDER INEQUALITIES IN AGRICULTURAL SYSTEMS

This chapter has attempted to define gender and examine the household as a site of inequalities. The chapter has considered the ways in which gender inequalities and other inequities (e.g., age, ethnicity) are embedded in agricultural systems, including access to productive resources and access to and involvement in agricultural science. In order to address inequalities at this scale, agricultural scientists need to understand some of these basic concepts and the ways in which these inequalities are formed and reproduced.

REFERENCES

Agarwal, B. (1994). "A Field of One's Own: Gender and Land Rights in South Asia." Cambridge University Press, Cambridge.

Agarwal, B. (1997). Relations: Within and beyond the household. *Fem. Econ.* **3,** 1–51.

Alwang, J., and Siegel, P. B. (1994). "Rural Poverty in Zambia: An Analysis of Causes and Policy Recommendations." The World Bank, Washington, DC.

Berti, P., Krasevec, J., and Fitzgerald, S. (2004). A review of the effectiveness of agriculture interventions in improving nutrition outcomes. *Public Health Nutr.* **7,** 599–609.

Bezner Kerr, R., Snapp, S., Chirwa, M., Shumba, L., and Msachi, R. (2007). Smallholder diversification with legumes in southern Africa: A case study from Malawi. *Exp. Agric.* **43,** 437–453.

Cornwall, A. (2003). Whose voices? Whose choices? Reflections on gender and participatory development. *World Dev.* **31,** 1325–1342.

David, S. (1998). Intra-household processes and the adoption of hedgerow intercropping. *Agric. Hum. Values* **15,** 31–42.

Deere, C. D., and Leon, M. (2003). The gender asset gap: Land in Latin America. *World Dev.* **31,** 925–947.

Devereux, S. (2002). The Malawi famine of 2002: Causes, consequences and policy lessons. *IDS Bull.* **33,** 70–78.

Doss, C. (1999). "Twenty-five Years of Research on Women Farmers in Africa: Lessons and Implications for Agricultural Research Institutions." CIMMYT, Mexico City.

Doss, C. R. (2001). Designing agricultural technology for African women farmers: Lessons from 25 years of experience. *World Dev.* **29,** 2075–2092.

Due, J., Magayane, F., and Temu, A. (1997). Gender again: Views of female agricultural extension officers by smallholder farmers in Tanzania. *World Dev.* **25,** 713–725.

Due, J. M. (1988). Intra-household gender issues in farming systems in Tanzania, Zambia and Malawi. *In* "Gender Issues in Farming Systems Research and Extension" (S. V. Poats, M. Schmink, and A. Spring, eds.), pp. 331–344. Westview Press, Boulder, CO.

Dwyer, D., and Bruce, J. (1988). "A home Divided: Women and Income in the Third World." Stanford University Press, Stanford.

Ellis, F., Kutengule, M., and Nyasulu, A. (2003). Livelihoods and rural poverty reduction in Malawi. *World Dev.* **31,** 1495–1510.

FAO (1995). A synthesis report of the Africa Region: Women, agriculture and rural development. Prepared under the auspices of FAO's Programme of Assistance in Support of Rural Women in Preparation for the Fourth World Conference on Women. FAO, Rome.

Ferguson, A. E. (1994). Gendered science: A critique of agricultural development. *Am. Anthropol.* **43,** 540–549.

Garcia, Z., Nyberg, J., and Saadat, S. O. (2006). "Agriculture, Trade Negotiations and Gender." FAO, Rome.

Humphries, S., Gonzales, J., Jiminez, J., and Sierra, F. (2000). Searching for sustainable land use practices in Honduras: Lessons from a programme of participatory research with hillside farmers. *Agric. Res. Extens. Network, paper no.* **104**.

Jewitt, S. (2002). "Environment, Knowledge and Gender: Local Development in India's Jharkhand." Ashgate, Burlington, VT.

Kennedy, E., and Peters, P. (1993). Household food security and child nutrition: The interaction of income and gender of household head. *World Dev.* **20,** 1077–1085.

Khadiagala, L. S. (2001). The failure of popular justice in Uganda: Local councils and women's property rights. *Dev. Change* **32,** 55–76.

Lastarria-Cornhiel, S. (1997). Impact of privatization on gender and property rights in Africa. *World Dev.* **25,** 1317–1333.

Lynch, B. D. (1997). Women and irrigation in highland Peru. *In* "Women Working in the Environment" (C. E. Sachs, ed.), pp. 85–102 Taylor & Francis, Washington, DC.

Peters, P. (1996). "Failed magic or social context? Market liberalization and the rural poor in Malawi. Development Discussion Paper No. 562." Harvard Institute for International Development, Cambridge, MA.

Quisumbing, A. R., and Maluccio, J. A. (2000). "Intrahousehold Allocation and Gender Relations: New Empirical Evidence from Four Developing Countries." *IFPRI Discussion, paper no.* **84**.

Ramamurthy, P. (1997). Rural women and irrigation: Patriarchy, class, and the modernizing state in South India. *In* "Women Working in the Environment" (C. E. Sachs, ed.), pp.103–126. Taylor & Francis, Washington, DC.

Ravazi, S. (2006). "UNRISD Research and Policy Brief 4: Land Tenure Reform and Gender Equity." UNRISD, Geneva.

Saito, K. A., and Weidemann, C. J. (1990). "Agricultural Extension for Women Farmers in Africa." The World Bank, Washington, DC.

Smale, M., and Heisey, P. W. (1995). Maize of the ancestors and modern varieties: The microeconomics of high-yielding variety adoption in Malawi. *Econ. Dev. Cult. Change* **43,** 350–368 .

Snapp, S., Kanyama, P. G. Y., Kamanga, B., Gilbert, R., and Wellard, K. (2002). Farmer and researcher partnerships in Malawi: Developing soil fertility technologies for the near-term and far-term. *Exp. Agric.* **38,** 411–431.

Snapp, S. S., Mafongoya, P. L., and Waddington, S. (1998). Organic matter technologies for integrated nutrient management in smallholder cropping systems of southern Africa. *Agric. Ecosyst. Environ.* **71,** 185–200.

Von Braun, J., Puetz, D., and Webb, P. (1989). "Irrigation Technology and Commercialization of Rice in the Gambia: Effects on Income and Nutrition." International Food Policy Research Institute, Washington, DC.

Von Braun, J., and Webb, P. (1989). The impact of new crop technology on the agricultural division of labor in a West African setting. *Econ. Dev. Cult. Change* **37,** 513–534.

Whitehead, A., and Kabeer, N. (2001). Living with uncertainty: Gender, livelihoods and pro-poor growth in rural sub-Saharan Africa. *IDS Working Paper* **134,** 1–31.

Whitehead, A., and Tsikata, D. (2003). Policy discourses on women's land rights in sub-Saharan Africa: The implications of the return to the customary. *J. Agr. Change* **3,** 67–112.

INTERNET RESOURCES

	Description and comments
http://hdr.undp.org/	Annual Human Development Reports provide many useful statistics and insight into global inequalities for different measures of development.
http://www.jubileesouth.org/ http://www.jubileedebtcampaign.org.uk/ http://www.dropthedebt.org/ http://www.50years.org/	These sites provide good background material on debt and ways to advocate for debt reduction.
http://www.etcgroup.org	The Action Group on Erosion, Technology, and Concentration (ETC group) provides an excellent resource on the magnitude of corporate power and consolidation in agriculture in its Oligopoly, Inc. series of reports.
http://www.unctad.org/en/docs/gdsafrica20031_en.pdf	The United Nations Conference on Trade and Development (UNCTAD) has long been a world leader in providing information about and analyses of the problems associated with commodity-dominated export economies. UNCTAD's report "Economic Development in Africa: Trade Performance and Commodity Dependence" (2003b) is another very valuable resource on these issues.
http://www.fao.org/gender/multimed/videos.htm	Online videos about gender and agriculture by the FAO, recommend especially the first video
http://www.unicef.org/photoessays/37446.html	Online UNICEF photo essay about the "double dividend" of gender equity
http://www.unicef.org/sowc07/lifecycle/index.html	Description of different gender issues at different stages in a girl's life cycle
http://web.worldbank.org/WBSITE/EXTERNAL/TOPICS/EXTGENDER/0,,contentMDK:20206498~pagePK:210058~piPK:210062~theSitePK:336868,00.html	Case studies of gender and agriculture issues by the World Bank
http://www.fao.org/docrep/009/a0493e/a0493e00.htm#Contents	Report by the FAO about agriculture, gender, and trade issues
http://www.unrisd.org/80256B3C005BCCF9/httpNetITFramePDF?ReadForm&parentunid=64FF792CAE6DF527C1257108003F59AA&parentdoctype=brief&netitpath=80256B3C005BCCF9/(httpAuxPages)/64FF792CAE6DF527C1257108003F59AA/$file/RPB4e.pdf	Report by UNRISD about land tenure and gender issues

(*Continues*)

INTERNET RESOURCES (*Continued*)

	Description and comments
http://www.fao.org/gender/en/stats-e.htm	This site provides some good, if somewhat dated, statistics on gender inequalities in agriculture.
http://www.unep.org/geo2000/ov-e/0002.htm	This report by the United Nations Environment Program gives a good overview of the major global environmental issues of this century.
http://viacampesina.org/	The Web site of Via Campesina, a major rural social movement made up of small farmers from around the world
http://www.mstbrazil.org	The world's largest rural social movement
http://www.landcoalition.org/ http:www.icarrd.org http://www.landaction.org/ http://www.fmra.org	A series of excellent sites on land reform
www.focusweb.org www.wtowatch.org www.ifg.org www.twnside.org.sg www.foodfirst.org www.iatp.org	These sites provide information about and analyses of the global trading system and multilateral institutions regulating it.

CHAPTER 10

The Nature of Agricultural Innovation

Czech Conroy

Summary

This chapter discusses the nature of innovation and innovations and points out that innovations can be **social** (e.g., formation of farmer groups) or **institutional** (e.g., development of new linkages between producers and market traders), as well as **technological**. Innovations range from being simple or minor in nature to complex or major.

It goes on to describe a number of theories about what drives innovation, particularly in agriculture, and identifies four main types of drivers, namely: scarcity of factors of production, such as land and labor; market opportunities, such as increased demand and improved access, and including globalization; a conducive enabling environment, including characteristics such as policies, trade rules, and veterinary services; and new research outputs/technologies.

The chapter describes different models of innovation processes, starting with the simple "transfer of technology" type in which there is only a single source of innovations; i.e., researchers which are "transferred" in a linear fashion to extensionists and then to farmers. It notes that there has been a shift over time away from this type of model to more complex ones that tend to see innovation as an interactive process involving a variety of organizations and individuals who possess different types of knowledge.

Finally, it highlights a few key issues that need to be carefully addressed if agricultural innovation is to be supported more effectively in the future. Issues include the importance of involving multiple stakeholders and improving the enabling environment for innovation (including both policies and institutions). It also identifies some issues associated with operationalizing the innovation systems concept, such as: how many and which different stakeholders to involve initially in an innovation platform; at what geographical level/scale (e.g., sub-district, province) a platform should be formed; how specific/broad the mandate of an IP should be (e.g., one crop, one constraint); and how poor farmers' and women's views can be effectively represented.

WHAT IS AGRICULTURAL INNOVATION?

There are many definitions of innovations (the plural implies things) and innovation (the singular implies a process). A simple definition of innovation is the application of technical, organizational, or other forms of knowledge to achieve positive novel changes in a particular situation.

Traditionally, the focus in agricultural development has been on **technological innovations**, such as varieties or breeds, types of equipment, or methods of pest control. These can improve agricultural enterprises in various ways: for example, they can be growth-increasing, cost-reducing, quality-enhancing, risk-reducing, and shelf life-enhancing. However, as the context of agricultural development has changed, ideas of what constitutes innovation have changed, and so have approaches for investing in it. It is now increasingly recognized that **social and institutional innovations** can be as important as technical ones (as illustrated in Boxes 2.3–2.6, Chapter 2, and in Case Study 8.5, Chapter 8). These include (1) innovation among producers and (2) development of innovatory linkages/networks between producers and service providers. Social innovation among producers may be formal or informal and includes the development of cooperatives, farmer groups, and self-help groups (SHGs) (Box 10.1).

BOX 10.1 Farmer Groups—A Common Type of Social Innovation

The benefits of social innovations can be as great as, if not greater than, those of technological innovations. For example, the formation of groups of farmers can benefit farmers by:

1. improving their access to government research and extension services;
2. strengthening farmers' bargaining power with traders;
3. reducing transaction costs for input suppliers and output buyers;
4. economies of scale (e.g., from bulking up in output marketing or storage); and
5. facilitating savings and access to credit.

Examples of farmer groups can be found in Boxes 2.5 and 2.6 of Chapter 2.

Distinguishing characteristics of innovations include the following (The World Bank, 2006):

- Innovations are new creations of social and economic significance. They may be brand new, but they are more often new combinations of existing elements.
- Innovation can comprise radical improvements but usually consists of many small improvements and a continuous process of upgrading.
- These improvements may be of a technical, managerial, institutional (that is, the way things are usually done) or policy nature.
- Very often innovations involve a combination of technical, institutional, and other sorts of changes.

It is important to distinguish between "innovation" and the concept of "invention," as they are quite different:

> Invention culminates in the supply (creation) of knowledge, but innovation encompasses the factors affecting demand for and use of knowledge in novel and useful ways. The notion of novelty is fundamental to invention, but the notion of the process of creating local change, new to the user, is fundamental to innovation—specifically, the process by which organizations [or farmers?] master and implement the design and production of goods and services that are new to them irrespective of whether they are new to their competitors, their country, or the world . . . (Mytelka, 2000).

Innovations can also be classified according to the degree of change they represent from existing practices—e.g., *evolutionary/minor* versus *revolutionary/major*. Differences between *evolutionary/minor* and *revolutionary/major* are really gradual over a spectrum. They can be defined as follows:

Minor innovations are ones that require the same or similar amounts of each factor of production; e.g., new crop varieties or crop types whose input needs are broadly similar to those of conventional varieties.

Medium innovations require substantial increases in one or more inputs; e.g., a switch from traditional to cross-bred milch cattle may require a lot more fodder, including green fodder year round, and more veterinary inputs.

Major innovations involve not just the replacement of existing technologies with new ones, but *social* (e.g., formation of irrigation management groups) and/or *institutional* innovations (e.g., contract farming, new market linkages) as well. They may even involve a complete change in the production system (e.g., changing from a scavenging poultry system to a confined poultry system). For example, a switch from rainfed to irrigated agriculture could involve changes in crops and cropping systems, credit sources, and marketing arrangements.

Minor innovations, such as the adoption by farmers of a new rice variety (Maurya *et al.*, 1988), may be implemented spontaneously by farmers on a significant scale without external support or encouragement whereas major innovations may require extensive external support to producers of a technical, financial, social and/or institutional nature (e.g., the Kebkabiya donkey plough, referenced in Case Study 5, Chapter 8). Even then innovations may not be adopted on a sustained basis if they have design weaknesses (e.g., the Bangladesh poultry model—see Case Study 5, Chapter 8). Alternatively, some major innovations may only be made by *resource-rich* producers or processors who have access to the human, social, or financial assets (see Figure 2.1, Chapter 2) required for the innovation to take place. The innovation may bypass the *resource-poor* and could even leave them worse off (less competitive vis-à-vis resource-rich producers), unless pro-poor development agencies intervene to improve their access to the required assets; e.g., credit.

THEORIES ABOUT DRIVERS OF INNOVATION

Does innovation occur randomly over time and space, or are there certain factors or conditions that stimulate innovation of certain types? A number of theories have been developed that aim to explain what drives innovation, and these will now be described.

Science Push/Transfer of Technology Model

The dominant view during the last few decades has been that scientific research is the main driver of innovation, creating new knowledge and technology that can be transferred to (and adapted to) different situations. The science push/ToT model of innovation mirrored the belief that "basic science leads to applied science, which causes innovation and wealth." The policy implications of the science push model were simple—If you want more economic development, you fund more science. The people who would reproduce and use the technology were not seen as sources of innovations or ideas in their own right. In this model technological change is *exogenous* to the economic system, originating outside the agricultural systems that are expected to benefit from it. The adoption of innovations by farmers that were developed on the basis of this model has generally been disappointing, particularly in the case of resource-poor farmers.

Population Pressure Models

Work by Boserup (1965) and Binswanger and McIntire (1987) identified increasing population density as the main driver in the evolution of agricultural systems—from extensive hunter/gatherer systems to slash-and-burn systems to more intensive farming systems: population growth (and the consequent scarcity of land) provides the impetus for *endogenous* technological change. As agricultural land becomes scarcer, traditional practices like long fallow periods are abandoned, and intensification technologies (often labor intensive) tend to be applied on a large scale, resulting in average increased output per hectare. Boserup also saw population growth as ultimately leading to cheaper transport, easier marketing, and more specialization, which in turn would lead to the growth of local towns and more profitable agriculture—*provided there were no cheap imports of domestically produced agricultural goods*. This model only addresses part of the processes driving agricultural innovation and is not relevant to situations in which *labor* is the most scarce factor of agricultural production—situations that are in some ways more relevant today with the spread of HIV/AIDS and labor migration to urban centers.

Market Pull Models

Boserup's model of innovation attached little importance to the influence of external markets, but with increased market integration and globalization, it has become more and more obvious that markets and output prices can exert a major influence on agricultural innovation. Good product prices may provide an incentive to farmers to improve their production practices or their marketing arrangements, and also the cash needed to do so. Models that assume that the primary driver of innovation is access to markets for agricultural and livestock products can be described as "market pull" models of innovation. The importance of markets is illustrated by data on livestock ownership and livestock population trends, presented in Chapter 8, that show how ownership of buffaloes and cross-bred cows increased in parts of India during the last few decades due to the growth in demand for milk, attractive producer prices, and the development of milk marketing infrastructure.

There has been a trend in recent decades towards economic *globalization*; i.e., increased economic integration between countries and a higher share of gross domestic product (GDP) being traded. This has provided opportunities for farmers to export their products to international markets, and "changing patterns of competition . . . in global markets, changing trade rules and the need for continuous upgrading to comply with them . . ." have become major drivers of innovation (The World Bank, 2006). For example, globally, demand for dairy and meat products has been stimulated by "new hygiene and public health management requirements as well as greatly increased product differentiation (cheese, yoghurt, yogurt drinks, cream, fluid milk, cold meats, prepared meals, and myriad other products)" (The World Bank, 2006). (It is important to recognize, however, that globalization has been far more marked in Asia than in most of sub-Saharan Africa.)

For producers and processors involved with a given product in a particular country (A), globalization can be a double-edged sword, as it may expose them to increased competition from producers in another country (B), who may have a comparative advantage (e.g., due to economies of scale, more efficient marketing systems, or more suitable agroecological conditions) in producing the same or similar products. For example, producers of groundnuts in India in the 1990s, whose main market was the domestic market for groundnut oil for cooking, faced increasing competition from imports of cheaper palm oil from Malaysia and Indonesia that led to a major fall in the demand for and price of groundnuts. In Andhra Pradesh, Indian groundnut producers failed to identify viable alternative crops, suffered increasing hardship, and as a result many of them ended up committing suicide.

In some cases, there may be scope for governments, researchers processors, traders, and/or producers in country A to deal with the threat of cheaper imports by increasing the competitiveness of the industry through innovation (technological, social, or institutional). In the groundnut example, there was substantial potential for efficiency improvements in processing and marketing in India, but this did not happen, which was due at least in part to government regulations.

Induced Innovation Model/Theory

In the science push/ToT, population push, and market pull models, there are only one major driver and point of origin/stimulus for innovation. It is clear that none of these models adequately describes and explains the drivers of innovation, and that a more comprehensive model is required. The induced innovation model closely links the emergence of innovations with the prevailing economic conditions (as opposed to innovations occurring randomly). Earlier versions focused on production-related innovations and the effects of factor (of production) endowments on the direction of technical change. A change in factor endowments induces a change in the relative prices of the factors of production (land, labor capital), thus spurring firms and households to invent and implement technologies that are directed at economizing the use of a factor that has become relatively expensive; for example, labor shortages will induce labor-saving technologies. In nineteenth century North America the steel plough, invented by John Deere, a farmer, was one of several mechanical labor-saving innovations that "were of crucial importance to the westward expansion of U.S. agriculture" (Sunding and Zilberman, 2001).

The Hayami and Ruttan version of the model (1985) identifies both scarcity of factors of production *and* market opportunities as stimuli of innovation. In addition, the emergence of innovations requires technical feasibility and the new scientific knowledge that will provide the technical base for the new technology. The development of these innovations is induced by changes in relative factor and product prices—the scarcity of a factor of production (e.g., labor) may attract the attention of scientists, administrators, or inventors. In this respect, technological change is *endogenous* to the economic system, in contrast to the science push model. (However, Hayami and Ruttan accept that there is also an *exogenous* [supply] dimension to the innovation process as well, as a result of general scientific and technological advances.) The potential demand and the appropriate knowledge base are integrated with the right institutional setup, and together they provide the background for innovation activities.

Summary of Innovation Drivers

The above models identify a range of innovation drivers, of which the main ones are:

- scarcity of factors of production, such as land and labor;
- market opportunities (increased demand, improved access);
- enabling environment (policies, trade rules, veterinary services); and
- new research outputs/technologies.

PROCESSES OF INNOVATION

In the induced innovation model described earlier, not much attention was paid to the ways in which the need or demand for innovations is transmitted to researchers. However, proponents of the model did recognize the need for "effective interaction among farmers, public research institutions and private agricultural supply firms" and the desirability of farmers' being organized into farmers' associations to strengthen their influence (Hayami and Ruttan, 1985). Nowadays, much greater attention is being given to innovation processes and the nature of the relationships between different actors.

During the last 20–25 years or so there has been increasing recognition that innovation process models that are (a) single source and (b) linear, such as the ToT model, are incomplete and overly simplistic; and that this kind of process seldom produces technologies that become widely adopted by farmers in less developed countries (LDCs). There have been many reasons for the failure of the ToT approach: new technologies often involved factor inputs (cash, labor, or land) that poor farmers simply did not have access to in the required amounts; and "receiving environments differ from those in which technologies have been developed, being more complex, more diverse, less controllable and more risk-prone" (Chambers, 1997). Perhaps the underlying problem has been the lack of communication and interaction between researchers and intended users.

The general failure of the ToT model to benefit poor farmers provided a major stimulus to the development of other models of innovation processes that are more complex and nonlinear. These models tend to see innovation as an interactive process involving organizations and individuals who possess different types of knowledge. Some of them also recognize the importance of the "particular social, political, policy, economic and institutional context" within which the process takes place (The World Bank, 2006).

Some of these models also have the development of innovation capacity as a major objective, whereas in the linear transfer of technology model, the primary objective was often seen as maximizing the number of adopters of new technologies. This is in recognition of the fact that circumstances change, and farmers need to be able to respond effectively to those changes with appropriate innovations: technologies or marketing arrangements that were ideal at one time may no longer be relevant 15 years later.

Farming Systems Research (FSR) Model

FSR was widely promoted in the early and mid-1980s, when FSR units were created in many NARIs, particularly in Sub-Saharan Africa (SSA). It complements the formal agricultural research system that plays the central role in the ToT model, and shares its premise that all technological innovations originate from formal research systems and organizations. However, it differs from the ToT model in that it places a strong emphasis on the importance of taking into account farmers' resources, circumstances, and objectives when developing technologies (and when setting research priorities), rather than assuming that circumstances should be modified to make them conducive to effective use of the technology. In this model farmers are consulted from an early stage in the research process and may even be asked to make plots available for experiments to test particular technologies. In the typology of participation described in Table 10.1, FSR generally follows a consultative mode.

Multiple Sources of Innovation Model

In this model "innovations come from anywhere in geographic space, from any research and extension institution, and from any instant in historical time" (Biggs, 1991). In particular, it recognizes that many ideas originate from farmers themselves, and that their development need not necessarily involve formal research systems at all. It emphasizes the non-linearity of the innovation process, in contrast to the ToT model.

This model recognizes that fact that before public and private sector research and extension services came into existence, farmers were the primary source of invention, innovation, and the spread of technical information; and that in many countries there is a wealth of local knowledge about agricultural production practices. During the last decade or two there have been numerous initiatives to collect and record indigenous technical knowledge (ITK) and sometimes to validate it.

The Participatory Technology Development (PTD) Model

The PTD model (related terms are "farmer participatory research" and "participatory innovation development") is a subtype of Biggs's model in that it explicitly recognizes that innovations may draw on local knowledge and/or formal science (Conroy, 2005; Farrington and Martin, 1987; Okali *et al.*, 1994). It also advocates that farmers, scientists, and development nongovernmental organizations (NGOs) undertake research together, typically in the form of on-farm experiments. It acknowledges that many rural people have valuable ITK and also that they conduct their own research, but usually in a more idiosyncratic way than do "scientific" researchers. Having less time and fewer resources but more local knowledge than full-time researchers, the majority of local people are more likely to engage in an *adaptive type* of research; i.e., trying out ideas and technologies borrowed from others and seeing how they work for themselves, making adjustments along the way.

Unlike the FSR model, the PTD model attaches importance to recognizing and understanding farmers' own capacity to experiment, adapt, and innovate and treating them as equal partners in the process. This is because technologies that are suitable for one farmer or livestock keeper may need modification before they are suitable for another (Pretty, 1995), and other farmers, rather than researchers, may be the best people to identify what modifications are required. In addition, there may be scope for farmers to improve a technology or use it more profitably, in ways that researchers might not think of themselves.

A valuable and widely used classification system for different types, or modes, of farmer participation in on-farm research, developed by Biggs in 1989, is summarized in Table 10.1. The degree of farmer involvement in decision making varies from mode to mode and increases in the modes to the right-hand side. These four modes are really different points on a continuum. There are no clear dividing lines between them, and a project may gradually move from one mode to another during its lifetime: indeed, it is common for projects to begin in the consultative mode and to shift to the collaborative mode as researchers and farmers (or livestock keepers) develop (a) a common understanding of experimental objectives and the best ways of achieving them and (b) a relationship based on trust.

TABLE 10.1 Four Different Modes of Farmer Participation in On-Farm Research

	Contract	Consultative	Collaborative	Collegiate
Type of relationship	Farmers' land and services are hired or borrowed; e.g., researcher contracts with farmers to provide specific types of land.	There is a doctor-patient relationship. Researchers consult farmers, diagnose their problems, and try to find solutions.	Researchers and farmers are roughly equal partners in the research process and continuously collaborate in activities.	Researchers actively encourage and support farmers' own research and experiments.

Source: Biggs (1989).

The traditional mode, in which the researchers are dominant and farmers are least involved, is the *contract* mode. The contract mode involves formal experimentation in specific on-farm situations, but the farmers' views are not actively sought by the researchers. The *consultative* mode, classically exemplified by applications of the farming systems research approach of the early to mid-1980s, includes "diagnosing farmers' practices and problems, planning an experimental programme, testing technological alternatives in farmers' fields and developing and extending recommendations" (Tripp, 1991). In this mode, it is the researchers who provide the solutions, plan the experiments, and finally recommend what is best practice.

In the *collaborative* mode, the ideas for interventions to be tested may also come from farmers or other knowledgeable people in the locality, and are the product of discussions between the researchers and farmers or livestock keepers. In the case of the *collegiate* mode, it is the farmers themselves who play the lead role in identifying what the content of the experiments will be, and the manner in which they will be conducted. PTD roughly corresponds to the *collaborative* and *collegiate* modes, although in the early stages it may be necessary to operate in the *consultative* mode.

Learning Selection Model

This is analogous in some ways to Darwin's theory of natural selection, which consists of three mechanisms (Douthwaite, 2001):

- novelty generation (random genetic mutations and sexual recombination of differing genetic material);
- selection (beneficial changes to achieve better survival and breeding rates); and
- diffusion and promulgation (the spread of beneficial differences to other areas).

Douthwaite argues that, when applied to innovation, these three mechanisms should be seen as part of a single learning process that is interactive and iterative, involving a number of different stakeholders: the originator(s) of the original idea, a product champion, the manufacturers, and the users. This model originated from his research on harvesting and post-harvesting technologies, whose main finding was that the successful technologies were the ones that manufacturers and users had modified the most. (Conversely, engineers and designers alone were singularly unable to develop machine designs that people adopted.) He uses the term "co-development" to describe this multistakeholder process.

Applying the model to agricultural innovation, a technology (depicted as a cogwheel in Fig. 10.1) begins as a "plausible promise" that motivates key stakeholders to develop it. The technology is then modified and gradually increases in fitness (in the biological sense, meaning improvements in the likelihood that technology will be adopted and promulgated) as it becomes "meshed in" to existing systems through the adaptation and learning that takes place. The meshing in is represented in Figure 10.1 by the move from a single cogwheel to three interlocked ones. The increase in knowledge is represented by the increases in the size of the cogwheels. Learning selection, which is based on an experiential learning cycle, is shown inside the black box, and is responsible for the evolution.

Innovation Systems (IS) Model

In this model innovation is seen as being "neither research nor science and technology, but rather the application of knowledge (of all types) to achieve desired social and/or economic outcomes" (Hall *et al.*, 2005). An innovation system can

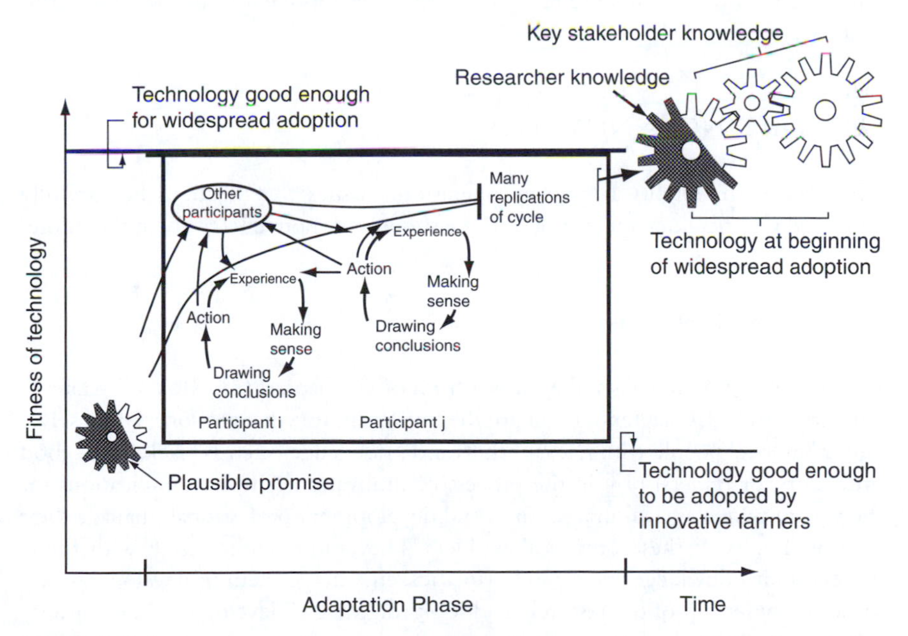

FIGURE 10.1 Learning selection is shown inside the black box and is responsible for the evolution.

be defined as "a network of organizations, enterprises, and individuals focused on bringing new products, new processes and new forms of organization into economic use, together with the institutions and policies that affect their behaviour and performance" (The World Bank, 2006).

The IS concept embraces not only the science suppliers but the *totality* of actors needed for innovation to take place, and the *interaction* of actors involved in innovation. In particular, it highlights the contribution of the *private sector* to innovation .... "It extends beyond the creation of knowledge to encompass the factors affecting demand for and use of knowledge in novel and useful ways" (The World Bank, 2006). The IS concept is derived from direct observations of industrial countries and sectors with strong records of innovation, and has been used predominantly to explain patterns of past economic performance in developed countries (Freeman, 1987). Nevertheless, it is closely related in some ways to the multiple sources of innovation model, which preceded it in the literature on agricultural innovation in LDCs.

The IS model is more holistic than the previous models of innovation processes in that it gives more emphasis to private sector service providers and also emphasizes the important influence of the enabling environment (both policies and institutions) on innovation. However, it does not necessarily give the same level of emphasis to the involvement of resource-poor farmers in the innovation process. The IS concept has been applied to agriculture in developing countries only recently, but has been enthusiastically embraced by international donors and others who are searching for a more effective approach to innovation in this context.

SUPPORTING INNOVATION

This section highlights a few key operational issues that need to be carefully addressed if agricultural innovation is to be supported more effectively in the future.

Involving Multiple Stakeholders

Recent innovation models and examples from other chapters (e.g., Box 3, Chapter 2, and Case Study 4, Chapter 8) highlight the important role that producers themselves and other local people with relevant skills and knowledge (such as blacksmiths, food processors, or traders) play in the process of innovation. This has implications for the way in which agricultural research and development professionals manage their relationships with these other stakeholders. They must work closely with them; respect their knowledge, views, and priorities; encourage them to develop a strong sense of ownership of the research; and look for ways of developing their capacity where necessary.

Growing recognition of the need to involve a variety of different stakeholders in the innovation process has recently led to the promotion of multi-stakeholder innovation platforms (Hall *et al.*, 2005) or learning alliances (Lundy *et al.*, 2005).

Operational Issues Associated with the IS Concept

The "operational aspects of the [innovation systems] concept remain largely unexplored" in relation to agriculture in developing countries (The World Bank, 2006). As agricultural research and development agencies consider how best to operationalize the concept, issues that need to be addressed include:

- How many, and which, different stakeholders should be involved initially in an innovation platform?
- At what geographical level/scale (e.g. subdistrict, province) should it be formed?
- How specific/broad should the mandate of an innovation platform (IP) be (e.g., one crop, one constraint)?
- How can poor farmers' and women's views be effectively represented in the deliberations of innovation platforms, so that the process is demand-led?
- How is it best to connect with and influence policy makers?
- What human capacity development and re-skilling are required to enable researchers and extension workers to implement the IS approach? (See Chapter 2 for discussion of this.)

Proponents of the IS concept have advised that, while it is important to intensify the level of interaction between key stakeholder groups, this does not necessarily mean that "everybody should be partners with everybody else in a mechanistic way" (Hall *et al.*, 2005).

Experiential Learning

Douthwaite's learning selection model, the PTD model, and the case studies in Chapter 8 (livestock) highlight the importance of hands-on learning by farmers and equipment manufacturers—learning by doing. This is consistent with theories of adult learning (Kolb, 1984), and the Farmer Field School approach is based on this (Gallagher, 2003). Interactive, iterative processes involving potential users may be a necessary condition of effective innovation, but, in contrast, traditional approaches to agricultural extension tend to treat farmers as passive recipients of technical knowledge. This may be effective in making farmers aware of technologies without leading to adoption of them, which may be another reason why there has been limited adoption of many technologies emanating from the formal research system.

The Enabling Environment

There is a danger that multistakeholder innovation platforms will be promoted without addressing key higher-level constraints to effective pro-poor innovation. The experience of initiatives and approaches intended to enhance agricultural innovation processes has shown that such constraints can greatly inhibit positive change. The enabling environment, which includes institutions, policies and laws, is more difficult to change, however, and hence tends to be overlooked in donor-sponsored initiatives to enhance agricultural innovation processes.

In the innovation system literature *institutions* are defined as "the sets of common habits, routines, practices, rules or laws that regulate the relationships and interactions between individuals and groups" (Hall *et al.*, 2005). For example, projects or initiatives that aimed to support NARIs to adopt participatory innovation processes have commonly found that the processes are not sustained much beyond the end of the initiative. Inhibiting factors have included entrenched habits, practices, and rules, such as a widespread attitude that participatory on-farm research is messy and does not conform to scientific and statistical norms, and the basing of career progression among NARI staff primarily on numbers of journal articles published rather than demonstrated benefits of their work to poor farmers.

The importance of *policies* was identified by practitioners of FSR, as well as more recently by proponents of IS. However, the IS concept stresses the need to look beyond research policy and to ensure that a wide range of policies is in place to address the "incentives, triggers and support structures needed to stimulate and sustain creativity" (Hall *et al.*, 2005).

REFERENCES

Biggs, S. (1989). "Resource-Poor Farmer Participation in Research: A Synthesis of Experiences from Nine Agricultural Research Systems." OFCOR Comparative Study Paper No. 3. ISNAR, The Hague.

Binswanger, H. P., and McIntire, J. (1987). Behavioural and material determinants of production relations in land-abundant tropical agriculture. *Econ. Dev. Cult. Change* **36**(1), 73–99.

Biggs, S. D. (1991). Multiple sources of innovation model of agricultural research and technology promotion. *World Dev.* **18**(11), 1481–1499.

Boserup, E. (1965). *The Conditions of Agricultural Growth: The Economics of Agrarian Change under Population Pressure.* Allen and Unwin, London.

Chambers, R. (1997). *Whose Reality Counts? Putting the First Last.* Intermediate Technology Publications, London.

Conroy, C. (2005). *Participatory Livestock Research: A Guide.* ITDG Publishing, Bourton-on-Dunsmore, UK.

Douthwaite, B. (2001). *Enabling Innovation: A Practical Guide to Understanding and Fostering Technological Change.* Zed Books, London and New York.

Farrington, J., and Martin, A. (1987). "Farmer Participatory Research: A Review of Concepts and Practices." Agricultural Administration Research and Extension Network Discussion Paper 19. Overseas Development Institute, London.

Freeman, C. (1987). *Technology and Economic Performance: Lessons from Japan*. Pinter, London, UK.

Gallagher, K. (2003). Fundamental Elements of a Farmer Field School. *LEISA Magazine*, March 2003.

Hall, A., Mytelka, L., and Oyeyinka, B. (2005). "Innovation Systems: Implications for Agricultural Policy and Practice." ILAC Brief 2. July 2005.

Hayami, Y., and Ruttan, V. W. (1985). *Agricultural Development: An International Perspective*. Johns Hopkins University Press, Baltimore, USA.

Kolb, D. A. (1984). *Experiential Learning*. Prentice Hall, Englewood Cliffs, NJ, USA.

Lundy, M., Gottret, M. V., and Ashby, J. (2005). "Learning Alliances: An Approach for Building Multi-Stakeholder Innovation Systems." ILAC Brief 8. August 2005.

Maurya, D. M., Bottrall, A., and Farrington, J. (1989). Improved livelihoods, genetic diversity and farmer participation: A strategy for rice-breeding in rainfed areas of India. *Exp. Agric.* **24**(3), 311–320.

Mytelka, L. (2000). Local systems of innovation in a globalised world economy. *Industry Innovation* **7**(1), 15–32.

Okali, C., Sumberg, J., and Farrington, J. (1994). *Farmer Participatory Research: Rhetoric and Reality*. Intermediate Technology Publications, London.

Pretty, J. (1995). *Regenerating Agriculture: Policies and Practice for Sustainability and Self-Reliance*. Earthscan, London.

Sunding, D., and Zilberman, D. (2001). The agricultural innovation process: Research and technology adoption in a changing agricultural sector. *In* "Handbook of Agricultural Economics" (B. L. Gardner and G. C. Rausser, eds.), Vol. 1. Elsevier Science, Amsterdam, pp. 207–261.

The World Bank (2006). *Enhancing Agricultural Innovation: How to Go beyond the Strengthening of Research Systems*. The World Bank, Washington, DC.

Tripp, R. (1991). The farming systems research movement and on-farm research. *In* "Planned Change in Farming Systems: Progress in On-Farm Research" (R. Tripp, ed.). John Wiley, Chichester, pp. 3–17.

CHAPTER 11

Outreach to Support Rural Innovation

Vicki Morrone

Summary

The mission of extension educators is undergoing a rapid transformation from focusing primarily on information dissemination to one of supporting local capacity to generate knowledge and facilitate rural development. Essential to this change process is engagement with communities and farmers in order to build ownership and move beyond transfer of technologies. A range of actors is involved, including private industry advisors and market-linked information, non-governmental organizations (NGOs), farmer organizations, and extension educators. Principles of innovative outreach are explored here, including building relevance and local capacity, models for scaling up, and sustainable development. Decentralized extension systems are described that integrate supply, research findings, and demand for information through priorities set by farmers. Farmer field school approaches are being used to support ecological literacy and farmer experimentation on a broad scale. Communication campaigns using radio drama have shown that farmer decision making around pest and crop management can be improved. These approaches focus on farmers as equal partners to ensure that change comes from within and takes on characteristics that reflect local priorities and is part of a sustainable process.

INTRODUCTION

Outreach is the topic of this chapter, where *outreach* is defined as "a way to extend information to new groups." Agricultural development requires resources, technology, processes, and information that are relevant to the lives and livelihood of smallholder farmers and community members. Until recently, the brokering and dissemination of information has been the critical role played by extension advisors. Today, extension staff from governmental organizations, private organizations, and NGOs are expected to act as facilitators and outreach advocates to catalyze innovations, develop local institutions, and generate local knowledge. Including community players in this process requires outreach advisors to take on the role of negotiators in order to assist in local priority setting, resolution of conflicting agendas, and implementing changes that are endorsed by the communities at stake.

Historically, extension systems have focused on agricultural development through extending technological and credit packages in support of production goals. In specific locations and socioeconomic contexts, extension advisors have been tremendously successful at working with farmers and their organizations to support successful technology uptake and rural innovation. An example in Fig. 11.1 illustrates the excitement of woman farmers in Zimbabwe who benefited from extension advisors' demonstration of crop response and the organization of revolving credit

FIGURE 11.1 Zimbabwean participants in a revolving credit club.

clubs to support access to inputs such as fertilizer and hybrid maize seed. Working collectively secured a lower cost through bulk pricing and overcame the transport barriers faced by many smallholder farmers.

It is possible to point to many remarkable local successes, where farmers' innovations and technologies have been promoted by extension. However, the challenge of how to "scale up" outreach and reach a wider audience remains. Overall, minimal agricultural change has occurred in the rainfed agroecosystems of the tropics. At the same time, reductions in public funding for agricultural research and extension are occurring throughout sub-Saharan Africa, Latin America, and Southeast Asia. These regions involve tremendous complexity of natural and social resources with limited infrastructure. In many cases this problematic context is compounded by a lack of transparency, limited accountability, a centralized bureaucracy, and instability in the economic and political environments.

Dissemination of targeted technologies has benefited some stakeholders and contributed to the remarkable yield gains associated with the Green Revolution, but it has been inadequate to address the complex realities of smallholder families and community goals for environmental services, such as the protection of soil and water quality. To find long-term solutions will require reform in how extension is conceptualized (Rivera and Alex, 2004). Throughout the developing world new models for community-led extension and learning are emerging, including examples highlighted in this chapter from Southeast Asia and sub-Saharan Africa.

Central principles in extension that will be explored in this chapter include:

1. **improving relevance** through demand-driven and decentralized extension;
2. **facilitation** of human and organizational capacity for innovation;
3. **scaling up** to reach more people; and
4. **sustainability** of outreach.

RELEVANCE OF EXTENSION

There is a growing consensus on the need for more demand-driven extension services. This is presented as the first principle for effective outreach, where locally defined priorities guide extension programming. As illustrated in Table 11.1, there are different approaches that can be used to improve the client orientation and integration of demand and supply in extension. Articulating local demand can be problematic, as farmers are often dispersed and numerous in comparison to support services. This highlights the role that institutional innovations[1] can play. Demand-driven extension examples include those built around farmer organizations, decentralized extension services that rely on local contracts, and market-linked extension.

[1]"*Innovation is the lifeblood of any organization.*" Quote from Don Sheelen, reported in Wikipedia definition of *innovation for organizations.*

TABLE 11.1 Fundamental Principles and Examples of How to Achieve More Relevant Outreach

Guiding principles	Extension models
Farmer-led extension - Farmers as empowered learners - Respect for indigenous and local knowledge - Farmer organizations to identify common needs	- Farmer field schools (see Case Study, Text Box 11.4) - Decentralized, demand-driven extension - Participatory action research and extension - Farmer organizations (e.g., the National Smallholder Farmer Association of Malawi)
Experiential learning - Biological principles taught, not set recommendations - Knowledge generation through action learning	- Farmer field schools - Primary and secondary agroecological curricula - Farmer research groups (CIALs) - Participation in mother-baby participatory research and extension[a]
Facilitation - Extension as facilitators and knowledge brokers	- Farmer field schools - Market-linked value chains
Iterative learning - Building working relationships among researchers, extension, NGOs, farmers, and other stakeholders - Transparent and joint priority setting - Systematic evaluation and reflection steps	- Participatory action research and extension - Farmer research groups (CIALs) - Demand-driven extension from farmer organizations and local contracts for information - Community roundtable discussions
Sustainability of extension - Ongoing support from public and private sectors jointly - Integration of demand and supply of extension information - Inclusion in country-wide goal-oriented program discussions	- Demand-driven extension from farmer organizations and local contracts for information - Market-linked value chains (see Case Study Text Box 11.2) - Local private providers of technical information - National Agricultural Advisory Services (NAADS) development of country-wide service (see Case Study, Text Box 11.3)

[a]See Snapp and Heong, 2003, for a description of this farmer-centered form of participatory action research.

Extension advisors are seen here as facilitators of a learning process, shifting away from a focus on communicating technical outputs (Table 11.1). Facilitators work to support local learning, build quality relationships, and develop an iterative process that has "built in" evaluation and reflection steps. The foundation for this extension approach is a respect for farmers as co-learners in the change process who are interested in learning about principles rather than set recommendations. An outreach process to address information or technology gaps needs to be approached from this "inside out" view. The end users must be offered a chance to provide input into the development of the program. Sustainability of the development process requires attention throughout the process of local knowledge generation and integration of extension services supply and demand, as articulated by community expectations and preferences.

A client-oriented extension approach has been criticized by those who contend that this process is time- and resource-consuming and difficult to implement. Issues have also been raised about equity, as questions arise about who is setting locally defined priorities. It is challenging to assure access to all community members and to negotiate the diverse agenda of multiple stakeholders. (See Chapter 9 for further insights into equity and access issues.) There are often differences in perspective amongst community members, NGO staff (who may be acting as facilitators for farmer organizations), extension advisors, and other educators. Some community members may value longer term returns, such as soil improvement, while other community members—perhaps at the edge of survival—may necessarily focus on immediate returns. Concerns about shifts in labor requirements may vary with agricultural responsibilities and priorities, depending on gender and age group. Respect for local knowledge, culture, and traditions is of paramount importance when facilitating negotiations for a community-led extension plan.

There are proven methods to promote technologies to target audiences, as illustrated in Table 11.2. Historically, the training and visit system was promoted by the World Bank and was shown to require considerable resources, but at the same time, the system was an effective means to promote technology transfer of relatively simple technologies such as fertilizer use on hybrid rice. For it to be successful, it required an environment of stable markets, infrastructure, and social capital, with strong linkages among research, extension, and farmers (Benor and Harrison, 1977). These conditions rarely exist in developing countries.

There are emerging examples of alternatives to traditional extension, including privatized or public–private hybrid models and decentralization of extension services (see Fig. 11.2). These have been tried in different permutations in Bolivia, New Zealand, and Uganda (Bently *et al.*, 2003). Concerns have been raised about the potential of market-driven extension systems not to prioritize the environment or pay sufficient attention to sustainable production techniques (Hall *et al.*, 1999).

ENHANCING CLIENT ORIENTATION AT DIFFERENT SCALES

Regardless of the scale of operation, extension must address local aspirations. This is a difficult task for a district-wide program, given the diversity of people, expectations, and resources. This is one reason that NGOs, and NGO partnerships with public and private sectors have emerged as globally significant. Many nonprofit organizations target local communities and have achieved notable successes in rural development. This includes areas where public institutions have failed to deliver

TABLE 11.2 Innovative Outreach Approaches to Support Agricultural Knowledge Dissemination

Agricultural knowledge system	Strengths	Challenges
Market information access through cell phones	Dissemination of real-time information on market prices	• Requires infrastructure, mobile phones, and human resources • Depends on quality of market information collected and interpreted
Entertainment: radio programming, drama clubs, songs	• Entertainment value • Successful means of engaging audience and addressing controversy • Comprehensive coverage (geographic and economic)	• Farmer access to radio • Sponsorship for program and timing of broadcast • Requires gifted script writers and actors • Entertainment media and programming needs will vary with audience targeted
On-site analyses of crop and soil health using soil test kits and plant color guides	• Rapid information on soil properties and crop health • Farmer empowerment through field diagnosis • Generates interest in remedies/ recommendations that are relevant	• Analyses vary in accuracy • Requires investment in calibration and training
Media campaigns: posters/ billboards, printed cloth, or T-shirts	• Can target heavily trafficked areas (markets) • Reaches a large audience over time	• Requires skill in message preparation • Language and style will vary depending on audience
Vouchers for inputs, with technical information	• Increases access to small amounts of inputs • Facilitates experimentation	• Requires careful implementation to support and *not* suppress local businesses

effective technical support. There are challenges inherent in NGO-led development, including a potentially narrow scope of operation or changeable priorities. For example, a specific group may be targeted by an NGO (e.g., children under five or widows), or the focus may shift from one technology to another (e.g., from agroforestry to nutrition), and the NGO may only work in one region or community.

A number of innovative examples are explored here, addressing the conundrum of how to meet local needs at different scales. These include integrated projects that grow over time, as well as more targeted, market-linked approaches (Table 11.3). An example of the former is a Malawi community fuel-wood tree project that linked indirectly to the private sector, while focusing on community-level natural resource management (Box 11.1).The tree project's goal was to offer communities a way to produce trees to reduce dependency on indigenous forests, to provide a renewable source of firewood, and to offer a means for families to earn income through tree

FIGURE 11.2 Private extension crop advisors work with university staff to support paprika production.

products and wood sales. This project was integrated over time into the national extension system, a process facilitated by joint planning from the onset. Links to local businesses can be built through solicitation of financial support and management advice from a business community service organization, as described in the case study. In this example, a cash crop with environmental impact was linked through a business organization to support community-led, sustainable resource management.

The importance of a facilitator role was shown for extension educators in this Malawi example (Box 11.1). As is often the case in natural resource management issues, conflict arose. The selection of tree species for village nurseries brought out different agendas amongst the participants. For example, some village chiefs were interested in promoting a single species for production of poles for sale. Other community members were interested in multiple species to address a range of needs, from fuel-wood to poles, soil fertility, and fruit production. Facilitation of community discussions and an evaluation process undertaken each year were strengths of the project that enhanced long-term success. This demonstrates how extension can support scaling up through attention to conflict resolution and communication.

The value chain[2] approach is another innovative means to form private–public partnerships. This has been highly successful in linking specific groups of smallholder farmers to market opportunities. In Zambia, extension from private and public sectors has supported smallholder production of paprika as an export crop (paprika is used as a natural food colorant); see Fig. 11.2. Another case study comes from a seed-producing area of southern Malawi (Box 11.2). In this example, farmer groups were

[2] *Value chain* is a systematic approach to integrating every step from field to market, through production and processing to sale of a product.

TABLE 11.3 Innovations in Extension to Enhance Demand-Driven Provision of Technical Services

Scale	Approach	Demand articulation	Facilitators	Supply	Case studies
Village	Enabling rural innovation at local scale	Farmer groups formed farmer field schools, supported by NGOs such as Africare; market plan produced and technical gaps identified	Farmer group leaders, NGOs, extension	Local, regional, and international technical providers; farmer groups now produce potatoes for fast-food chain	Nyabyumba farmers' group of Kabale district, Uganda (Kaaria *et al.*, 2006)
Province	Farmer organizations co-learning with technical advisors	Farmer organizations (e.g., irrigation committees, women's credit groups) identify opportunities and service providers	Farmer organization leaders	Seek services from government and NGO technical staff, interact frequently	Irrigation districts, Mali (Pesche, 2004)
Region	Planning for services	Village groups create "demand-for-services" plans, clustered into larger plans from different villages	Extension agents; government planners integrate local demand for services plans	Government extension and environmental protection unit; local NGOs if government technical expertise not available	Azad Jammu and Kashmir region, Pakistan (Qamar, 2004)
Countrywide	National Agricultural Advisory and Development Services	Farmer fora and farmer field school groups form proposals on technical gaps and services needed; prioritized proposals developed at regional meetings	Farmer Fora (often NGO facilitated priority setting at province level; criteria emphasize market crops)	Contracts awarded for technical services by NGOs, extension	Eastern Uganda (Friis-Hansen, 2004)

BOX 11.1 A Case Example—Malawi's Fuel-Wood Tree Nurseries

In Malawi, as in many developing countries, rural families depend on wood for cooking and heating. Deforestation and soil erosion are of growing concern in the populated areas of southern and central Malawi. In 1996, the University of Malawi's Department of Rural Development initiated a survey to assess local requirements for tree production to meet increasing demands for firewood in the growing capital city of Lilongwe. Interest was high, and clubs were formed to support fuel-wood tree production. Linkages with extension and local businesses were established from the beginning of the project, including financial support from the Tobacco Exporters Association of Malawi (TEAM). The tobacco industry relies on wood to cure burley tobacco; thus, there is a consciousness of the need for trees in the countryside from the environmental and economic perspective.

University lecturers managed the program initially and hired students, who learned real-world experience by working with the farmers' clubs and with extension educators of the National Agroforestry team. The clubs received education through training sessions and farmer-to-farmer visits for information on performance of tree species and seedling establishment techniques. TEAM provided long-term support for the project, including funding for the local collection of seed from superior trees. This supported growth from 3 farmer clubs in year 1, to 17 clubs by year 3. Participating households each planted an average of 100 trees, and over time the project has become integrated into the country's extension program, offering Malawians throughout the country the opportunity to learn how to establish their own wood lots.

An assessment of impact has shown unexpected benefits from educational opportunities associated with the project. These include experiential training in participatory extension for agricultural university undergraduates who participated through internships and expanded primary education for children who collected seed and were able to fund school fees. Environmental education on the role of trees in conservation of biodiversity, soil, and water has also expanded at the community level.

supported through NGO, researcher, and private enterprise collaboration. Education in production and business techniques, on-farm research, and market information were all required to assist farmers in addressing this market niche (Box 11.2).

There are prominent examples of extension successes, but these often remain localized. It is vital to acknowledge that agricultural information remains unavailable to the majority of farmers who face severe cash constraints and are unable to pay for advisory services. Production of cash crops for local, regional, and international markets is one pathway that has been used to generate funds to support technical advice. For farmers with limited market access, and those with few

BOX 11.2 Local Seed Production Has Markedly Improved the Livelihoods of Farmers in Chingale, Malawi

Through markedly enhancing the quality and quantity of grain legumes produced, the area became known as a producer of seed-quality grain, which was marketed to government and NGOs. This success built on educational efforts of the NGO World Vision, which initiated training in farmer empowerment and business education in the late 1990s. Partnerships were developed with national and international agricultural research institutes and private sector grain legume marketers. Farmers gained knowledge in seed production techniques and access to new varieties. By 2003, Malawi farmer groups marketed 30 to 50 metric tons of seed annually, allowing these entrepreneurial farmers to build modest houses and upgrade thatched houses with tin roofs.

Source: Setimela *et al.*, 2004.

resources to invest in cash crops, a publicly-funded mechanism such as a voucher system may be needed to ensure access to extension among the poorest farmers. Coordinated action and cooperation among farmers may be the most important means to aggregate demand and improve extension services to distant or poorly resourced areas (Leeuwis and van den Ban, 2004).

Examples of ways farmers have organized across a district are illustrated here through an example from an irrigation catchment in Mali, West Africa (Table 11.3). Success was possible at this site due to a multidisciplinary group of extension providers working closely with farmer groups over an extended period of time. Flexibility was crucial for the extension educators, as the knowledge gaps that farmers identified changed over time. This was often a source of frustration, as farmer groups met repeatedly with technical service providers yet found that the technologies promoted were rarely appropriate or cost effective. Farmer organizations reported that researchers were not very responsive and continued with research that was perceived locally as irrelevant (Pesche, 2004).

DEMAND-DRIVEN MODELS

Centralized and bureaucratic approaches to extension have been critiqued as lacking in flexibility and in responsiveness to farmers. Experiments in new approaches to extension based around decentralization and a "demand-driven" structure are underway in various countries. The village extension system of Laos is a case in point. As described in a interview with Mr. Sisanonh,[3] the director of Central

[3]Reported in the *BeraterInnen News*, January 2004. See http://www.agridea-international.ch/fileadmin/10_International/PDF/RDN/RDN_2004/BN_1–04_The_Village_Extension_System.pdf.

Extension and Training Development in Laos, it is the responsibility of village leaders to organize extension and choose experienced farmers to be extension workers who are supervised by village authorities. These village extension workers address the topics of greatest interest to local households, whether those interests are livestock, crops, or an agricultural specialty. The Laos government extension service responsibilities are to provide technical backstopping and to facilitate networking of village extension workers, in order to promote the spread of innovations.

Another example comes from Uganda, where over half the country is involved in a bold initiative to reorganize extension; see Box 11.3.

A central aspect of the Ugandan NAADS demand-driven extension system is the aggregation and expression of demand by local people. What is the process by which a farmer's demand is articulated, who facilitates it, and how are the identified knowledge gaps prioritized? Government extension, NGO facilitators, and farmer

BOX 11.3 Demand-Driven Advisory Services in Uganda

Uganda has launched institutional reform in provisioning of services to limited resource farmers. Local farmer groups are involved in contracting out information requirements, with the goal of integrating demand and supply. This radical reorganization of extension is called the National Agricultural Advisory and Development Services (NAADS). A recent review was conducted in Soroti, one of the first districts where NAADS was initiated. The number of farmers reached by NAADS was impressive, and many had made significant educational and economic gains. The study also highlighted the importance of earlier investments in farmer education in the district, most notably farmer field schools. NAADS was shown to have been most successful in meeting the needs of market-linked production systems.

The guiding principles of NAADS focus on developing an iterative, demand-driven process. This requires a commitment to change throughout the extension service. Information is sourced from private and public advisors, whoever can best fulfill a local contract for extension services. Priorities generated by farmer forum groups are prioritized at parish levels and then at subcounty level. A list of guiding principles developed by the government of Uganda NAADS program can be viewed at http://www.naads.or.ug/publications.php.

NAADS has undertaken an ambitious experiment in decentralizing and "building in" a demand component to extension services. Success has been variable but, overall, highly significant. This has lead to recent efforts to broaden the NAADS approach and try it in other countries.

organizations are the key players involved in facilitating the articulation of farmers' demands and priorities. There can be conflicting expressions of farmer demand, where local priorities interpreted by one group may differ from those interpreted by another group. Furthermore, a process for regional aggregate demand requires prioritization of multiple local farmer group requests for information. In Uganda, this has led to dissatisfaction among some groups that have found the prioritization process to be slow, cumbersome, and market driven (see Box 11.3). Critiques of this process focus on how to speed up the priority setting process and improve flexibility. Discussion in national and international forums has also centered on the prioritization of cash crops and livestock, which may have led to the bypass of some subsistence cropping systems (Obaa *et al.*, 2005). This transparent process of critique and lively discussion at many levels shows an exciting commitment of Uganda NADDS to a process of continual improvement.

Implementing a change from top-down to decentralized extension at a national level is a bold experiment. Changes in extension structure have been implemented in several countries to enhance accountability, but the flow of resources (financial and technical) to local areas has not always followed the decentralization of priority setting. Resistance to change within an institution is to be expected. In Uganda a sequential process was followed: initially, changes in extension structure were implemented in trailblazing districts, to allow for modifications and growing acceptance before larger scale implementation. A similar progressive model has been followed for implementing the Village Extension System in Laos.

FACILITATION OF LOCAL CAPACITY

Active Learning

Building an environment for active learning is one of the new pillars of extension. For this to occur, it is important to fully respect the local knowledge of stakeholders and to realize that all participants face multiple time demands. Maintaining this perspective will support the building of quality relationships with all actors, from start to finish. Experiential learning builds on indigenous knowledge, promotes science-based education, and engages participants in testing research questions. This reinforces principles and engages farmers in discovery.

Farmer field schools (FFSs) are a form of extension promoted around the world to support discovery learning. Farmers are encouraged to observe the natural world and to conduct experiments in their own fields. Insect and plant interactions were the initial focus of FFSs, with farmers participating in weekly educational sessions and testing principles through field studies[4]. Active, query-based learning about

[4]More information on farmer field schools is available at many web sites, including the Food and Agriculture Organization and http://www.farmerfieldschool.net/document_en/FFS_GUIDe.doc.

agroecology is highly suited to the knowledge-intensive nature of adapting farm management technologies to local conditions. Sustainable agricultural practices in particular are thought to be supported by enhanced understanding of biology (see Chapter 4).

Initially focusing on integrated pest management, the growth of the FFS movement has led in surprising directions, with curriculum as diverse as living soils, livestock health, and learning about the scientific method. From Southeast Asia there are exciting examples of synergistic collaboration between farmer field schools and primary education. The future of agriculture is in the hands of the next generation. Providing experiential learning opportunities in farm settings is an important investment to reach the hearts and minds of future farmers (see Box 11.4).

A typical FFS involves a significant time commitment on the part of farmers and a talented facilitator who has strong biological training, as well as understanding of active learning techniques. This level of educational investment is not available through many extension systems, but where it is, there is evidence that the return to investment is multifold and long-term. In Uganda the experience with NAADS advisory services has been successful in regions where FFSs were

BOX 11.4 New Directions for Farmer Field Schools

FFSs train farmers in biological principles and empower farmers to design integrated crop management strategies that are relevant to their farming practices. An exciting new development in FFS involves linkages with primary education in Southeast Asia (see http://www.communityipm.org/downloads.html). The rural ecology and agricultural livelihoods (REAL) program was initiated in Cambodia, Thailand, Vietnam, Bangladesh, Laos, Indonesia, and the Philippines, through cooperation of Education and Agricultural Ministries. At the core of the REAL approach is the student-centered study of local agriculture and rural livelihoods through real-world experiments. Information collected in rural environments is used as a basis for integrated, activity-based learning about mathematics, science, art, and culture. It facilitates intergenerational learning and creates a bridge between community members, teachers, parents, and children. If ecological literacy is at the heart of a relevant curriculum, as claimed by educational theorists such as David Orr (1992), then a REAL approach shows how ecological literacy can be implemented. As explained by Chutima, a fifth grader from Central Thailand, participating in an Integrated Pest Management (IPM) program in her school allowed her to gain knowledge about the life cycles of insects and the skills required to follow environmentally sound practices in vegetable production (Bartlett and Jatiker, 2004). Most importantly, she learned how to learn.

initiated first. This provided a large number of educated farmers who were effective at identifying knowledge gaps and participating with NAADS in agricultural development (Friis-Hansen, 2004).

AGRICULTURAL KNOWLEDGE AND INFORMATION SYSTEMS

A model for agricultural knowledge generation and dissemination has been proposed that is referred to as the agricultural knowledge and information systems (AKIS) paradigm. This places farmers, NGOs, and the private sector at the center of a triangle that has three nodes: University, Research, and Extension (Fig. 11.3).

This AKIS model has been criticized because of its somewhat linear conceptualization of delivering technology to three sectors from the three nodes. Collaborative learning and promotion of local innovation is not well represented in this model. The three core institutions of education, technology generation, and extension advisory staff need to be interwoven and interlinked to promote farmer-centered learning and the continuous exchange of information among all parties.

A participatory research and extension process is at the foundation of AKIS to integrate stakeholders in identifying problems, prioritizing, and charting a way forward. This type of interaction is essential to create buy-in and build bridges among development organizations (e.g., NGOs, extension, and researchers), community governance structures (e.g., village councils headed by village chiefs), and other parties such as agricultural industry or regulatory institutions.

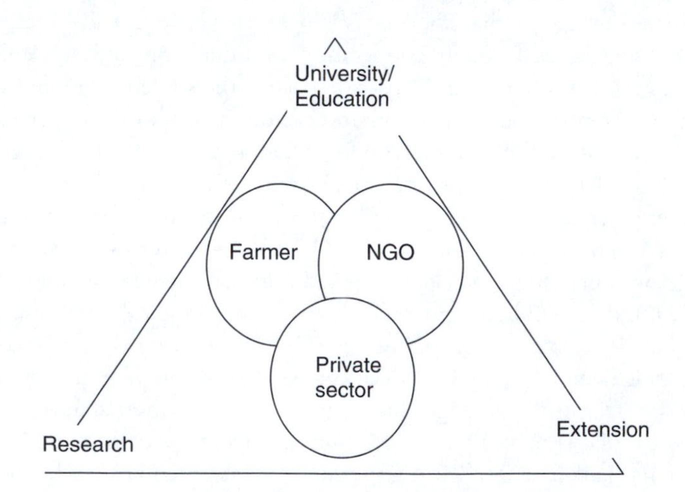

FIGURE 11.3 Model for the AKIS paradigm.

Central to this vision of participatory AKIS is that extension plays the role of facilitator. This shifts extension staff from acting as advisors on technical issues to catalysts, helping in conflict resolution and assisting communities in identifying key priorities, opportunities, and methods to resolve problems. End users of AKIS should be considered broadly and include individuals and farm families, cultivators, livestock producers, traders, seed buyers, sellers, and consumers.

An iterative process of learning is important where there are built-in evaluation steps that include midcourse corrections. This attention to process will promote co-learning and enhance the relevance of outcomes. It is essential that all team members, and notably farmers, have "buy-in" and are part of identifying priorities, not just implementing development activities. This practice will strengthen the chance for success, as participants will have ownership in the process.

There are many resources available today to support participatory research and extension. A recently published and notable collection is available on the web from CIP-UPWARD. This three-volume source book, *Participatory Research and Development for Sustainable Agriculture and Natural Resource Management*, sets out theory and case studies from 30 countries. See http://www.cip-upward.org/main/Publications/Publications.asp.

The UPWARD project grew out of a commitment to assist extension advisors in understanding process dynamics and to move beyond perfunctory implementation of a participatory methodology. Some observers contend that extension should play a facilitator role by focusing on enhanced communication between and among farmers and other stakeholders (Hagmann *et al.*, 1999). Others have emphasized the centrality of rural innovation, seeing extension as the catalyst that brings together different players and supports new ventures and entrepreneurship. In the synthesis presented in the UPWARD source book, four elements were identified as crucial for participatory agricultural development:

1. **Assessment and diagnosis**: problem diagnosis and participatory assessment of resources, needs, and opportunities;
2. **Experimenting with technology options**: joint agenda setting, testing a range of options, on-farm research, and collaborative evaluation;
3. **Facilitating local innovation**: institutional and policy innovations, negotiation and conflict management, supporting community organizations, local capacity development; and
4. **Dissemination and scaling up**: document findings, promote networking and horizontal linkages for information flow, outreach to a broad audience.

The initial points of assessment and experimentation are addressed in Chapter 2 of this book, while facilitating local innovation is the major topic in Chapter 10. The final element, dissemination and scaling up, is the theme of the next section of this chapter.

SCALING UP

A primary goal of agricultural advisors is the generation of knowledge to serve a wide audience, but this raises the key question of how to scale from a small number of beneficiaries to many? Or from a farmer-led perspective, how can we support local change at many locations? There are trade-offs here, as participatory engagement and extension in one area requires time and resources and may preclude working in other topic areas. Farmer organizations can play a key role in enhancing the effectiveness of extension efforts and reaching a broad audience.

Farmer groups are becoming more effective at demanding relevant technical advice, as explored earlier in this chapter. There are also lessons to be learned from countrywide literacy campaigns and farmer-led movements. (See examples in Snapp and Heong, 2003.) Here we present an effective extension campaign from Southeast Asia that improved farmer understanding and pest management in rice. Millions of farmers in Vietnam and Thailand were reached with a message that challenged current practice in pesticide application and encouraged experimentation (Box 11.5). This led to substantial changes in pest management practice: the number of sprays used by 2 million farmers was reduced from 3.4 sprays per season to 1.6.

This experience illustrates the effectiveness of a campaign based on in-depth understanding of the audience, with a carefully targeted message that engaged farmers and stimulated experimentation. Table 11.2 presents other innovative approaches to information dissemination, exploring attributes and challenges associated with each approach.

The learning cycle described in the Box 11.5 continues today. Members of the team have broadened their scope beyond entomology to consider fertilizer and seeding rates. Research on integrated management of rice indicated that cost savings of about $85.00 US per hectare were possible through judicious, coordinated reductions. This outcome was the genesis for a campaign called the "Ba Giam Ba Tang" or "Three Reductions." Farmers across the Mekong Delta in Vietnam and other regions are testing for themselves the economics involved in reducing rates for three inputs simultaneously. The pioneer of this multidisciplinary effort, IRRI entomologist Dr. K.L. Heong, pointed out in a recent interview that, "We should be training extension workers to communicate more effectively, to deliver correct information to farmers and to motivate them to evaluate it objectively. We can't afford to leave pesticide education to those who profit by spreading misinformation about these chemicals." (See http://www.irri.org/media/press/press.asp?id=79)

SUSTAINABILITY

Long-term effectiveness is the ultimate criterion for a successful extension effort. Public sector investments in extension continue to dwindle, so sustainability depends upon the extent to which advisory services can be sourced and supported locally.

BOX 11.5 Shifting Farmer Beliefs and Pest Management Practice

High rates of pesticide use in rice production are a significant concern in Southeast Asia. This motivated a team of entomologists, extension advisors, and public media specialists to improve farmer understanding of pest dynamics. Initially, the project surveyed Vietnamese farmers to document local perceptions about insect damage and management strategies used to control pests. These practices were compared to biological findings on pest thresholds for damage and rice–insect interactions. A contradiction was exposed between the view of entomologists that early leaf damage had minimal impact on yield and a common belief of Asian rice farmers that insecticide sprays should be used to control highly visible leaf damage. This led to costly early season spray applications and to secondary pest problems arising from indiscriminant killing of insects early in the season, including leaf folders, whorl maggots, grasshoppers, and beetles.

Radios are a primary source of information and entertainment in villages around the globe. So, a large scale media campaign of radio drama was launched to expose these contradictions. Actors used scenarios to challenge farmers to test the following "rule of thumb" on a small area of their field: "Spraying for leaf feeder control in the first 30 days after transplanting (or 40 days after sowing) is not necessary." Farmers were encouraged to try an experiment that compared normal practice to a 500 square meter area of a rice field where no insecticide was applied the first 30 days after transplanting. In follow-up workshops farmers shared their results with local communities, extension field staff, and researchers. Over 85% of farmers who participated found the yields of the two plots were identical. The farmer's own experiment helped him or her to resolve the conflicting information, and beliefs changed. Before participating in the experiment 68% of the farmers surveyed applied insecticides in the first 30 days. Thirty-one months later, excessive insecticide use dropped by 53%.

Source: Escalada *et al.*, 1999.

Farmer organizations are one means to enhance sustainability, as such associations provide institutional continuity and, in some cases, financial support. Another way is through local entrepreneurs who provide technical knowhow or act as an information brokering service. Examples include individuals who obtain training or develop a technical innovation and then offer services to other farmers for a small fee. This is illustrated through a case study of mud stove construction in Niger (Box 11.6).

Sustainable extension requires integrated social and technological development. An example comes from West Africa, where thousands of farmers have adopted "micro-dosing" of fertilizer in sorghum and millet (Fig. 11.4). This technology targets fertilizer to planting hills at very low rates, 4 to 10 kg per ha^{-1}. It was shown to

BOX 11.6 Mud Stove Project

Niger is an arid country located in West Africa, in the Sahelian region. Firewood is the primary source of fuel for cooking in the rural areas, and collection takes up to five hours per day per family. Wood stoves that use three rocks to support the pot save up to 50% of wood consumption compared to an open fire. The stoves are made from local mud, which is also used to construct homes. Stoves were constructed by the families with guidance from a Peace Corps volunteer and a local teenager who assisted with the project. This young man saw an unmet need as families adopted the mud stove but required technical advice in order to seal the stove so it would last over the rainy season. Following construction, he would conduct a use assessment and then offer to return to seal the stove with cement. He charged a small fee to seal the stove, offering a needed service and earning some income. He also followed up with a post-survey and checked in with the family to assure the stove was satisfactory. This provided district-wide sustainable support for the continuance of mud stove construction.

be technically feasible through a network of on-farm trials carried out across five countries (Buerkert *et al.*, 2001). However, no farmer uptake occurred until extension and NGOs became involved in institutional innovations, most notably the "warrantage" credit inventory system (Fig. 11.5). These village-based grain banks support credit access and more reliable economic returns through sponsoring sales when grain prices are high. This has been critical to farmer adoption. What remains to be seen is how rapidly this social and technical "integrated innovation" will spread. Farmers continually assess the socioeconomic returns to agricultural technologies, and adopt, adapt, or disadopt technologies as incentives change.

Agricultural change is not only influenced by climate and biophysical resources, but, most importantly, by the economic and social context. These complexities require extension educators and development agents to consider how to engage with policymakers and other key actors in the social environment.

CONCLUSION

The actors and institutions involved in outreach are undergoing rapid change. Historically, agricultural information often followed a technology transfer mode, with a linear flow of information from researchers to farmers. This was not effective at meeting the needs of poor farmers, nor did it take into account the complexity at local levels. In contrast, the emerging outreach models described here emphasize relevance and local capacity building. Notably, these are decentralized and

FIGURE 11.4 Targeting microdoses of 4 kg P ha^{-1} fertilizer to planting hills has greatly enhanced this farmer's millet crop.

FIGURE 11.5 The inventory credit system "warrantage" has proven to be a successful institutional innovation in West Africa that reduces risk, enhances returns, and ensures credit access.

demand-driven extension systems, where priority setting involves village leaders, farmer organizations, and market linkages. The role of extension advisors is thus to act as facilitators and advocates, to catalyze innovations, and to develop local capacity and institutions. This requires the interaction of many stakeholders, including farmers, extension educators, traders, private industry, NGO advisors, entrepreneurs, and researchers. Furthermore, this broad vision of extension encompasses engagement with governmental regulators, policymakers, and members of the media. Agricultural change requires the courage to try new models and modes of operation.

RESOURCES AND REFERENCES

Agricultural Research and Extension Network. Back issues of AGREN are a valuable extension resource and available at http://www.odi.org.uk/agren/.

Anandajayasekeram, P., Davis, K., and Workneh, S. In press. Farmer field Schools: An alternative to existing extension systems? Experience from Eastern and Southern Africa. *J. of Int. Agr. Ext. Ed.* **14–1,** 81–93.

Bartlett, A., and Jatiker, M. (2004). Growing up in the REAL world. *LEISA Magazine*, June 2004, 9–11. http://www.leisa.info/

Benor, D., and Harrison, J. Q. (1977). "Agricultural extension: The training and visit system." The World Bank, Washington DC.

Bently, J. W., Boa, E., van Mele, P., Almanza, J., Vasquez, D., and Eguino, S. (2003). Going Public: A new extension method. *Int. J. Agric. Sustainability* **1,** 108–123.

Esbern-Hansen. (2004). "Uganda case study: Demand driven advisory services." Paper presented at Neuchatal meeting in Arhus.

Friis-Hansen. (2004). "Demand driven advisory service as a pathway out of poverty: Experience from NAADS in Soroti district, Uganda." Paper presented at the Neuchatel meeting in Århus.

Hall, H. M., Morriss, S., and Kuiper, D. (1999). Privatization of agricultural extension in New Zealand: Implications for the environment and sustainable agriculture. *J. Sustain. Agr.* **14,** 59–71.

Kaaria, S., Abenakyo, A., Alum, W., Asiimwe, F., Best, R., Barigye, J., Chisike, C., Delve, R., Gracious, D., Kahiu, I., Kankwatsa, P., Kaganzi, E., Muzira, R., Nalukwago, G., Njuki, J., Sanginga, P., and Sangole, N. (2006). "Enabling rural innovation: Empowering farmers to take advantage of market opportunities and improve livelihoods." Paper presented at the Innovation Africa Symposium, Kampala. CIAT.

Leeuwis, C., and van den Ban, A. (2004). "Communication for rural innovation. Rethinking agricultural extension." Blackwell Science, Oxford, UK.

Obaa, B., Mutimba, J., and Semana, A. R. (2005). "Prioritizing farmer's extension needs in a publicly-funded contract system of extension: A case study from Mujono District, Uganda." Agricultural Research and Extension Network (AGREN) Paper No. 147, http://www.odi.org.uk/agren/papers/agrenpaper_147.pdf.

Orr, D. W. (1992). "Ecological literacy: Education and the transition to a postmodern world." State University of New York Press, Albany, NY, USA.

Pesche, D. (2004). "The role of farmers' organizations in management of agricultural services: Demand as a learning process." CIRAD Neuchatel Initiative Annual Meeting, Aarhus, Denmark.

Peterson, W. (1997). The context of extension in agricultural and rural development. Chapter 3. *In* "Improving agricultural extension. A reference manual." FAO. http://www.fao.org/docrep/W5830E/w5830e05.htm

Praneetvatakul, S., Waibel, H., and Meenakanit, L. (2007). Farmer field schools in Thailand: History, economics and policy. A Publication of the Pesticide Policy Project, Special Issue Publication Series, No. 12, Hannover, 89.

Qamar (2004). "Demand-for-services planning by villagers: A case study from Pakistan." CIRAD Neuchatel Initiative Annual Meeting, Aarhus, Denmark.

Rivera, W., and Alex, G. (eds.) (2004). "Extension reform for rural development: Vol. 5. National strategy and reform process." Agricultural and Rural Development Discussion Paper 12. The World Bank. http://lnweb18.worldbank.org/ESSD/ardext.nsf/11ByDocName/ExtensionReformforRuralDevelopmentVolume5NationalStrategyandReformProcess/$FILE/Extension_Reform_V5_final.pdf.

Setimela, P. S., Monyo, E., and Banziger, M. (eds.) (2004). "Successful community-based seed production strategies." CIMMYT, Mexico, DF.

Snapp, S. S., Blackie, M. J., and Donovan, C. (2003). Realigning research and extension services: Experiences from southern Africa. *Food Policy* **28,** 349–363.

Snapp, S. S., and Heong, K. L. (2003). Scaling up: Participatory research and extension to reach more farmers. *In* "Uniting science and participation: Managing natural resources for sustainable livelihoods." (B. Pound, S. S. Snapp, C. McDougal, and A. Braun, eds.). Earthscan, U.K., and IRDC, Canada, pp. 67–87.

Swanson, B., Bentz, R., and Sofranko, A. J. (1998). Improving agricultural extension: A reference manual. Food and Agricultural Organization, Rome, Italy. http://www.fao.org/docrep/w5830e/w5830e00.htm#contents.

The Laos Village Extension System. http://www.i-p-k.ch/subsites/Report_LEAP/Consider_VES.htm, http://www.laoex.org/websites.htm.

CHAPTER 12

Tying It Together: Global, Regional, and Local Integrations

Malcolm Blackie

THE POVERTY OF BIG IDEAS

Any discussion of poverty quickly turns to the numbers, which are frightening. As Easterly (2006) noted so graphically, 850 million people do not have enough to eat; 10 million children die from preventable diseases. We know how to produce food—obesity is a problem of increasing severity in the developed world. The diseases killing all those children have well-known causes—the children do not need to die. Surely it is just a straightforward process of getting the answers out to those who need them. Data on which this chapter is developed come largely from Africa. This is deliberate. It is not my intention to ignore the very real dimensions of the poverty that exists in the rest of the world; rather it is an explicit acknowledgment of the limits of my own knowledge. Most of my field work has been done in Africa, and understanding poverty needs a solid base of experience in the real world. I can—and do—use the literature to widen the discussion beyond Africa. However, the farmers I have worked with provide the human face to the tale (of both hope and of concern), which I shall develop in this final chapter. Furthermore, of the three major flash points of poverty in the world, Africa is the one where the signals—child mortality, malnutrition levels, and life expectancy[1]—are all pointing the wrong way. Success in turning round the trends in Africa is fundamental to "making poverty history."

Because the numbers are huge, those who want effectively to address the evident poverty crisis that affects so much of the human race must, if their efforts are to have any impact, think big as well. Big thoughts need big plans to carry them into action. The past several years have seen a plethora of plans: the United Nations came up with the Millenium Development Goals (MDGs); Tony Blair, the British Prime Minister of the time, noting that poverty in Africa was a "scar on the conscience of the world," set up the Africa Commission to provide a coherent set of guidelines for creating change on that continent; and Bob Geldorf, through his "Make Poverty History" campaign, set the agenda for the 2005 G8 conference at Gleneagles in Scotland.

The record of such plans is unpromising. After World War II, the Marshall Plan for Europe helped rebuild a devastated continent and set the scene for future (largely) peaceful developments in the region. However, in the developing world, progress is unspectacular (and dismal in Africa), despite enormous and costly programs (Easterly, 2006). This chapter explores further Easterly's (2006) proposition that

> It doesn't make sense to have the goal that your cow will win the Kentucky Derby. No amount of training will create a Derby-winning race cow. It makes much more sense to ask, "What useful things can a cow do".... [D]ecades of experience show aid agencies to be cows, not racehorses.

Building on Hope

Using real-world contemporary examples from the troubled continent of Africa, I show that significant, widespread change is possible, affordable, and reliable when we use the skills and knowledge of the poor (linked to high-quality analysis and

[1]As well as a range of others (to which not everyone subscribes), including population increase, poor governance, low productivity, poor trade performance and rising aid dependency, increasing polarity of wealth, and crippling external debt.

science) to help them solve their own problems rather than imposing our "big plans" on them. The moving words of Bishop Tengatenga of Malawi, who bravely and forthrightly spoke at the 2005 "Malawi after Gleneagles Conference" in Edinburgh, say it all:

> It is difficult to believe in your own self worth if all the time you are told you are failing. The poor struggle every day to survive; recognise what they are doing, the obstacles in their way, and give them a hand of friendship and encouragement. Build–don't destroy–their confidence and they will repay a hundredfold. That is the help they need.

While there is desperate poverty in the developing world, there is also a powerful urge to succeed. There have been notable successes in creating change for the better that are often overlooked in the bleak landscape of failure that is so frequently perceived as the reality in poor countries. The need for further investment in both human capital and infrastructure is evident; nevertheless there are opportunities to exploit more effectively the human capital and infrastructure that exist. A strategy that builds on the best and is directed by farmers' needs and informed by the commercial, social, and ecological environments of the continent can provide gains, not only for the better-off producers, but also for the poor and excluded.

SOME HISTORY

A mere 40 years ago, the focus of poverty discussion was Asia, not sub-Saharan Africa. The developed world then, as today, struggled with the moral dilemma of huge food surpluses while populations starved in Asia. It was evident then, as today, that food aid could provide, at best, temporary relief in emergency situations—food aid could not be a viable long-term development policy. In the 1960s, the concept of triage among Asian countries (the emerging problems of sub-Saharan Africa, curiously, did not register) was being discussed actively (see Paddock and Paddock, 1967). In a carefully documented presentation, it was argued that a population–food collision was inevitable. With science unable to offer sufficient increases in agricultural production, the hungry nations of the 1960s would inevitably become the starving nations of the 1970s. Their solution, which was presented in the name of reason, U.S. self-interest, and true humanitarianism—was a famine-disaster version of the military medical "triage" system. The United States needed to divide the developing nations into three categories:

- those so hopelessly headed for or in the grip of famine (whether because of overpopulation, agricultural insufficiency, or political ineptness) that aid would be a waste; these "can't-be-saved nations" would be ignored and left to their fate (these included India, Egypt, and Haiti)
- those who could survive without aid, "the walking wounded" (Gambia and Libya)
- those who could be saved with help (Pakistan and Tunisia).

The deeply disturbing Paddock scenario did not come about. The reason is largely because they [in common with Meadows *et al.* (1972)] underestimated the power of science. In India alone, wheat production increased sixfold from the 1960s to today, rice yields more than doubled, and the percentage of undernourished people fell from 38% to 21% (Mukherjee, 1987). This was the celebrated green revolution. The model focused on improving the productivity of cereals through the development of crop varieties that could exploit intensive cropping systems. It was a child of its time—capital intensive, hierarchical, and based on the Schultz hypothesis of disruptive, rapid change to make things happen in what were perceived as deeply conservative peasant societies (Schultz, 1964).

However, while there have been many and obvious gains from the green revolution, it was not a universal solution, most notably in Africa. The green revolution did not miss Africa; it failed to take root. The Schultzian "big bang" approach, relying on the Asian combination of improved seeds and enhanced fertility management, was not sufficient to bring about needed change. In many (although not all) African countries, agriculture is the mainstay of the economy, with the majority of the population relying on cropping to feed themselves and their family—and to earn a little cash to buy clothes, pay school fees, and purchase medicines (Conroy *et al.*, 2006). However, the land is degraded, soils are exhausted, and rainfall is erratic. Crop yields are low. Perhaps farmers could put part of their land down to cash crops and use the income to buy the inputs they need to improve their food security. High transport costs make it difficult to compete in agricultural export markets, and the terms of trade for many primary commodities continue to deteriorate. Rising international and local fertilizer prices make this essential input unprofitable to use except on very high value crops, which are typically too risky for the poor to grow.

As a result, the poor (some 80%+ in the case of Malawi) become poorer, their food supplies dwindle, and national growth stalls. At the same time, the AIDS pandemic causes suffering and death in almost every family. African farmers know about improved varieties and are desperate for access to fertilizer. However, they typically face a dreadful series of choices based on technologies that are incomplete, often uneconomical, and do not provide a reliable and effective road from poverty (Fig. 12.1). The options offered to them as a route out of poverty are deeply flawed, and, unsurprisingly, the poor (who may be illiterate but are not stupid) reject them firmly. The green revolution fails to take off.

The stalling of the green revolution, particularly in Africa, has led to a decline in support for agricultural research and for interventions that address agricultural productivity. However, there is an alternative approach that builds on a powerful partnership of scientists, farming communities, and development agencies (both private and public). I will term this a "green evolution." The green evolution encourages the efficient and swift transformation of agricultural production through harnessing the best of skills in a collaborative, "learning by doing" manner in which all feel ownership and pride. Existing structures are improved and enhanced to

FIGURE 12.1 Smallholder farm families such as the one pictured here face difficult choices.

build change through an evolutionary rather than a revolutionary approach. This is cost-effective, brings the best of developing country and international expertise together in a problem-solving format, and can (as shown later in the chapter) be scaled up rapidly to reach the poor quickly and effectively.

A green evolution strategy is based around the highly efficient use of the right inputs used in the right way. This creates broad-based opportunities for the poor to benefit directly from effective access to the improved seed, fertilizers, and other critical inputs that are the foundations of the essential growth in productivity. Efficiency and consistency are the guiding principles to developing a productive, commercialized, and profitable agricultural sector, with broad-based participation, and specifically involving the poor and vulnerable in creating realistic and profitable options for change.

A GREEN EVOLUTION STRATEGY

The green evolution, in common with evolution in nature, is efficient in selecting the best and encourages partnership and collaboration. Unlike the hierarchical and prescriptive nature of the "big bang" revolution, an evolutionary strategy uses multiple channels and players and allows choices to emerge and be tested, with the best being adopted. It fits comfortably into the increasingly practiced participatory framework for development that facilitates the empowerment of the poor and disadvantaged. Such a strategy, with a foundation of good science, directed by farmers' needs and informed by the commercial, social, and ecological environments of developing countries, can provide gains, not only for the better-off producers but also for the poor and excluded.

Poor farmers in Africa do not want to be poor. The reason so many are living on the "edge of survival" is that too many of their traditional approaches to agricultural production are breaking down. Economic growth has been insufficient to offer alternative means of employment for the rural poor. Profits from farming at low levels of productivity have been too small to allow farmers to reinvest in their farms and maintain productivity at acceptable levels (Blackie, 1994; Eicher, 1990). Meanwhile, continual increases in population have depleted both the available resource base and the social entitlements that hitherto provided a state of equilibrium in rural areas of Africa (Lele, 1989). Finally, until recently,[2] increases in agricultural productivity in developed regions of the world (facilitated by both science and subsidy) have pushed world agricultural commodity prices down, making it increasingly difficult for marginal land farmers to operate profitably within existing technical and economic parameters (Sachs, 2005).

For many of the rural poor on the continent, a severe lack of cash constrains the feasible options. In Malawi (admittedly at the extreme), the average cash income of a farmer is $0.10/day (Conroy *et al.*, 2006). The farmer has so many demands on her very limited resources of cash and labor that she needs to know, as far as it is possible, that any investment she makes in her farming enterprises will repay the labor or cash adequately and reliably. If she has access to sufficient productive land, she may grow enough to feed herself and her family, providing her health is good and the weather favorable. However, the start of the rains brings diarrhea and malaria. Often, illness in herself or her children will result in her planting her crop late. With a poor rainy season her crop may fail. Too often she will be unable to produce enough food for her family's needs and will seek work or food elsewhere—often planting, weeding or fertilizing a neighbor's crop—which means that her own is left unplanted, unweeded, and unfertilized until later in the season. Late planting and poor weeding mean a poor harvest and once again she finds herself without food before the next crop comes in. This is the downward spiral that creates much of Africa's poverty (Kumwenda *et al.*, 1996). Add in the devastating AIDS pandemic (which in no small part is both driven and exacerbated by poverty; see Conroy *et al.*, 2006) and farm households can find themselves enmeshed in a poverty trap with no evident escape.

DeVries and Toennissen (2001) set the scene graphically:

> It is that of a single mother whose primary means of income is a one hectare plot of unimproved land on an eroded hillside.... From each harvest she must provide for virtually all the needs of her family throughout the year, including clothing, health care, education costs and housing. Because she can afford few purchased inputs, the yield potential of her farm is low...perhaps 2000 kilograms of produce....

[2]See Box 12.1 for a recent, more pessimistic perspective on world food price trends.

BOX 12.1 Nuns Mug Orphan! Soon We'll All Be Fighting for Food

NUNS mug disabled orphan for bag of crisps. As a headline, it certainly grabs the attention. This is the title of the latest note from the iconoclastic fund managers at Bedlam Asset Management, and it relates to a phenomenon that is sweeping the world—rising food prices. Merrill Lynch has also coined an eye-catching term for the process: agflation. The prices of rice, wheat, corn, barley, cattle, and pork are all up by more than 30% since March 2005. In America, the annual rise in the producer-price index for finished consumer foods has picked up from a little over 1.5% last year to almost 4%. Food prices are rising faster, relative to other measures of core producer-price inflation, than at any time since the early 1980s.

At one level, the rising prosperity of India and China is ensuring higher demand. In the normal course of things, this would not be a problem. Farmers should easily be able to respond to higher prices by increasing output, particularly by bringing marginal land back into cultivation. Reports of increased corn plantings have pushed corn prices down so far this year, but wheat is substantially more expensive. Even so, markets may not be as responsive as they used to be.

The primary reason is climate change. The Australian drought is a clear case of unusual weather conditions having a direct impact on food production. But the indirect effects are probably more important, as land used for food production is diverted into use for ethanol, to provide an alternative to gasoline. In turn, land used to produce soybeans is diverted to higher-priced corn (until recently, this was forcing up soybean prices). And as animal feed grows more expensive, so do the prices of beef, pork, and chicken. Tyson Foods, an American agri-giant, said that higher feed costs led to a 41% drop in poultry profits in the first quarter; the company will try to pass on those costs (processed chicken prices are up nearly 30% year-on-year).

Government interference in the system does not help. American consumers would be better off if Brazilian sugar, rather than Midwestern corn, were used to produce ethanol; high tariffs make that difficult. Across the Atlantic, grants from the European Union are designed to reduce the area of land under cultivation.

At the very least, higher food prices (rather like oil) will reduce the ability of consumers to spend money on other goods. Other things being equal, higher food prices should cause a poorer trade-off between headline inflation and growth. Agflation may also cause political problems around the globe. According to Bedlam, food accounts for 33% of China's consumer-price index and 45% of India's. Perhaps we are unlikely to see nuns stealing food. But we could see governments fall—bread riots are one of history's most recurring phenomena. And governments may start scrapping with each other over natural resources. That can lead to geopolitical tensions and greater risk across all securities markets.

Source: ©The Economist Newspaper Limited, London, May 6th, 2007.

> In the course of a given season, innumerable threats to the crops appear....The impact of drought plus whatever combination of pests and diseases attacks the crop in a given year can often reduce the average harvest on her farm by perhaps 50–60%, to 1000 kilograms of produce. *At this level of productivity, the family is on the edge of survival.*

Those most in need of new livelihood options are the least able to pay for them. Furthermore, the advice they receive on the choices open to them is disgraceful—what the farmer needs is reliability and consistency of performance. A single mother, hoping to harvest a ton of rice on a hectare of depleted upland soil, can ill afford to lose 100 kg of her harvest to a crop pest or disease in a single season even if, under some conditions (which she may not be able to achieve), she can potentially get a higher yield from a new variety. She needs to move to a higher level of productivity but cannot afford the means to lift herself there. While group savings and credit schemes (such as Savings and Credit Cooperative Societies, Household Income Security Associations, and Self-Help Groups) can help poor families access inputs to get out of the poverty spiral, the effectiveness of such interventions is badly blunted when the inputs themselves are poorly tailored to the needs of the poor.

Much of the debate on poverty revolves around the low prices that farmers get for their produce. Remember first that the priority for the poor is to grow their own food. Because many of the rural poor do not even produce enough to feed themselves all year round, they buy food when supplies are short and prices high.[3] Poor people do not need expensive food. Thus an evident priority in the struggle against poverty is to bring food prices down. The costs of many improved technologies (seeds, fertilizer, improved livestock breeds) needed by smallholders, despite ongoing efforts at market development, will remain high. Low cash cost technologies (home-produced seed, household composts) often have a substantial cost in terms of labor, which is also a scarce resource in many poor households.

The advice given to many poor farmers for the use of essential inputs (both those purchased from outside and those which the farmer may generate from homestead resources, such as manures and home-saved seed), actually serves actively to discourage their use. In large part, this is because of inadequate incorporation of basic economic parameters into farmer recommendations (Blackie, 2005). The information provided frequently overlooks the obvious fact that an expensive input (whether in cash or labor terms) can be profitable if it is used efficiently. The knowledge the poor seek is how to make best use of the limited amounts that they are able to purchase. As a result, poverty alleviation and food security have to be arranged around low food prices and efficient production methods. With low food prices, the poor can use their limited cash to invest in better housing, education, and health care. With high food prices, they are further trapped in poverty, and the opportunities for livelihood diversification are few.

[3]This can be countered by warehouse receipt systems and grain banks, which aim to reduce the need for farmers to sell their crops immediately after harvest when prices are low and the need to buy in food on the open market when prices are high, but the reach of these schemes is limited.

The focus of green evolution is on quality and impact, facilitated through enhanced networking and coordination among the various sector stakeholders and international organizations. The best options are pulled together and then promoted through large-scale initiatives. The poor influence the choice of recommendations, while the private sector contributes toward sector needs, such as seed and market systems. The promotion of proven and well-validated research, using proven and novel (but justified) communication pathways, can have a rapid impact on poverty. Existing projects, which have known technical and social strengths, can efficiently add value to a carefully focused development initiative. This serves to strengthen farmer–extension–researcher–market policy linkages. The objective is to create multiagency, multidisciplinary buy ins and to build teams that work systematically and with strong national leadership, to develop solutions to pressing national problems.

There are examples of this happening (see Blackie, 2005). These studies suggest (based on the analysis of a limited number of case studies) that innovative partnerships, such as those described in the preceding paragraph or those more directly arising from market opportunities, are indeed making real impacts on poverty in the developing world. These partnerships encourage a coordinated, cost-effective, and efficient technology transfer process (through learning by doing at all levels) using the best of national and international expertise in a focused, problem-solving effort (Fig. 12.2). Local knowledge and expertise (at farmer, market/private sector, and researcher/policy maker level) can be tapped to link research, extension, markets, and national policy to improve living standards for rural people reliant on agriculture. The green evolution embodies the farmers' need for consistency and reliability, with the requirement for substantially enhanced efficiency over key areas of the farming system.

IMPLEMENTING THE DREAM

This is not just a fantasy; it is entirely possible. The first step is to build the teams, engaging a broad range of individuals and institutions, to focus on addressing the central theme of rural development in poor countries: the alleviation of poverty and the development of sustainable livelihoods for the poor and excluded. Innovation requires close and effective collaboration between "public good" research and the market. While increasing the demand-led component of the research agenda is important, this will not, on its own, act sufficiently fast to lift the technologically disconnected rural poor out of poverty. A fruitful interaction among academia, government, and industry, which has led to the technology explosion in the wealthy parts of the globe, is needed to produce a strong and effective partnership between national and international science, and between science and the user of science, who is typically the resource-poor smallholder.[4]

[4]This already exists for cash crops such as tobacco in Africa (and in non-African countries for a wide range of cash-earning commodities). The need now is to create conditions that make it happen for staple crops also (see Janssen, 2002).

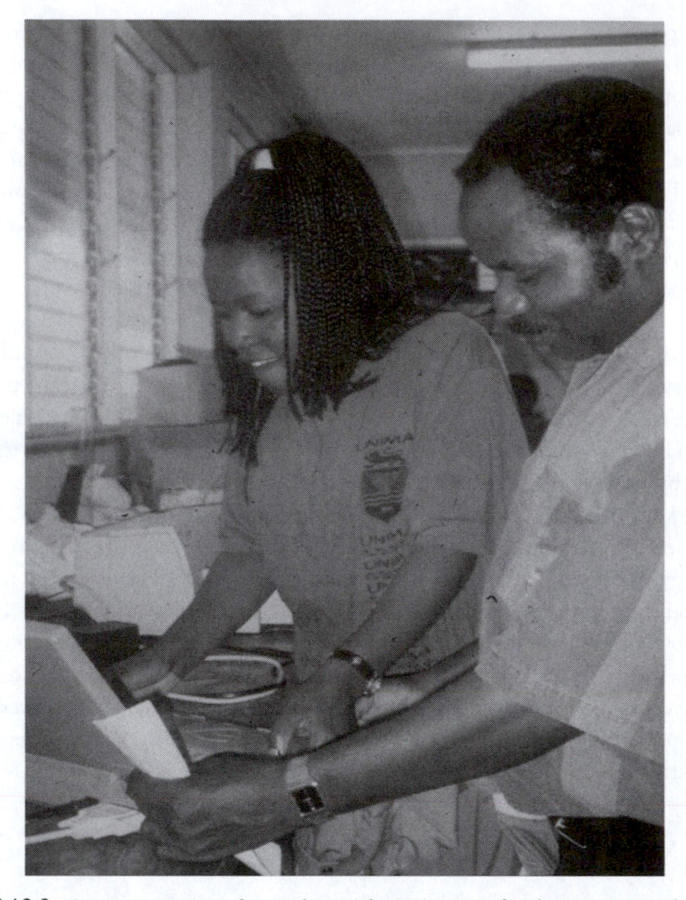

FIGURE 12.2 A new generation of researchers at the University of Malawi is meeting the challenge of conducting cutting-edge research through partnerships with farmers.

Pro-poor agricultural development needs to involve low-income farmers and consumers as active participants in setting priorities for, and in the implementation of, development initiatives. Led in southern and eastern Africa by efforts by Collinson (2002) in the 1970s and 1980s (building on work by a number of others, particularly in Asia and Latin America), farming systems research (FSR) was introduced to help guide technology development to address the priorities of the poor. FSR was based around a farm management-oriented informal survey process, supplemented by secondary data from key sources and informants (Collinson, 2002). Variations on this theme, with a broader, less directly agricultural focus, such as rapid rural appraisal and participatory rural appraisal, have been developed. Implicit in this is the central role of technology as a route out of poverty (the role of the market has, too often, been ignored or underemphasized). The scientist facilitates the development of ideas and helps define options rather than entering with already identified solutions. The overall theme is that of encouraging participants to take control of the process of change, thus empowering them to become more active partners in development.

Building Confidence, Trust, and Ownership

The key element in creating farmer involvement is building the trust and respect of the farmers. This requires a continuing exercise of discussing and coming to a consensus on options, together with obtaining routine and informed feedback on results. Some of the tools are already in use. Researchers, in particular, have been highly innovative in developing the necessary tools to meet the challenge of conducting participatory activities with many clients over an extended geographical area in a cost- and time-effective manner. Snapp, for example, developed the "mother and baby" trial design[5] (see Snapp *et al.*, 2003). The design comprises "mother" trials, which test a number of different technologies, and "baby" trials, which test a subset of three (or fewer) technologies, plus one control. The design makes it possible to collect quantitative data from mother trials managed by researchers and to cross-check them systematically with baby trials managed by the farmers. By facilitating hands-on experience for the farmers, the clustered mother and baby trials provided a relatively rapid approach to developing "best-bet" options (Fig. 12.3). The linked trial approach provides researchers with tools for quantifying feedback from farmers and helps generate new insights and priorities (Snapp *et al.*, 2003).

Another element necessary for enhancing impact and sustainability of change is building ownership of the process by the poor. Making farmers proud of their involvement in creating and contributing to change and their participation in the process an enjoyable and interesting experience is typically, often unintentionally, downplayed in many development programs. The benefits from correcting this neglect are considerable. A radio soap opera in Kenya, "*Tembea na majira*," incorporating development issues, has an audience of nine million with impressive percentages of people taking on board the development messages embedded in the story lines.

An impressive program in Tanzania that involves farmers learning to work as research teams to solve their own problems provides another example of technology-led innovation. Beans over large areas of eastern and southern Africa are widely grown by the poor. This fact was used as an access point for involving the poor in creating change within the community. Yields of the crop were well below their potential, in large part because of severe plant disease problems. The obvious start was to use new technology to address the diseases and, to this end, a bean-integrated pest management (IPM) project in Malawi (Dedza district), Tanzania (Mbeya,

[5] The terminology is, in fact, the farmers', who were delighted to have responsibility for their own trials.

FIGURE 12.3 Farmer experimentation with multipurpose legume intercrops such as the soybean and pigeon pea pictured here was fostered through participation in farmer-designed and -implemented "baby trials."

Mbozi, Lushoto, and Hai districts), Kenya (Kisii and Rachuonyo districts in Nyanza province), and Uganda (Kabale in southwestern Uganda) was developed to reach marginalized groups, particularly small-scale women farmers (Ward *et al.*, 2007).

However, the project went beyond technology. The team involved in design and implementation focused not only on improving bean production but also on making individuals proud of their farming and on encouraging the community to increase the value they place on farming. Scaling up was achieved through showing community groups how their knowledge could bring others benefit, who then spread the ideas further. The emphasis in the implementation was to build pride and interest in the new technologies and to help encourage the poor to engage with the program. Through humor, drama, and music, the implementation team made the interaction between farmers and outside actors enjoyable. Early adopting farmers, extension personnel, community development staff, and local leaders sensitized the rest of their community farmers to organize themselves into research groups for effective access to information and technologies. These farmer groups also served to boost farmer confidence as a forum of like-minded people. Farmers volunteered pieces of land for experimentation to inform others about what they were doing. This was especially appreciated by women who had little time for experimenting or participatory learning but were very keen to see how the crop responded.

The bean IPM project successfully made participating farmers realize that they could construct solutions to their production constraints. Encouraged by their experiences with beans, farmers started actively to seek improved services (such as quality seed, markets, credit, improved livestock, fertilizers, tree nurseries, irrigation facilities, soil and water conservation methods) (Blackie and Ward, 2005) and raised these issues openly with local officials and visitors. Farmers' groups have brought an important and dynamic component to the local innovation system. The project started in 1997 with no farmer groups, no activities, and no partners. By 2006, there were almost 80,000 farmers in almost 400 farmer groups and 11 constraints addressed with 13 different partners. Over 60% of the group members are women farmers who also play key roles in group leadership. Group members continue to be very keen to learn by doing and sharing knowledge, exchanging experiences, training other farmers, and reporting their own research results. Participating farmers and partners organized and implemented cross visits to other locations for different lessons independently from the project.

Building in the Market

Moving from subsistence (or below subsistence) to an agriculture that provides a healthy and sustainable life means providing opportunities for the poor to engage effectively in markets—both for needed inputs and for the production they generate. One typical problem the poor face is getting access to inputs; conventional markets struggle to serve a community of dispersed households living in poverty. Building ownership among the poor provides one route out. For example, local, reliable seed systems for "orphan" crops (which are important to the poor), such as sweet potatoes, groundnuts, and pigeon pea, can be developed among the poor themselves. In Uganda, the Kapchorwa Seed Potato Production Association reduced the devastating disease of bacterial wilt in Irish potatoes to below 1%. Improved sweet potatoes, with enhanced vitamin A, were disseminated rapidly in Central Uganda. In just 2 years (2001–2003), 36,000 farmers tripled sweet potato production, producing 34,000 tons with a value of over £1.2 million (Blackie and Ward, 2005).

FIPS-Africa, a "not-for-profit" company in Kenya, works with the private sector in that country—Athi River Mining (ARM), Monsanto, Western Seed Company, Lachlan Agriculture, and Kenya Seed Company—and with public sector agencies such as the Kenya Agriculture Research Institute (KARI). FIPS-Africa uses a participatory methodology to build demand for improved inputs at the farm level, not by imposing predefined options on farmers, but by helping farmers make informed choices as to what is best for them and then ensuring that these choices are available in convenient locations. Farmers are helped to evaluate a range of different technology options, including new seeds of maize, beans, and vegetables; fertilizer types; and weed control options. FIPS acts as an "honest broker," helping farmers choose and evaluate various technologies in an unbiased manner. It then builds market links by working through local stockists to ensure that needed inputs are available for sale in quantities the farmer can afford (not just those that are convenient for the supplier to deliver) and helping in the promotion of new options for farmers through a range of activities. These include collaboration with church and school groups, providing information at market days, and the provision of samples for farmers to experiment with on their own.

Evaluation data on FIPS show the approach to be remarkably successful in improving food security among participating farmers. Once their food crop is secure, they quickly start to use very small quantities of fertilizer (which they can buy at their local stockist) on high-value cash crops such as kale and cabbage for income generation. Income growth and livelihood diversification follow rapidly. Spillover to neighboring farmers shows that farmer-to-farmer advice carries the best of the new messages quickly among the poor. In an analysis of FIPS impact, food security among poor farmers with whom FIPS was working nearly doubled within 3 years from around 30% to 60%. However, in the same time period, food security among the clearly less well off who were not working with FIPS also doubled from nearly 15% to almost 30%. A simple modeling exercise suggested that, with realistic assumptions, adoption of improved technologies recommended by FIPS would be 84% among farmers (some 300,000 new adopters of improved technologies) in the three districts currently served by within 8 years (Blackie *et al.*, 2006).

Building Partnerships with the Wider Community

An African green evolution is both viable and successful. Examples that follow provide reason for such optimism. These are by no measure a comprehensive review of achievements in transforming African agriculture. Rather they highlight, across a range of very different commodities, from food crops to high-quality export production, the power of linking farmer knowledge and interest with the best of modern science. Each example has influenced the lives of tens of thousands of poor families in Africa—so it can be done. The common theme running through them all is that through linking farmers, researchers, and policy makers it is possible to bring the power of new technology positively and effectively into the process of poverty alleviation. In each case, the influence of leadership, teamwork, and effective networking is evident. The need to learn from, and build on, these successes is urgent. Millions of Africans are food insecure and too many of the young die from nutrition-related illnesses every year (Conroy *et al.*, 2006).

THE TANZANIA FERTILIZER PARTNERSHIP (TFP)

In 2007, in both Malawi and Tanzania, formal public/private partnerships are being formed to bring these concepts to scale. The TFP is a collaboration of Tanzanian central and local government agencies, an international consortium of private and public sector actors, and a number of national and local organizations, operations, and projects that are participating in different aspects of the work (including FIPS, which is scaling out its operations from Kenya to the region). The role of partners is based on the capacity of each to deliver specific operational output in the field or to provide necessary support and inputs. The partnership runs with minimal formality and procedure. It has evolved from a primarily fertilizer-marketing focus to broader small farmer productivity and marketing activity addressing prioritized constraints along the entire length of the value chain. It aims to do "business unusual". Formalities are included only as and when they are seen to be necessary and useful. Initially, TFP included Yara International, Rabobank, NORAD, and Norfund. The government of Tanzania joined in August 2006. The Rockefeller Foundation has provided technical support from an early stage. There are currently no rules of membership. The principles of self-selection and operational relevance are the critical concerns, with overall coordination through the Agricultural Council of Tanzania. A similar partnership is being developed for Malawi.

RESCUING AFRICA'S CASSAVA FARMERS

Cassava crop protection successes are an interesting mixture of farmer-led enterprises and focused scientific endeavors. Cassava was introduced to Africa by Portuguese traders to trading stations in the Congo in the mid-1500s. The crop was attractive to farmers because of its drought tolerance, known resistance to locusts, low labor requirements, and capacity to survive in low-fertility soils (Gabre-Madhin and Haggeblade, 2001; Jones, 1957). It supplanted yams in some locations and cereals in others, spreading across Central Africa (Jones, 1957). Introduced into East Africa after 1800, cassava spread west into the interior from Zanzibar and Mozambique. It is now a major African staple food and, in particular, is an important source of household food security for many of the continent's poor (Gabre-Madhin and Haggeblade, 2001).

In the 1920s and 1930s, cassava mosaic virus, spread by a white fly, threatened this increasingly important food security crop in Ghana, Nigeria, Cameroon, Central African Republic, Tanganyika, and Madagascar (Jones, 1957). Farmers responded immediately by replacing affected plants with cuttings from unaffected varieties. This theme was taken up by colonial agricultural research stations in Tanzania, Kenya, Madagascar, and Ghana, which introduced cassava breeding into their programs for the first time (Cours *et al.*, 1997). The result, after a decade of intensive research, was a series of new resistant varieties that spread rapidly and largely replaced the affected "local" varieties (Gabre-Madhin and Haggeblade, 2001).

In the early 1970s, two imported pests—the cassava mealy bug in the Democratic Republic of Congo (then Zaire) and the cassava green mite in Uganda—threatened the crop. Lacking natural predators, both spread rapidly across the continent. The mealy bug, the more voracious of the two, caused crop losses of 80% as it ate its way across the continent at over 300 km per year. By the early 1980s, the mealy bug had infested the entire African cassava belt, where it threatened the principal food source of over 200 million Africans (Herren and Neuenschwander, 1991). A decade of collaborative work by international and national research institutes led to the identification of a natural predator of the mealy bug. The International Institute for Tropical Agriculture mounted a mass rearing and distribution program in collaboration with African NARSs. First released in 1981, the predator wasp had, by 1988, largely controlled the mealy bug threat throughout Africa (Gabre-Madhin and Haggeblade, 2001). A rather more challenging program used to identify a suitable predator for the cassava green mite has also proved successful.

DISEASE PREVENTION IN UGANDA'S STAPLE: BANANAS

Bananas in the Central Highlands of Africa owe their importance as a food crop to skillful farmer plant selection over about the last 800 years. The crop, an introduction, like cassava (but by Arab traders), was well suited to the climate in what are now Uganda, Rwanda, Burundi, and eastern Congo. Farmers liked the crop because of high calorie yields per hectare and its ability to protect the soil from erosion (Gabre-Madhin and Haggeblade, 2001). Uganda farmers cultivate some 60 different cultivars, the largest pool of genetic diversity anywhere in the world, despite difficulties of undertaking crop improvement with a vegetatively propagated crop (de Langhe *et al.*, 1996; Reader, 1997).

In recent times, while the banana remains an established staple, it is increasingly threatened by pests and fungal disease, and farmers have not been able to develop varieties sufficiently quickly to meet these new challenges. Tissue culture methods have been introduced to promote rapid and sterile multiplication of pathogen-free planting material. The KARI, in conjunction with a local private biotechnology company, has begun to produce *in vitro* banana plants commercially. These have been shown roughly to double both yield and income under farmer conditions (Qaim, 1999). This farmer/scientist collaboration has supported the development of a highly suitable food security crop that currently accounts for over one-fourth of caloric consumption in countries such as Rwanda and Uganda. A commercial tissue culture laboratory is now established in Uganda, and tissue culture plants produced by a South African company have been used in national trials in Uganda.

DIVERSIFYING INTO EXPORT HORTICULTURE

Kenya, with a high value tourist industry, developed a local quality vegetable production capacity in the 1950s. The rehabilitation of previously ecologically declining areas such as Machakos bears testimony to the positive effects of this industry on smallholders with access to markets associated with the expanding tourist industry in Kenya (Tiffen *et al.*, 1994). In 1957, private traders in Kenya began expanding this trade into the export of off-season vegetables and tropical and temperate fruits. After 1970, this trade expanded steadily as a result of growing demand in Europe, improved technologies and marketing systems for fresh vegetable distribution there, and substantial increases in air-freight space from Nairobi to Europe, a by-product of Kenya's booming tourist industry (Gabre-Madhin and Haggeblade, 2001).

Steadily increasing production quality standards, particularly in Europe, have led to a marked expansion in the considered use of pest control methods among the 500,000 smallholder vegetable farmers who today supply about 75% of all vegetables and 60% of all fruits under contract to exporters (Noor, 1996). The value of horticultural exports has risen from $13 million in 1970 to $155 million in 1999. Uganda, Zimbabwe, and Zambia have all entered this market in recent years.

SMALLHOLDER COTTON SUCCESSES IN ZIMBABWE

In Zimbabwe, before the 1960s, virtually no cotton was grown by smallholders (Blackie, 1986). By 1980, some 42,000 Zimbabwean smallholders produced nearly a third of the national cotton crop. A few years later the number of registered smallholder cotton growers had doubled and they were producing consistently more than half the national cotton crop. By 2000 (a record year), over 80% of national cotton production was produced by smallholders. Not only were smallholders growing more cotton than their large-scale counterparts, typically they were producing a higher quality lint through careful picking and sorting before delivery. Cotton had become the biggest smallholder cash crop in Zimbabwe. This rapid uptake of a new cash crop came about through the work of Melville Reid, one of the most innovative extension workers involved with Zimbabwean smallholders. Through careful discussion with both farmers and research colleagues, Reid made cotton an attractive crop by removing the obstacles facing smallholders. He devised a low-cost cotton production system suited to family labor and cash availability of the typical smallholder household. He arranged training courses for farmers and farm advisors and ran regular field days to promote the crop. He also worked with the marketing agency to create a "smallholder-friendly" marketing system.

THE MALAWI STARTER PACK: BREAKING THE CYCLE OF POVERTY

In response to Malawi's serious food crises of the late 1980s, in a remarkably few years, Centro Internacional de Mejoramiento de Maiz y Trigo (CIMMYT) Mexico and Malawian scientists produced new varieties of flinty, high-yielding hybrid maize well suited to Malawi's needs. Companion agronomic research promised to reduce the need for commercial fertilizer and improve soil fertility. It identified crop rotations and complementary agroforestry cultures that both economized on purchased inputs and improved diets. Once new hybrids were developed and complementary crops were identified, a Rockefeller Foundation-inspired institutional innovation, the Maize Productivity Task Force (MPTF), organized 5 years of extensive farmer trials, nearly 2000 a year, to identify, for each of Malawi's major agroclimatic zones, the most economically efficient package of practices; the "best bet" for that region. By 1998, the MPTF could recommend with confidence these improved systems to farmers.

However, few Malawians had the needed cash to purchase even minimal amounts of the new seeds and fertilizer that could help them break the cycle of poverty. This gave rise to the concept of giving a starter pack of the new inputs to all farmers. The universal starter pack was designed to use the promise of best-bet technology to jump-start maize production for all smallholders. This would simultaneously improve the food security of all food-deficit smallholder households and sharply increase the marketed surplus available to urban consumers (bringing food prices and inflation down sharply).

All smallholders in Malawi were given a small package containing enough hybrid seed and the economically viable recommended quantity and type of fertilizer sufficient to plant 0.1 ha of land. Each household would gain sufficient extra maize to feed itself for a month in the food shortage season. All of the inputs in the starter pack generated incremental production. They did not displace commercial purchases, as the poor could not afford even these small amounts. There were evident rewards to good husbandry, especially to timeliness of planting, fertilizing, and weeding, which provided a strong incentive and reward for using the inputs well. It provided a nationally implemented, but individually operated, technology testing and demonstration program for a small part of each farm, facilitating experimentation by farmers of promising, but not yet widely adopted, technologies. The program was intended to be developed, refined, and adapted in future years to "fast track" further technology choices into the smallholder sector, thus diversifying farming systems and increasing smallholder incomes.

Best-bet inputs and practices were incorporated into the starter packs and were distributed to 2.8 million smallholder farmers in 1998 and 1999, together with a carefully developed extension message to assist farmers in the use of the pack (Blackie, 1998; Mann, 1998). Evaluation data showed that the starter packs raised maize production on average by about 125 to 150 kg per household (significantly more than was estimated). Production in each of those 2 years was approximately 2.5 million tons, 500,000 tons higher than ever before or since; 67% higher than the 20-year average. In terms of cost-effectiveness, the starter pack program performed extremely well compared to alternative food crisis prevention measures, such as general fertilizer price subsidies, and relief interventions, such as subsidized commercial food imports and food aid (Levy, 2005).

In the years following 1999, as a donor-imposed "exit strategy," the universal starter pack—based around scientifically verified best-bets technologies—was changed to a targeted inputs program aimed at alleviating short-term poverty among the poorest. The link with science was lost, the skills and experience of the MPTF were allowed to dissipate, and the program quickly lost direction. Implementation became an annual struggle between donors and the Malawi government as to the numbers of packs to be distributed (MPTF argued persuasively that it was impossible to distinguish the desperately poor from the rest in a country where poverty is so pervasive) and the pack contents. Packs included inappropriate fertilizer and seed (both maize and legumes) of dubious quality. Unsurprisingly, Malawi has steadily slipped back into national food insecurity. As importantly, the uncertainty around pack content and distribution has served actively to disrupt the commercial market in seed and fertilizer.

LOOKING BEYOND THE FARM GATE

The story we have developed so far is founded on building from the bottom up: the poor are actively engaged in finding avenues through which they can change their own lives. However, they cannot work alone—there are forces beyond their control that can outweigh what they can do for themselves. For a comprehensive analysis of the factors that trap the poor in poverty, see, for example, Conroy *et al.* (2006). Here, partly because of space (and also possibly more honestly) because the best of what I know is at the farmer level, not in the more macro arenas of trade and agricultural policy, I turn (too briefly) to look at the issues beyond the "farm gate."

Trade and Subsidies

One of the (few) things the major developed countries seem to agree on is that subsidies to poor people (in poor countries) are unsustainable and need to be eliminated. I will not rehearse the arguments here—suffice it to say that, as commonly implemented, most subsidies to agriculture in the developing world are a spectacularly inefficient way of helping poor people. However, this is only half the story: developed countries cheerfully subsidize their own agriculture by huge amounts. This matters since the outcome is not only to support rich country farmers but also to take income directly from the poor in poor countries. In 2001, agricultural subsidies in Organization for Economic Cooperation and Development countries were about $311 billion. Compare this to development assistance to all developing countries in the same year of $12 billion.

The most effective way we could help the poor in poor countries would be to allow them to trade freely with the developed world. (Pingali *et al.*, 2006). Brazil and India, among others, have shown how effectively they can compete in the production (and processing) of cotton, sugar, soybeans, maize, and many other products—only to find that their goods are shut out of the markets where consumers have the cash to purchase them. These market distortions mean that farm commodity prices in the developed world no longer reflect supply and demand. This leads, in the first instance, inevitably to a growth in production (beyond that which the national market can absorb). As a result, the surplus is dumped on the international market—depressing prices in markets open to poor farmers. Space does not permit discussion of the full complexities of farm subsidies either in the developed or in the developing world. Hypothetically, there are three effects that agricultural support measures of advanced countries may have on agricultural production in developing countries (Herrmann, 2006):

- **None**: because advanced countries produce and support different things from developing countries—temperate agricultural products as opposed to tropical ones.
- **Negative**: the elimination of agricultural support in advanced countries means the cost of food imports in developing countries rises as international food prices increase.
- **Positive**: the elimination of agricultural support in advanced countries increases international prices to which developing country farmers respond.

Herrmann (2006) showed clearly that negative effects dominate. Trade is a key driver of economic growth. Developing countries, particularly Asia, have used trade to break into new markets and transform their economies. However, in Africa, the last 3 decades have had a collapse in the continent's share of world trade from around 6% in 1980 to 2% in 2002 (Commission for Africa, 2005). The trade barriers imposed by the rich nations are "politically antiquated, economically illiterate, environmentally destructive and ethically indefensible" (Commission for Africa, 2005, p. 49) but little effort is being implemented to do away with them (see, e.g., the case of cotton [Oxfam 2004]).

The costs are substantial, as Oxfam (2004) illustrated, using the case of sugar subsidies—a product that developing countries are especially good at producing. The European Union (EU) is the world's second-biggest sugar exporter, yet the cost of producing a kilogram of sugar in the EU is more than six times higher than in Brazil. In addition, Oxfam (2004) claimed that the EU subsidy is not just the $1.5 billion of subsidies to farmers; it also includes a further $0.63 million of "hidden subsidies" that go to large EU sugar refiners. On the Oxfam analysis, Brazil loses around $500 million a year and Thailand about $151 million, even though these two countries are the most efficient sugar producers in the world. Less efficient, and poorer, African countries lose out as well: Mozambique lost some $38 million in 2004, as much as it spends on agriculture and rural development, while sugar subsidies cost Ethiopia what it spends on HIV/AIDS programs.

Disease

Any discussion of poverty is incomplete without consideration of disease. Take the effects of HIV/AIDS—a pandemic of increasing severity across the developing world. Governance, macroeconomic management, economic policy, health, HIV/AIDS, agricultural collapse, and hunger are all linked. Poor governance and macroeconomic stability deter investment and undermine growth, while the AIDS pandemic undermines the capacity to implement programs in poverty alleviation. Food crises exacerbate malnutrition and fuel the AIDS pandemic as people are forced into high-risk sexual behavior as a survival strategy. The threat of food shortages creates macroeconomic difficulties, as scarce foreign exchange is diverted to purchase and import food reserves, diverting resources from investment in development programs. External and internal debt rises inexorably. Recall Easterly's (2006) analysis that opened this chapter. World poverty need not persist and could be halved within the coming decade. Billions more people could enjoy the fruits of the global economy. Tens of millions of lives can be saved. Practical solutions exist, the framework is established, and the cost is affordable.

Central to making it affordable is using aid as a development tool and not as a political weapon to help the powerful. Dr. Banda in Malawi was able to finance his draconian and exploitative policies by posing as an anticommunist. Jonas Savimbi in Angola was able to finance a dreadful war of attrition against his compatriots using a similar strategy. The outcome has been too many murderous regimes—some deliberately destructive such as Samuel Doe in Liberia and some that have simply served to impoverish their peoples and drive them further into poverty, such as Daniel arap Moi in Kenya. The evidence is surely incontrovertible that this is a spectacularly bad way to promote change for the better in the developing world.

An international development assistance regime that does what it claims to do would not prop up the arap Mois and Samuel Does of this world, but rather provide space for the many talented and concerned individuals in poor countries, who are too often sidelined at present, to begin to influence development policy (as they so ably did during the 2002 famine in Malawi[6]—for the benefit of all (Conroy *et al.*, 2006). Rich countries could then know that their aid investments were, indeed, creating change for the better. Poor countries would no longer stagger from crisis to crisis, but be able to put in place thoughtful, long-term strategies for development. The only losers would be the tyrants, people smugglers, and war mongers of rich and poor countries alike.

Aid built on a genuine sense of solidarity and mutual trust will put the economics of the poor world on durable developmental paths that go beyond ending hunger (Mkandawire, 2005).

[6]A remarkable collaboration of individuals across the public, private, and voluntary sectors mobilized, at very short notice, a feeding and rehabilitation program in Malawi to provide emergency aid to nearly 3.5 million Malawians in an efficient and timely manner.

As Anne Conroy wrote so passionately in the closing chapter of Conroy *et al.* (2006), 20 years ago, President Julius Nyerere asked the governments of the West "Should we really starve our children to pay our debts?" It appears the (silent) answer was "yes." "The heaviest burden of a decade of reckless borrowing is falling not on the military or on those who conceived the years of waste, but on the poor who have to do without necessities" (Peter Adamson quoted in Lewis [2005]).

Let us return to Malawi. Malawi owes some $3.1 billion, of which 82% is owed to multilateral creditors (the World Bank, the International Monetary Fund, and the African Development Bank), 17.5% to bilateral creditors, and 0.5% to commercial creditors. Debt service totaled $112 million in 2004. Yet 5 million Malawians are in need of humanitarian assistance today, and there is a major gap between the resources pledged by the international community and requirements for both food aid (to keep people alive) and any substantial agricultural recovery program (to help Malawians pull themselves out of poverty). The costs of servicing Malawi's external debt will be malnutrition and famine unless someone mobilizes the resources to provide smallholder households with sufficient seed and fertilizer to increase productivity.

Stephen Lewis wrote angrily in June 2007:

> Everyone is aware of the solemn promises that were made at Gleneagles in July of 2005. They followed in the wake of Tony Blair's Commission on Africa, with all of the attendant triumphalism, and it seemed to promise a new dawn for the African continent.... Fast forward, then, to 2007 and the G8 Summit just completed in Germany. In the weeks prior to the Summit itself, quite predictably a number of groups and institutions took stock of the extent to which the promises at Gleneagles had been honoured. Every single assessment found a staggering shortfall... What actually happened in Germany is deeply, deeply troubling, and it's worthy of every piece of scorn that can be heaped upon it. The G8 communiqué is deficient in so many ways: fundamentally, it's intellectually dishonest and riddled with arithmetic sleight-of-hand.

Lewis concludes: "It's a terrible thing we do to the uprooted and disinherited of the earth. Together, we must bring it to an end." He says it all.

CONCLUDING COMMENTS

Ending poverty will require honest delivery of the commitments made at multiple high-level meetings—Rio, Monterey, and Gleneagles—of new money provided to meet the needs of the poor in the developing world. Thandike Mkandwire is forthright:

> ... by the mid-1990s, "institutional reforms"—or "good governance", as this was popularly known in donor circles—became the new mantra in the policy world. A wave of institutional reforms swept across the African continent. Already by the beginning of the millennium, there were increasing doubts about the "institutional fix" and the institutionalists began to lose ground. While many countries had, under the aegis of the international financial institutions, introduced major institutional reforms, the economic recovery

> remained elusive. This prompted the new question, "Why is it that even when countries adopt the recommended polices and the right institutions, economic growth does not take place?" One response to this new question is that "institutions do rule," but the institutions peddled by the international financial institutions were the wrong ones, partly because of "monocropping" through the one-size-fits-all institutional design, and "monotasking" that insisted that all institutions should be harnessed to the protection of property rights.
>
> These institutions differed radically from not only those behind the East Asia miracle and China but also from those of any successful case of development in modern times. Indeed, in the successful "late industrializers" many of the institutions being pushed as prerequisites for development never served the functions attributed to them and they were assiduously avoided in all strategies of "catching up."

Mkandawire is right. National ownership and leadership are fundamental to coherent progress on development and poverty reduction. Enlightened political leadership at national and international levels is needed now more than ever before. The process has to be led by the nationals of the country—the "green evolution" concept can be developed to include faith communities; civil society and the technocrats who have all shown a remarkable ability to represent the needs and priorities of the poor. With imagination, effort, and hard work, change can come about. What is needed for those in the developing world is to be given the opportunity to express themselves as equals and not as supplicants. This will only happen when we transform values to genuinely respect the dignity and equality of all human beings.

REFERENCES

Blackie, M. (1986). Restructuring agricultural delivery systems in sub-Saharan Africa: Case studies from Zimbabwe. *In* "Accelerating Agricultural Growth in Sub-Saharan Africa" (J. Mellor, C. Delgado, and M. Blackie, eds.). Johns Hopkins, Baltimore, MD.

Blackie, M. (1994). Maize productivity for the 21st century: The African challenge. *Outlook Agric.* **23,** 189–196.

Blackie, M. (2005). "The possible dream: Food security and prosperity in Africa." Keynote address, 7th Annual African Crop Science Society Conference, Entebbe, Uganda, December 6–10, 2005.

Blackie, M., Thangata, P., Kelly, V., and Makumba, W. (2006). "Agricultural Sustainability in Malawi: Transforming Fertilizer Subsidies from a Short-Run Fix for Food Insecurity to an Instrument of Agricultural Development—Technical and Policy Considerations." Paper presented at the International Association of Agricultural Economists Conference, Gold Coast, Australia, August 12–18, 2006.

Blackie, M., and Ward, A. (2005). Breaking out of poverty: Lessons from harmonising research and policy in Malawi. *Aspects Appl. Biol.* **75,** 115–126.

Collinson, M. (2002). "A History of Farming Systems Research." Springer, The Netherlands.

Commission for Africa (2005). "Our Common Interest: Report of the Commission for Africa." London (http://commissionforafrica.org/english/report/thereport/finalreport.pdf) May 2005.

Conroy, A., Blackie, M., Whiteside, A., Malawezi, J., and Sachs, J. (2006). "Poverty, AIDS and Hunger in Malawi: Breaking out of Malawi's Poverty Trap." MacMillan Palgrave, London.

Cours, G., Fargette, D., Otim-Nape, G., and Thresh, J. (1997). The epidemic of cassava mosaic virus disease in Madagascar in the 1930's–1940's: Lessons for the current situation in Uganda. *Trop. Sci.* **37,** 238–248.

de Langhe, E., Swennen, R., and Vuysteke, D. (1996). Plantain in the early bantu world. *In* "The Growth of Farming Communities in Africa from the Equator Southwards" (J. Sutton, ed.). Azania, Nairobi.
DeVries, J., and Toennissen, G. (2001). "Securing the Harvest." CABI, Wallingford.
Easterley, W. (2006). "The White Man's Burden." Penguin, New York.
Eicher, C. (1990). Africa's food battles. *In* "Agricultural Development in the Third World" (C. Eicher and J. Staatz, eds.). Johns Hopkins, Baltimore, MD.
Gabre-Madhin, E., and Haggeblade, S. (2001). "Successess in African Agriculture: Results of an Expert Survery." IFPRI, Washington, DC.
Herren, H., and Neuenschwander, P. (1991). Biological control of cassava pests in Africa. *Annu. Rev. Entomol.* **36,** 257–283.
Herrmann, M. (2006). "Agricultural Support Measures of Advanced Countries and Food Insecurity in Developing Countries." Research Paper No. 2006/141, UNU-WIDER.
Janssen, W. (2002). "Institutional Innovations in Public Agricultural Research in Five Countries." ISNAR Briefing Paper No. 52. IPFRI, Washington, DC.
Jones, W. (1957). Manioc: An example of innovation in African economies. *Econ. Dev. Cult. Change* **5,** 97–117.
Kumwenda, J., Waddington, S., Snapp, S., Jones, R., and Blackie, M. (1996). Soil fertility management in the smallholder maize-based cropping systems of Africa. *In* "The Emerging Maize Revolution in Africa" (C. Eicher, and D. Byerlee, eds.). Lynne Reinner, Boulder, CO.
Lele, U. (1989). Managing agricultural development in Africa. *In* "Agricultural Development in the Third World" (C. Eicher, and J. Staatza, eds.). Johns Hopkins, Baltimore, MD.
Levy, S. (ed.) (2005). "Starter Packs: A Strategy to Fight Hunger in Developing Countries." CABI, Wallingford.
Lewis, S. (2005). "Race against Time." Anansi, Canada.
Lewis, S. (2007). "The G8 Betrayal of Africa." Stephen Lewis Foundation, June 15, 2007. http://www.stephenlewisfoundation.org/news_item.cfm?news=1900.
Meadows, D. L., Meadows, D. H., Randers, J., and Behrens, W. (1972). "The Limits to Growth." Potomac Associates, New York.
Mkandawire, T. (2005). "Malawi after Gleneagles: A Commission for Africa Case Study: Text of the Keynote Address at the Scotland Malawi Forum." Scottish Parliament, November 2005.
Mukherje, S. K. (1987). "Fertilisers in Developing Countries: Opportunities and Challenges." 15th Francis New Memorial Lecture.
Noor, M. (1996). "Successful Diffusion of Improved Cash Crop Technologies." Proceedings of the Workshop Developing African Agriculture: Achieving Greater Impact from Research Investment. Addis Ababa, Ethiopia. September 26–30, 1995.
Oxfam (2004). "Dumping on the World." Oxfam Briefing Paper No. 61. Oxfam International, Washington, DC.
Paddock, W., and Paddock, P. (1967). "Famine 1975—America's devision: Who Will Survive." Little Brown, Boston.
Pingali, P., Stamoulis, K., and Stringer, R. (2006). "Eradicating Extreme Poverty and Hunger: Towards a Coherent Policy Agenda." ESA Working Paper 61, Agriculture and Development Division. FAO, Rome.
Qaim M. (1999). A socioeconomic outlook on tissue culture technology in Kenyan banana production. *Biotechnol. Dev. Monit.* **40,** 18–22.
Reader, J. (1997). "Africa: A Biography of the Continent." New York: Vintage Books.
Sachs, J. (2005). "The End of Poverty: How We Can Make It Happen in Our Lifetime." Penguin, London.
Schultz, T. W. (1964). "Transforming Traditional Agriculture." Yale University Press, New Haven, CT.
Snapp, S., Blackie, M., and Donovan, C. (2003). Realigning research and extension to farmers' constraints and opportunities. *Food Policy* **28,** 349–363.

Tiffen, J., Mortimore, M., and Gichuki, F. (1994). "More People, Less Erosion." Oxford University Press, Oxford.

Ward, A., Minja, E., Blackie, M., and Edwards-Jones, G. (2007). "Beyond participation—building farmer conFIdence: Experience from Sub-Saharan Africa?" *Outlook Agric.* **36**.

INTERNET RESOURCES

http://www.odi.org.uk. The London based Overseas Development Institute. A very useful source of well-documented case studies and research reports on development issues.

http://www.eldis.org. An excellent broad-based source of summaries of information from a wide range of development agencies.

http://www.africanhunger.org/. The Partnership to Cut Poverty and Hunger in Africa. A very high-powered group and some excellent publications. You have to join to access the Web site fully and the cost is quite high unless your institution is paying. Lower costs apply to individuals and institutions in Africa.

http://www.scidev.net/. An international science and development network that publishes a wide range of challenging materials. There is a regular newsletter that always has information of interest and a very accessible database.

http://www.irinnews.org/. This is a UN-based Web site that provides news updates in areas and fields that you can specify for yourself. It is a useful tool for keeping up-to-date on information from the field.

http://www.globalpolicy.org/. Another UN Web site—the Global Policy Forum. This is a very good source of information on a wide range of topics.

http://www.economist.com/. You need to be a subscriber and the journal is not inexpensive, but there is a wealth of accessible and well-written materials. It is an excellent database and a very good source for teaching information and for closely argued oversights on topics of importance in development.

Index

B

C

D

E

H

I

K

L

M

N

T

U

V

W

Y

Z